PROCEEDINGS OF THE ANALYSIS CONFERENCE,
SINGAPORE 1986

NORTH-HOLLAND MATHEMATICS STUDIES 150

NORTH-HOLLAND – AMSTERDAM • NEW YORK • OXFORD • TOKYO

# PROCEEDINGS OF THE ANALYSIS CONFERENCE, SINGAPORE 1986

*Edited by*

Stephen T. L. CHOY
Judith P. JESUDASON
and
P. Y. LEE

*Department of Mathematics*
*National University of Singapore*

1988

NORTH-HOLLAND – AMSTERDAM • NEW YORK • OXFORD • TOKYO

ISBN: 0 444 70341 1

*Published by:*
ELSEVIER SCIENCE PUBLISHERS B.V.
P.O. BOX 1991
1000 BZ AMSTERDAM
THE NETHERLANDS

*Sole distributors for the U.S.A. and Canada:*
ELSEVIER SCIENCE PUBLISHING COMPANY, INC.
52 VANDERBILT AVENUE
NEW YORK, N.Y. 10017
U.S.A.

PRINTED IN THE NETHERLANDS

# PREFACE

It has become a tradition that one or two mathematical conferences be held annually in Singapore, and the second such conference of 1986 was a workshop and conference on analysis, held on the campus of the National University of Singapore from June 12 through June 21, 1986. This volume forms the proceedings of the workshop and conference, which emphasized mainly harmonic and functional analysis. The invited speakers were E. Hewitt, S. Igari, A. Miyachi, G. Pisier, and J. J. Uhl, Jr., and they have contributed a total of five papers to these proceedings. One other invited speaker, S. Gong, was unfortunately unable to attend the conference, due to unforeseen circumstances, but has kindly sent us a paper based on the address that he would have given. As there were more papers submitted than could be included in this volume, the remainder of the contributed papers were selected on the basis of referees' reports. A list of the talks given at the conference, as well as the names and affiliations of participants and contributors immediately follow this preface.

We would like to thank our colleagues, authors, referees, publisher, typists, and many others who have helped us in the preparation, editing, and production of these proceedings; in particular we are grateful to Madam Lum and Miss Tan for their expert typing of most of the manuscripts. Also, we would like to thank all of the organizations who gave financial support, including the Department of Mathematics, National University of Singapore, the Singapore Mathematical Society, the Singapore National Academy of Science, and the Southeast Asian Mathematical Society. We hope that this volume will serve as a useful record of our conference, and that the memories of ten days spent in Singapore it brings to the participants and interested readers are both as fruitful, and as pleasant, as those memories that we recall.

The Editors,
Singapore, August 1987

## Invited Addresses

E. Hewitt, University of Washington, *Marcel Riesz's theorem on conjugate Fourier transforms : a progress report I-III; Alfred Haar and his measure.*

S. Igari, Tôhoku University, *Application of an interpolation theorem for mixed normed spaces I : An estimate of Riesz-Bochner means of Fourier transforms; Application of an interpolation theorem for mixed normed spaces II : Restriction problem of Fourier transforms.*

A. Miyachi, Hitotsubashi University, *A factorization theorem in Hardy spaces; Boundedness of pseudo-differential operators with non-regular symbols; Estimates for pseudo-differential operators with exotic symbols.*

G. Pisier, Université de Paris VI, *Factorization through weak-*$L_p$ *and* $L_{p1}$ *and non-commutative generalizations.*

J. J. Uhl, Jr., University of Illinois at Urbana, *Differentiation in Banach spaces I, II; Geometry and Dunford-Pettis operators on* $L_1$.

## Short Communications

W. R. Bloom and J. F. Fournier : *Generalized Lipschitz spaces on Vilenkin groups.*

P. S. Bullen : *On the solution of* $\dot{y}_{ap} = f(x,y)$.

T. S. Chew : *A Denjoy-type definition of the nonlinear Henstock integral.*

M. T. Chien : *Perturbations of* $C^*$*-algebras.*

C. H. Chu and L. S. Liu : *A localized version of Choquet's theorem.*

S. Darmawijaya and P. Y. Lee : *The controlled convergence theorem for the approximately continuous integral of Burkill.*

C. S. Ding : *Absolutely Henstock integrable functions.*

J. L. Geluk : *Asymptotically balanced functions.*

B. Jeffries : *Pettis integral operators.*

C. H. Kan : *Extreme contractions from* $L_p$ *to* $L_q$, $p \leq 1 \leq q$.

C. W. Kim : *Shift invariant Markov measures.*

E. M. Lagare : *Approximations of integrals of Henstock integrable functions using uniformly regular matrices.*

H. C. Lai : *Translation invariant operators and multipliers of Banach-valued function space.*

P. Y. Lee : *A proof of the generalized dominated convergence theorem for Henstock integrals.*

D. J. Luo : *On limit cycle bifurcations.*

P. P. Narayanaswami : *The separable quotient problem for Frechet and (LF)-spaces.*

C. W. Onneweer : *Weak $L_p$-spaces and weighted norm inequalities for the Fourier transform on locally compact Vilenkin groups.*

G. Z. Ouyang : *Multipliers of Segal algebras.*

P. L. Papini : *Norm-one projections onto subspaces of sequence spaces.*

S. Pethe : *On linear positive operators generated by power series.*

W. Ricker : *Joint spectral subsets of $\mathbb{R}^m$ for commuting families of operators in Banach spaces.*

Rosihan Mohamed Ali : *Möbius transformations of convex mappings.*

S. Y. Shaw : *Uniform ergodic theorems for semigroups of operators on $L^\infty$ and similar spaces.*

I. H. Sheth : *Centroid operators.*

R. N. Siddiqi : *On density of Fourier coefficients of a function of Wiener's class.*

S. L. Tan : *The successive conjugate spaces of dual $C^*$-algebras.*

R. Taviri : *Fixed points of nonexpansive mappings in Banach spaces.*

H. J. Tu : *Some new applications of potential theory to conformal mappings.*

H. C. Wang : *On the Fourier transforms.*

S. L. Wang : *Weighted norm inequalities for some maximal functions.*

J. A. Ward : *A reflexivity condition for some homogeneous Banach spaces.*

B. E. Wu : *The second dual of Cesaro sequence spaces of a non-absolute type.*

D. Yost : *There can be no Lipschitz version of Michael's selection theorem.*

X. W. Zhou : *On a conjecture of band limited function extrapolation.*

## Workshop Lectures

S. Igari : *Interpolation of linear operators on product measure spaces, I-IV.*

A. Miyachi : *Singular integral operators and pseudo-differential operators, I-IV.*

G. Pisier : *Probabilistic and volume methods in the geometry of Banach spaces, I-III.*

## List of Participants and Contributors

IZZAH BTE. ABDULLAH, Universiti Kebangsaan Malaysia, Malaysia
ACHMAD ARIFIN, Institute Teknologi Bandung, Indonesia
RAVI P. AGARWAL, National University of Singapore, Singapore
AZLINA AHMAD, Universiti Kebangsaan Malaysia, Malaysia
SRICHAN ARVORN, Chiangmai University, Thailand
NACHLÉ ASMAR, California State University at Long Beach, U.S.A.
LYN BLOOM, WACAE (Nedlands Campus), Australia
WALTER R. BLOOM, Murdoch University, Australia
P. S. BULLEN, The University of British Columbia, Canada
SERGIO S. CAO, University of the Philippines, Philippines
CHAN CHUN-WAH, Hong Kong Polytechnic, Hong Kong
CHAN KAI MENG, National University of Singapore, Singapore
CHAN WAI-KIT, Hong Kong University, Hong Kong
SHAO-CHIEN CHANG, Brock University, Canada
TSU-KUNG CHANG, National Tsing Hua University, Taiwan
LOUIS H. Y. CHEN, National University of Singapore, Singapore
CHENG KAI NAH, National University of Singapore, Singapore
CHEW TUAN SENG, National University of Singapore, Singapore
MAO-TING CHIEN, Soochow University, Taiwan
CHONG CHI TAT, National University of Singapore, Singapore
STEPHEN T. L. CHOY, National University of Singapore, Singapore
C. H. CHU, University of London, United Kingdom
SOEPARNA DARMAWIJAYA, Institut Teknologi Bandung, Indonesia
DING CHUANSONG, Northwestern Teacher's College, People's Republic of China
DANIEL FLATH, National University of Singapore, Singapore
J. F. FOURNIER, University of British Columbia, Canada
J. L. GELUK, Erasmus University, Netherlands
S. GONG, The Chinese University of Science & Technology, People's Republic of China
R. C. GUPTA, National University of Singapore, Singapore
RENATO GUZZARDI, Università della Calabria, Italy
EDWIN HEWITT, University of Washington, U.S.A.
HU KAIYUAN, National University of Singapore, Singapore
S. IGARI, Tôhoku University, Japan
BRIAN JEFFERIES, Macquarie University, Australia
JUDITH P. JESUDASON, National University of Singapore, Singapore

KAN CHARN HUEN, National University of Singapore, Singapore
CHOO-WHAN KIM, Simon Fraser University, Canada
YATI KRISNANGKURA, Srinakharinwirot University, Thailand
EMMANUEL M. LAGARE, Mindanao State University, Philippines
HANG-CHIN LAI, National Tsing Hua University, Taiwan
LEE PENG YEE, National University of Singapore, Singapore
LI SHI XIONG, Anhui University, People's Republic of China
LIM SUAT KHOH, National University of Singapore, Singapore
LIU LIANG SHEN, Zhongshan University, People's Republic of China
LIU YU-QIANG, South China Normal University, Peoples's Republic of China
LOU JIANN HUA, National University of Singapore, Singapore
LUO DINGJUN, Nanjing University, People's Republic of China
SIDNEY S. MITCHELL, Chulalongkorn University, Thailand
A. MIYACHI, Hitotsubashi University, Japan
TARA R. NANDA, National University of Singapore, Singapore
P. P. NARAYANASWAMI, Memorial University of Newfoundland, Canada
NG BOON YIAN, Universiti Malaya, Malaysia
NG K. F., Chinese University, Hong Kong
NG PENG NUNG, National University of Singapore, Singapore
NURUL MUCHLISAH, Universitas Hasanuddin, Indonesia
ONG BOON HUA, Universiti Sains Malaysia, Malaysia
C. W. ONNEWEER, University of New Mexico, U. S. A.
OUYANG GUANGZHONG, Fudan University, People's Republic of China
ROONGNAPA PAKDEESUSUK, Chiangmai University, Thailand
PIER LUIGI PAPINI, University of Bologna, Italy
SHARADCHANDRA PETHE, Universiti Malaya, Malaysia
MINOS PETRAKIS, University of Illinois at Urbana, U.S.A
GILLES PISIER, Université de Paris VI, France
ROGER POH KHENG SIONG, National University of Singapore, Singapore
QUEK TONG SENG, National University of Singapore, Singapore
W. J. RICKER, Macquarie University, Australia
I. ROBERTS, Darwin Institute of Technology, Australia
ROSIHAN MOHAMED ALI, Universiti Sains Malaysia, Malaysia
AHAMAD SHABIR SAARI, Universiti Kebangsaan Malaysia, Malaysia
SEN-YEN SHAW, National Central University, Taiwan
ICHHALAL HARILAL SHETH, Gujarat University, India
R. N. SIDDIQI, Kuwait University, Kuwait
BAMBANG SUDIJONO, Universitas Gajah Mada, Indonesia
TAN SIN LENG, Universiti Malaya, Malaysia
ABU OSMAN BIN MD TAP, Universiti Kebangsaan Malaysia, Malaysia
RAKA TAVIRI, University of Papua New Guinea, Papua New Guinea

TU HONGJI, Fuzhou University, People's Republic of China
PETER C. T. TUNG, National University of Singapore, Singapore
J.J. UHL JR., University of Illinois at Urbana, U.S.A.
HWAI-CHIUAN WANG, National Tsing Hua University, Taiwan
WANG SILEI, Hangzhou University, People's Republic of China
JO WARD, Murdoch University, Australia
S. J. WILSON, National University of Singapore, Singapore
WU BO-ER, South China Normal University, People's Republic of China
XU FENG, Northeast Normal University, People's Republic of China
DAUD YAHAYA, Universiti Malaya, Malaysia
LEONARD Y. H. YAP, National University of Singapore, Singapore
DAVID YOST, Australian National University, Australia
ZHANG WENYAO, Liaoning Normal University, People's Republic of China
ZHENG XUE AN, Anhui University, People's Republic of China
ZHOU XING WEI, Nankai University, People's Republic of China
ZOU CHENZU, Jilin University, People's Republic of China

## CONTENTS

Proceedings of the Analysis Conference, Singapore 1986
S.T.L. Choy, J.P. Jesudason, P.Y. Lee (Editors)



# MARCEL RIESZ'S THEOREM ON CONJUGATE FOURIER SERIES AND ITS DESCENDANTS

by

Nakhlé Asmar and Edwin Hewitt[1]

*Marcel Riesz centum annos ex anno natalis sui dedicatus*
*Opera imperitura reliquit*

## §0. Notation.

Throughout this paper we adhere to the following notation: $\mathbb{N}$ denotes the positive integers; $\mathbb{Z}^+$ the nonnegative integers; $\mathbb{Z}$ the integers; $\mathbb{Q}$ the rational numbers; $\mathbb{R}$ the real numbers; $\mathbb{C}$ the complex numbers; $\mathbb{D}$ the set $\{z \in \mathbb{C} : |z| < 1\}$. The symbol $G$ will always denote a locally compact Abelian group, with one or another property imposed on it. The symbol $X$ will always denote the character group (dual group) of $G$. Haar measure (chosen arbitrarily) on $G$ will be denoted by $\mu$. All notation and terminology not otherwise explained are as in Hewitt and Ross [1979, 1970][2]. We suppose that the reader is at home with the basic facts of Fourier transforms on locally compact Abelian groups.

For a real number $p > 1$, $p'$ means $\frac{p}{p-1}$.

## §1. Marcel Riesz's original theorem.

**(1.1) Early history.** Conjugate Fourier series, despite their intrinsic interest and many applications, are for many analysts a little mysterious. Some explanation may therefore be appropriate. The earliest study that we have found of conjugate Fourier series is a Jugendarbeit of Alfred Tauber [1891][3]. His description is that followed to the present day. Let

$$F(z) = \sum_{\nu=0}^{\infty}(a_\nu - ib_\nu)z^\nu \quad (a_\nu, b_\nu \text{ real, } b_o = 0) \tag{1}$$

be a power series convergent in $\mathbf{D}$. (The minus sign in (1) is merely a convenience.) Write $z = r\exp(it)$ with $0 \le r < 1$ and $-\pi < t \le \pi$. The definition (1) becomes

$$\begin{aligned} F(z) &= F(r\exp(it)) \\ &= \sum_{\nu=0}^{\infty} r^\nu(a_\nu \cos \nu t + b_\nu \sin \nu t) + i\sum_{\nu=1}^{\infty} r^\nu(-b_\nu \cos \nu t + a_\nu \sin \nu t) \\ &= \varphi(r,t) + i\psi(r,t). \end{aligned} \tag{2}$$

Tauber *loc. cit.* and a host of later authors have been concerned with the relation between the real-valued harmonic function $\varphi$ and its conjugate real-valued harmonic function $\psi$. For later use, we will write (1) and (2) a bit differently.

For $n \in \mathbf{N}$, let $c_n = \frac{1}{2}(a_n - ib_n)$ and let $c_{-n} = \frac{1}{2}(a_n + ib_n)$. Let $c_o = (a_o + ib_o) = a_o$. It is plain that

$$F(z) = F(r\exp(it)) = \sum_{\nu=-\infty}^{\infty} r^{|\nu|}c_\nu \exp(i\nu t) + \sum_{\nu=-\infty}^{\infty} \operatorname{sgn}\nu\, r^{|\nu|}c_\nu \exp(i\nu t), \tag{3}$$

where $\operatorname{sgn}\nu = \frac{\nu}{|\nu|}$ for $\nu \neq 0$ and $\operatorname{sgn}0 = 0$.

To apply (3) to Fourier series, we now consider a complex-valued function $f$ in $\mathfrak{L}_1(-\pi,\pi)$. The *Fourier transform* $\hat{f}$ of $f$ is the function on $\mathbf{Z}$ such that

$$\hat{f}(n) = \frac{1}{2\pi}\int_{-\pi}^{\pi} f(u)\exp(-inu)du \quad (n \in \mathbf{Z}), \tag{4}$$

and the *Fourier series of f* is the formal infinite series

$$\sum_{\nu=-\infty}^{\infty} \hat{f}(\nu)\exp(i\nu t). \tag{5}$$

The *conjugate Fourier series of f* is the formal infinite series

$$-i\sum_{\nu=-\infty}^{\infty} \operatorname{sgn}\nu\, \hat{f}(\nu)\exp(i\nu t). \tag{6}$$

The function $f$ is real-valued if and only if $\hat{f}(-n) = \overline{\hat{f}(n)}$ for all $n \in \mathbb{Z}^+$. In this case we go back to the infinite series (1) and make the following special choice of its coefficients:

$$a_n = \hat{f}(n) + \hat{f}(-n) \tag{7}$$

and

$$b_n = i(\hat{f}(n) - \hat{f}(-n)). \tag{8}$$

We now write the two infinite series in (3), quite formally, with $r = 1$. This gives us the two formal series (5) and (6), and these turn into

$$\tfrac{1}{2}a_o + \sum_{\nu=1}^{\infty}(a_\nu \cos \nu t + b_\nu \sin \nu t) \tag{9}$$

and

$$\sum_{\nu=1}^{\infty}(-b_\nu \cos \nu t + a_\nu \sin t), \tag{10}$$

respectively. From now on, the coefficients $a_n$ and $b_n$ in series of the form (9) and (10) will be as they are defined in (7) and (8) respectively, that is, they will be the real forms of the Fourier transform of $f$.

Already in 1891 much was known about the convergence of the Fourier series (9), and it was natural to find conditions under which the conjugate series (10) converges. The first successful effort in this direction was made by Pringsheim [1900]. His results are now only of historical interest, since for example he did not have the powerful instrument provided by the Lebesgue integral.

Fatou's thesis [1906] contains the seed of much of the study of conjugate series. Again we go back to the two infinite series in (3), with $c_\nu = \hat{f}(\nu)$, where $f$ is an arbitrary complex-valued function in $\mathfrak{L}_1(-\pi, \pi)$. If $f$ is not real-valued, we will have $c_{-\nu} \neq \overline{c_\nu}$ for some integer $\nu$. This need not prevent us from writing the infinite series in (3), and the Riemann-Lebesgue lemma implies that the radius of convergence of the power series in $r$ appearing on the right side of (3) is not less than 1. We write the first series in (3) as

$$f(r,t) = \sum_{\nu=-\infty}^{\infty} r^{|\nu|}\hat{f}(\nu)\exp(i\nu t). \tag{11}$$

An easy calculation shows that

$$\begin{aligned} f(r,t) &= \frac{1}{2\pi}\int_{-\pi}^{\pi} f(t-u)\frac{1-r^2}{1+r^2-2r\cos u}du \\ &= \frac{1}{2\pi}\int_{-\pi}^{\pi} f(t-u)\Gamma(r,u)du. \end{aligned} \tag{12}$$

We write the second series in (3) as

$$\tilde{f}(r,t) = -i \sum_{\nu=-\infty}^{\infty} \operatorname{sgn}\nu\, r^{|\nu|} \hat{f}(\nu) \exp(i\nu t).^{4} \tag{13}$$

Another easy calculation yields

$$\begin{aligned} \tilde{f}(r,t) &= \frac{1}{2\pi} \int_{-\pi}^{\pi} f(t-u) \frac{2r \sin u}{1 + r^2 - 2r \cos u} du \\ &= \frac{1}{2\pi} \int_{-\pi}^{\pi} f(t-u) Q(r,u) du. \end{aligned} \tag{14}$$

The integrals appearing in (12) and (14) are very convenient in studying $\lim_{r\uparrow 1} f(r,t)$ and $\lim_{r\uparrow 1} \tilde{f}(r,t)$. However, the generalizations to locally compact Abelian groups that we have in mind *require* the form (13) of $\tilde{f}(r,t)$. So we introduce it here.

For continuous $f$, Fatou [1906], p. 360, showed that

$$\lim_{r\uparrow 1} \tilde{f}(r,t) = \tilde{f}(t) \tag{15}$$

exists for a fixed $t$ if and only if the Cauchy principal value

$$\lim_{\varepsilon\downarrow 0} \tfrac{-1}{2\pi} \int_{\varepsilon}^{\pi} [f(t+u) - f(t-u)] \operatorname{ctn} \tfrac{1}{2} u\, du \tag{16}$$

exists, and then the limits in (15) and (16) are equal. On p. 363, he showed that if $f$ satisfies a Lipschitz condition of order $\alpha < 1$, then $\tilde{f}(t)$ exists for all $t$ and is also in Lip$\alpha$.

Privalov [1916$^{(1)}$] extended this result. If $f$ is of class Lip1, then $\tilde{f}(t)$ exists for all $t$ and is of class Lip$\beta$ for all $\beta < 1$. He gave an example showing that $\tilde{f}$ need not be of class Lip1.

W. H. Young [1911] showed that if $f$ is real-valued and has finite variation on $[-\pi, \pi]$, then the series (10) converges at a given point $t$ if and only if the limit (16) exists for this value of $t$, and then (16) is the sum of (10).

W. H. Young and G. C. Young [1913] made a thoroughgoing study of the Riesz-Fischer theorem. One of their theorems is the following (p. 57). Let $\sigma_n(f(t)$ be the $n^{\text{th}}$ Cesarò mean of the partial sums $S_k f(t)$ of the Fourier series (9) of a real-valued function $f$ in $\mathfrak{L}_1(-\pi, \pi)$. For $1 < p < \infty$, $f$ belongs to $\mathfrak{L}_p(-\pi, \pi)$ if and only if the $\mathfrak{L}_p$ norms of the functions $\sigma_n f$ are bounded over all $n$. (See for example Zygmund [1959, Vol. I], Ch. IV, p. 145, Theorem (5.7).) On page 58, the Youngs ask if this characterization remains true when the means $\sigma_n f$ are replaced by the partial sums $S_n f$ of the Fourier series. Should this be the case, the authors point out a remarkable consequence. For $f \in \mathfrak{L}_p(-\pi, \pi)$ and $g \in \mathfrak{L}_{p'}(-\pi, \pi)$, the equality

$$\frac{1}{2\pi} \int_{-\pi}^{\pi} f(t) \overline{g(t)} dt = \sum_{\nu=-\infty}^{\infty} \hat{f}(\nu) \overline{\hat{g}(\nu)} \tag{17}$$

holds, the series on the right side of (17) being convergent. (We have recast the Youngs' statement in complex form.) It was already known to the Youngs that the series on the right side of (17) is Cesarò summable to the left side of (17). The convergence of the series on the right side of (17) the Youngs regarded as so unlikely that they rejected the conjecture that the $\mathfrak{L}_p$ norms of the partial sums $S_n f$ are bounded in $n$.

Fejér [1913] used the conjugate Fourier series to compute $\lim_{\varepsilon \downarrow 0}[f(t+\varepsilon) - f(t-\varepsilon)]$. His result was later completed by Lukácz [1920]. See Zygmund [1959, Vol. I], Ch. II, p. 60, Theorem (8.13).

We come now to a decisive step in the theory of conjugate Fourier series.

**(1.2) Theorem (Privalov [1917, 1919]).** *Let $f$ be in $\mathfrak{L}_1(-\pi, \pi)$, and let $\tilde{f}(r,t)$ be as in (1.1.13) and (1.1.14). Then the limit*

$$(i) \qquad \lim_{r \uparrow 1} \tilde{f}(r,t) = \tilde{f}(t)$$

*exists and is finite for almost all $t$ in $[-\pi, \pi]$. The value of $(i)$ where it exists is the expression (1.1.16).*

**(1.3) Subsequent history of Privalov's theorem.** The Comptes Rendus note Privalov [1917] contains a complete statement of the first assertion of Theorem (1.2), as well as a sketch of the proof. It attracted scant attention. The only citation we have found is in a preliminary announcement by Hardy and Littlewood [1924], who describe it as follows.

*The subject matter of these notes[5] is extremely interesting, but the indications of demonstrations are insufficient.*

In view of the great importance of Theorem (1.2), and the efforts that Hardy and Littlewood later lavished on conjugate Fourier series [1925], we think that the two English savants should have exerted their great powers to fill in the trifling gaps in Privalov's note [1917].

Privalov [1919] is a book dealing with boundary values of analytic functions defined in various domains. It was apparently unknown in the West for some time. The earliest reference we have found is Lichtenstein [1924/1926]. Privalov [1941] revised his book [1919] and added much new material. We have found no non-Soviet references to this work. After Privalov's untimely death in 1941 at the age of 50, his friends undertook a revision of Privalov [1941], which appeared as Privalov [1950]. This work is widely available and widely cited. A complete proof of Theorem (1.2) appears there; indeed it is merely the overture in a long work of great profundity. Many other proofs of Theorem (1.2) are in existence. See Zygmund [1959, Vol. I], Ch. IV, §3, pp. 131–136 and the notes to this §, p. 377. The proof in Ch. VII, §1, pp. 252–253, is very like Privalov's.

It was observed early on that the conjugate Fourier series of an integrable function need not be the Fourier series of an integrable function, even if the conjugate series converges everywhere. Kolmogorov showed that something close holds.

**(1.4) Theorem (Kolmogorov [1925]).** *Let $f$ be in $\mathfrak{L}_1(-\pi,\pi)$ and let $\varepsilon$ be a real number such that $0<\varepsilon<1$. The function $\tilde{f}$ is in $\mathfrak{L}_{1-\varepsilon}(-\pi,\pi)$, and there is a constant $C$ such that*

$$(i) \qquad \|\tilde{f}\|_{1-\varepsilon} < \tfrac{C}{\varepsilon}\|f\|_1.$$

Privalov kept his interest in special conditions for the convergence of the conjugate Fourier series under various conditions: see Privalov [1923, 1925].

**(1.5)** Finally we are ready to describe Marcel's Riesz's fundamental contribution. If $f$ is in $\mathfrak{L}_2(-\pi,\pi)$, the Riesz-Fischer theorem shows that $\tilde{f}$ exists as an $\mathfrak{L}_2$-limit: Privalov's Theorem (1.2) is not needed. It is clear that

$$(1) \qquad \widehat{\tilde{f}}(\nu) = -i\ \mathrm{sgn}\nu \hat{f}(\nu), \quad \nu\in\mathbb{Z},$$

and that

$$(2) \qquad \|\tilde{f}\|_2 \le \|f\|_2,$$

with equality holding in (2) if and only if $\hat{f}(0)=0$. One may ask if something like this holds for all $p>1$: *viz.*, if $f$ is in $\mathfrak{L}_p(-\pi,\pi)$, does $f$ also belong to $\mathfrak{L}_p(-\pi,\pi)$, does (1) hold, and does something resembling (2) hold?

In the early 1920's, Marcel Riesz provided affirmative answers to all three of these questions. Lars Gårding [1970] gives an interesting account of exchanges between Riesz and Hardy. In 1923, Riesz wrote to Hardy that he had affirmative answers. We now quote from Gårding.

> *Hardy wrote back "some months ago you said 'j'ai démontré que 2 séries trig. conjuguées sont toujours en même temps les séries de Fourier de fonctions de classe $L_p(p>1)$...' I want the proof. Both I and my student Titchmarsh have tried in vain to prove it ..." And in the next letter "Very many thanks - you supply all that is essential. I have sent on your letter to Titchmarsh. Most elegant and beautiful. Of course p. 2 is the real point. It is amazing that none of us should have seen it before (even for $p=4$!)."*

Directly thereafter, Riesz [1924] published a Comptes Rendus note stating his theorems but giving no proofs. Hilb and Riesz [1924] stated some theorems about $\tilde{f}$ for $f$ in $\mathfrak{L}_p(-\pi,\pi)$, citing a "demnächst erscheinenden Arbeit" of M. Riesz. Actually four years passed before Riesz published the paper. E. W. Hobson [1926], pp. 610–614 and p. 698, cited Riesz's work without furnishing a proof and drew some interesting consequences. Titchmarsh [1926] published a theorem very like Riesz's and also – ***verbatim*** – Theorem (1.8) ***infra***. He notes Riesz's priority but fails to mention that he had seen Riesz's 1923 letter to Hardy. One may judge this as ungracious.[6] In 1928, Riesz published the whole story.

**(1.6) Theorem (M.Riesz [1928]). Let $p$ be a real number such that $1 < p < \infty$, and let $f$ be a function in $\mathfrak{L}_p(-\pi, \pi)$.** *The function $\tilde{f}$ is also in $\mathfrak{L}_p(-\pi, \pi)$, and there is a constant $M_p$ such that*

(*i*) $$\|\tilde{f}\|_p \leq M_p \|f\|_p{}^7.$$

*Furthermore, we have*

(*ii*) $$\hat{\tilde{f}}(\nu) = -i \operatorname{sgn}\nu \hat{f}(\nu), \quad \nu \in \mathbb{Z}.$$

**(1.7)** Riesz did not write out (1.6.ii) explicitly, though he indicated it plainly. A stronger version of this identity is due to Titchmarsh [1929], *viz.*: if $f$ is in $\mathfrak{L}_1(-\pi, \pi)$ and if $\tilde{f}$ is also in $\mathfrak{L}_1(-\pi, \pi)$, then (1.6.ii) holds. Something similar appears in Smirnov [1929].

**(1.8)** As corollaries to Theorem (1.6), Riesz proved that

(*i*) *the $\mathfrak{L}_p$ norms of the partial sums $S_n f$ of the Fourier series of $f$ are bounded over all $n$,*

and

(*ii*) $$\lim_{n\to\infty} \|f - S_n f\|_p = 0.$$

From (*i*) he proved that the series on the right side of (1.1.17) converges for all $f \in \mathfrak{L}_p(-\pi, \pi)$ and $g \in \mathfrak{L}_{p'}(-\pi, \pi)$, and that its sum is the left side of (1.1.17). He gave an example to show that the series on the right side of (1.1.17) may converge only conditionally if $p \neq 2$. This settled decisively the question raised by the Youngs [1913], already described in §1. Though Riesz did not spell it out, one can show from (*i*) that a function $(c_\nu)_{\nu=-\infty}^{\infty}$ on $\mathbb{Z}$ is the Fourier transform of a function in $\mathfrak{L}_p(-\pi, \pi)$ if and only if the $\mathfrak{L}_p$ norms of the functions $\sum_{\nu=-n}^{n} c_\nu \exp(i\nu t)$ are bounded over $n$. (See Hobson [1926], pp. 610–614.)

There is yet more in Riesz [1928].

**(1.9) Theorem (Riesz [1928]).** *Let $p$ be a real number such that $1 < p < \infty$, and let $f$ be a function in $\mathfrak{L}_p(\mathbb{R})$. The Cauchy principal value*

(*i*) $$Hf(x) = \frac{1}{\pi}\int_{-\infty}^{\infty} \frac{f(y)}{x-y} dy = \lim_{\varepsilon \downarrow 0} \frac{1}{\pi} \int_{|x-y|>\varepsilon} \frac{f(y)}{x-y} dy$$

*exists for almost all $x \in \mathbb{R}$. The function $Hf$ belongs to $\mathfrak{L}_p(\mathbb{R})$, and an inequality*

(*ii*) $$\|Hf\|_p \leq M_p \|f\|_p$$

*obtains, the constant $M_p$ being the same as in (1.6.i).*

**(1.10)** The function $Hf$ is of course the *Hilbert transform*[8] *of $f$*. We shall have constant recourse to it in the sequel.

Zygmund soon contributed to the subject.

**(1.11) Theorem (Zygmund [1929]).** *Suppose that* $|f|\log^+|f|$ *is in* $\mathfrak{L}_1(-\pi,\pi)$[9]. *Then* $\tilde{f}$ *is in* $\mathfrak{L}_1(-\pi,\pi)$ *and there are constants* $B$ *and* $C$ *such that*

$$(i) \qquad \|\tilde{f}\|_1 \le B\int_{-\pi}^{\pi} |f(t)|\log^+|f(t)|dt + C.$$

Zygmund cites, and may well have been inspired by, Riesz [1928]. Pichorides [1972] comments on the smallests values of $B$ and $C$ in Theorem (1.11) and also on the smallest value of $C$ in (1.4.i).

A great deal more has been published on Hilbert transforms and other singular integrals. This enormous and still expanding theme is not part of our modest endeavor.

## §2. The Bochner-Helson theorem.

Bochner [1939][10] achieved a far-reaching generalization of Riesz's Theorem (1.6), which can be formulated, as Helson [1959] recognized, in terms of orders on groups.

**(2.1) Definitions.** Let $X$ be an Abelian group, written additively. A subset $P$ of $X$ is called *an order in (or on) $X$* if:

(1) $P + P = \{\chi + \psi : \chi, \psi \in P\} = P$;

(2) $P \cap (-P) = P \cap \{-\chi : \chi \in P\} = \{0\}$;

(3) $P \cup (-P) = X$.

We write $\chi \leq \psi$ if $\psi - \chi \in P$. The relation $\leq$ is a complete order on $X$, $P$ being the set of nonnegative elements. With $P$ we associate the function $\text{sgn}_P$ on $X$. This function is defined by:

$$\text{sgn}_P(\chi) = \begin{cases} 1 & \text{if } \chi \in P\backslash\{0\}; \\ 0 & \text{if } \chi = 0; \\ -1 & \text{if } \chi \in -P\backslash\{0\}. \end{cases} \tag{4}$$

**(2.2) Remarks.** An Abelian group with an order is $\{0\}$ or is torsion-free and infinite. Every torsion-free Abelian group admits at least two orders, and may admit a great many. For a detailed discussion of orders and their Haar measurability if $X$ has a locally compact topology, see Hewitt and Koshi [1983]. We shall have occasion *infra* to use some results from this paper.

**(2.3) Definitions.** Let $X$ be an Abelian group with an order $P$, and let $G$ be the (compact and connected) character group of $X$. Let $\alpha$ be a complex-valued function on $X$ that vanishes except for a finite number of points in $X$. For tradition's sake, we write $\alpha_\chi$ for the value of $\alpha$ at the element $\chi$ of $X$. A function

$$t \mapsto \sum_{\chi \in \mathbf{X}} \alpha_\chi \chi(t) = f(t) \tag{1}$$

is called a *trigonometric polynomial on $G$*. The trigonometric polynomial

$$-i \sum_{\chi \in \mathbf{X}} \text{sgn}_P(\chi)\alpha_\chi \chi = \tilde{f} \tag{2}$$

is called *the conjugate trigonometric polynomial to $f$*.[11]

**(2.4) Theorem (Bochner [1939]).** *Notation is an in (2.1) and (2.3). Let $p$ be a real number such that $1 < p < \infty$. There is a constant $A_p$ such that*

$$\|\tilde{f}\|_p \leq A_p \|f\|_p, \tag{i}$$

*for all trigonometric polynomials $f$ on $G$.*

We will see in the sequel that $A_p$ is exactly the constant $M_p$ in the original Theorem (1.6) of Marcel Riesz.

**(2.5) Notation and remarks.** Let $G$ be a locally compact Abelian group (not necessarily compact) with character group $X$. Recall that $\mu$ denotes a Haar measure on $G$. For a real number $p \geq 1$, we write $\mathfrak{L}_p(G)$ for the space of all complex-valued Haar-measurable functions $g$ on $G$ for which the norm

$$(1) \qquad \|g\|_p = \left[\int_G |g(t)|^p d\mu(t)\right]^{1/p}$$

is finite. For $g \in \mathfrak{L}_1(G)$, its *Fourier transform* is the function $\hat{g}$ on $X$ such that

$$(2) \qquad \hat{g}(\chi) = \int_G g(t)\overline{\chi(t)}d\mu(t), \quad \chi \in X.$$

If $G$ is compact (and only if $G$ is compact), all trigonometric polynomials $f$ are in $\mathfrak{L}_1(G)$. Well-known orthogonality relations show that for a trigonometric polynomial $f$ as in (2.3.1), we have

$$(3) \qquad \hat{f}(\chi) = \alpha_\chi, \quad \chi \in X.$$

If $X$ has an order $P$, so that we can define the conjugate trigonometric polynomial $\tilde{f}$ as in (2.3.2), it is clear from (3) that

$$(4) \qquad \widehat{\tilde{f}}(\chi) = -i \operatorname{sgn}_P(\chi)\hat{f}(\chi), \quad \chi \in X.$$

**(2.6) Theorem (Bochner [1939], Helson [1958]).** *Let $G$ be a compact Abelian group with ordered character group $X$, as in (2.3). Let $p$ be a real number such that $1 < p < \infty$. Let $f$ be a function in $\mathfrak{L}_p(G)$. There is a function $\tilde{f}$ in $\mathfrak{L}_p(G)$ such that*

$$(i) \qquad \widehat{\tilde{f}}(\chi) = -i \operatorname{sgn}_P(\chi)\hat{f}(\chi), \quad \chi \in X.$$

*The function $\tilde{f}$ satisfies the inequality*

$$(ii) \qquad \|\tilde{f}\|_p \leq A_p\|f\|_p,$$

*where $A_p$ is as in (2.4). The mapping $f \mapsto \tilde{f}$ is thus a bounded linear transformation of $\mathfrak{L}_p(G)$ into itself with norm $A_p$.*

**(2.7) Remarks.** Theorem (2.4) is essentially though not explicitly in Bochner [1939]. Helson's contribution was in formulating Bochner's conditions in terms of orders on Abelian groups (Helson [1959]). The paper Helson [1959] is not concerned with Theorem (2.4), but rather with the $\mathfrak{L}\log^+\mathfrak{L}$ case. The paper Helson [1958] deals with the $\mathfrak{L}_p$ case, but only for some special groups.

Theorem (2.4) is by no means trivial to prove. A succinct treatment appears in Rudin [1962], Ch. 8, pp. 216–220. The proof is as in M. Riesz [1928] plus some abstract functional analysis. Theorem (2.6) follows readily from Theorem (2.4), the density of the set of trigonmetric polynomials in $\mathfrak{L}_p(G)$, and the identity (2.5.4).

The identity (2.6.1) plainly defines $\tilde{f}$ uniquely, since the Fourier transform of a function identifies the function. We may therefore call $\tilde{f}$ *THE conjugate function of* $f$. Let $G$ be a noncompact locally compact Abelian group whose dual group $X$ admits a Haar-measurable order $P$. For $1 < p \leq 2$, and $f \in L_p(G)$, suppose that there is a function $\tilde{f}$ also in $L_p(G)$ such that

$$\widehat{\tilde{f}}(x) = -i \ \mathrm{sgn}_P \hat{f}(\chi)$$

for almost all $\chi \in X$. Again we will call $\tilde{f}$ the *conjugate function of* $f$.

**(2.8) Connections with M. Riesz's theorem.** Consider the multiplicative group $\mathbb{T} = \{\exp(it) \in \mathbb{C} : -\pi < t \leq \pi\}$. Its character group consists of all functions $\exp(it) \mapsto \exp(int)$ for $n \in \mathbb{Z}$ and so is isomorphic with the additive group $\mathbb{Z}$. The group $\mathbb{Z}$ admits exactly two orders. Let $P$ be the order in $\mathbb{Z}$ that contains the integer 1. Bochner's Theorem (2.6) can be interpreted as a generalization of M. Riesz's theorem (1.6), except of course that the conjugate function $\tilde{f}$ in Theorem (1.6) is explicitly known from Privalov's Theorem (1.2), while in Theorem (2.6) it is known only as the $\mathfrak{L}_p$ limit of a certain sequence of unspecified trigonometric polynomials.

The point of view of Theorem (1.6) differs markedly from that of Theorem (2.6). The auxiliary property (1.6.ii) of the conjugate function $\tilde{f}$ becomes the *definiens* in Theorem (2.6), while the conjugate function $\tilde{f}$ becomes the *definiendum.*

**(2.9) More about the Hilbert transform.** Riesz in [1928] does not take up the Fourier transform of the Hilbert transform $Hf$ of $f$ in $\mathfrak{L}_p(\mathbb{R})$ for $1 < p \leq 2$. He could have, since Titchmarsh [1924] had already proved that the Fourier transform $\hat{f}$ exists for $f \in \mathfrak{L}_p(\mathbb{R})$ if $1 < p < 2$. For $p = 2$, this fact is of course Plancherel's classical theorem. For $1 < p \leq 2$ and $f \in \mathfrak{L}_p(\mathbb{R})$, we know that

$$\widehat{Hf}(t) = -i \ \mathrm{sgn}\ t\hat{f}(t) \tag{1}$$

for almost all $t \in \mathbb{R}$. (Here of course $\mathbb{R}$ is regarded as its own character group.) The first mention that we have found of (1) is Titchmarsh [1937], Ch. V, p. 120, formula (5.1.8). Now, the additive group $\mathbb{R}$ admits exactly two Lebesgue (= Haar) measurable orders (see Hewitt and Koshi [1983]). One of these orders is the set $\{t \in \mathbb{R} : t \geq 0\}$. Thus Theorem (1.9) provides a perfect analogue of the identity (1.6.ii). The Hilbert transform $Hf$ behaves just like the conjugate function $\tilde{f}$ for the two measurable orders of $\mathbb{R}$, for all values of $p$ for which the Fourier transform of $Hf$ is defined.

Rudin [1962], Ch. 8, p. 226, Theorem 8.7.11, has proved an analogue for (1) for half-spaces in $\mathbb{R}^k$ $(k \in \{2,3,\ldots\})$ but does not construct the Hilbert transform.

**§3. Themes of this essay.**

**(3.1)** We emphasize that Theorem (2.6) is a pure existence theorem. It offers no way of computing $\tilde{f}$ from $f$ pointwise $\mu$-almost everywhere. (The same is true of Plancherel's theorem, though pointwise methods are well known here.) Thus we have two problems.

**(3.2)** Suppose that we are given some specific discrete Abelian group $X$ containing an order $P$. As usual, we write $G$ for the (compact, connected) character group of $X$. Suppose that $f$ belongs to a space of functions on $G$ that contains all of the spaces $\mathfrak{L}_p(G)$ for $p > 1$: for example, $\mathfrak{L}_1(G)$ or $\mathfrak{L}\log^+\mathfrak{L}(G)$. Is there a way of constructing a function $\tilde{f}$ on $G$ for which (2.6.i) holds?

**(3.3)** Suppose that (3.2) has an affirmative answer in some particular case. Do analogues of Kolmogorov's Theorem (1.4) and Zygmund's Theorem (1.11) hold?

**(3.4)** Hewitt and Ritter [1983] have given reasonably complete answers to (3.2) and (3.3) for all noncyclic subgroups of the additive group $\mathbb{Q}$. Each of these groups contains exactly two orders, one of them being the nonnegative rational numbers in the group. They succeeded in constructing $\tilde{f}$ only for $f$ in $\mathfrak{L}\log^+\mathfrak{L}$, and so could not address Kolmogorov's Theorem (1.4). All else goes through without a hitch, although the constructions and computations are formidable. To our knowledge, this paper is the only published construction of $\tilde{f}$ for any compact Abelian group other than $\mathbb{T}$.

**(3.5)** Now suppose that $G$ is a noncompact locally compact Abelian group with (nondiscrete) character group $X$ and that $X$ contains a Haar-measurable order $P$. Is there an analogue $Hf$ of the Hilbert transform (1.9.i) defined at least for $f \in \mathfrak{L}_p(G)$, $1 < p \leq 2$, such that

$$\widehat{Hf}(\chi) = -i \operatorname{sgn}_P(\chi)\hat{f}(\chi) \tag{1}$$

for $\theta$-almost all $\chi \in X$?

**(3.6)** If $Hf$ exists, can it be explicitly computed?

In the remainder of this essay, we explore the questions (3.2)–(3.6).

## §4. Orders on $\mathbb{Z}^a$.

In the present section we classify the orders on $\mathbb{Z}^a$ $(a = 2, 3, \ldots)$. Our goal is to obtain analogue of Hewitt and Koshi [1983], Theorem (3.8), for orders in $\mathbb{Z}^a$ and hence obtain a complete description of orders on $\mathbb{Z}^a$ as does the theorem of Hewitt and Koshi [1983] for nondense orders in $\mathbb{R}^a$.

**(4.1) Definition.** (a) Let $X$ be a torsion-free Abelian group. Let $A$ be a subset of $X$. We say that $A$ is ***positively independent*** (*over* $\mathbb{Z}$) if, whenever $a_1, a_2, \ldots, a_\ell$ are in $A, n_1, n_2, \ldots, n_\ell$ are nonnegative integers, and

$$\sum_{j=1}^{\ell} n_j a_j = 0,$$

we have $n_1 = n_2 = \ldots = n_\ell = 0$.

(b) An order $P$ in $X$ is said to be *Archimedean* if whenever $x$ and $y$ are in $P\backslash\{0\}$, there exists a positive integer $n$ such that $nx > y$, which is to say $nx - y \in P\backslash\{0\}$.

**(4.2) Theorem.** *Let $P$ be an order in $\mathbb{Z}^a$.*

(*i*) *The order $P$ is Archimedean if and only if there is a vector $\boldsymbol{\alpha} = (\alpha_1, \alpha_2, \ldots, \alpha_a)$ in $\mathbb{R}^a$ such that the set $\{\alpha_1, \alpha_2, \ldots, \alpha_a\}$ is linearly independent over $\mathbb{Q}$ and*

$$P = \{\mathbf{x} \in \mathbb{Z}^a : \boldsymbol{\alpha} \cdot \mathbf{x} \geq 0\}.$$

(*ii*) *Let $\mathbf{v} = (v_1, v_2, \ldots, v_a)$ be a nonzero vector in $\mathbb{R}^a$ such that the set $\{v_1, v_2, \ldots, v_a\}$ is linearly dependent over $\mathbb{Q}$. Then there is a non-Archimedean order $P_1$ such that $\{\mathbf{x} \in \mathbb{Z}^a : \mathbf{v} \cdot \mathbf{x} > 0\} \subsetneqq P_1 \subsetneqq \{\mathbf{x} \in \mathbb{Z}^a : \mathbf{v} \cdot \mathbf{x} \geq 0\}$.*

**Proof. Ad (i).** Use Theorem (8.1.2.c) of Rudin [1962], p. 194.

**Ad (ii).** Let $\mathbf{x}_1, \mathbf{x}_2, \ldots, \mathbf{x}_\nu$ be any finite sequence of elements of $\mathbb{Z}^a$ such that $\mathbf{v} \cdot \mathbf{x}_\ell > 0$ for $\ell = 1, \ldots \nu$. Let $\alpha_1, \alpha_2, \ldots, \alpha_\nu$ be nonnegative integers. Suppose that $\sum_{\ell=1}^{\nu} \alpha_\ell \mathbf{x}_\ell = 0$. We then have

$$\mathbf{v} \cdot \sum_{\ell=1}^{\nu} \alpha_\ell \mathbf{x}_\ell = \sum_{\ell=1}^{\nu} \alpha_\ell \mathbf{v} \cdot \mathbf{x}_\ell = 0. \tag{1}$$

Since $\mathbf{v} \cdot \mathbf{x}_\ell > 0$ for $\ell = 1, 2, \ldots, \nu$, (1) implies that $\alpha_1 = \alpha_2 = \ldots = \alpha_\nu = 0$. That is, the set $\{\mathbf{x} \in \mathbb{Z}^a : \mathbf{v} \cdot \mathbf{x} > 0\}$ is a positively independent subset of $\mathbb{Z}^a$. Apply Lemma (2.3) and Theorem (2.5) of Hewitt and Koshi [1983] to obtain an order $P_1$ on $\mathbb{Z}^a$ such that

$$\{\mathbf{x} \in \mathbb{Z}^a : \mathbf{v} \cdot \mathbf{x} > 0\} \subsetneqq P_1. \tag{2}$$

Note that

$$\mathbb{Z}^a = \{\mathbf{x} \in \mathbb{Z}^a : \mathbf{v} \cdot \mathbf{x} > 0\} \cup \{\mathbf{x} \in \mathbb{Z}^a : \mathbf{v} \cdot \mathbf{x} = 0\} \cup \{\mathbf{x} \in \mathbb{Z}^a : \mathbf{v} \cdot \mathbf{x} < 0\}. \tag{3}$$

Since $P_1$ is an order, (2) shows that

$$\{\mathbf{x} \in \mathbb{Z}^a : \mathbf{x} \cdot \mathbf{v} < 0\} \cap P = \emptyset. \tag{4}$$

The equalities (3) and (2) show that

$$\begin{aligned} P_1 &\subset \{\mathbf{x} \in \mathbb{Z}^a : \mathbf{x} \cdot \mathbf{v} > 0\} \cup \{\mathbf{x} \in \mathbb{Z}^a : \mathbf{x} \cdot \mathbf{v} = 0\} \\ &= \{\mathbf{x} \in \mathbb{Z}^a : \mathbf{x} \cdot \mathbf{v} \geq 0\}. \end{aligned}$$

Since the set $\{\mathbf{v}_1, \mathbf{v}_2, \ldots, \mathbf{v}_a\}$ is linearly dependent over $\mathbb{Q}$, the set $\{\mathbf{x} \in \mathbb{Z}^a : \mathbf{x} \cdot \mathbf{v} = 0\}$ is a nonzero subgroup of $\mathbb{Z}^a$ which must contain nonzero elements of $P_1$. Hence $P_1$ is not a set described by $(i)$, and so $P_1$ is a non-Archimedean order. □

**(4.3) Notation and Remarks.** (a) From here on, when dealing with an infinite Abelian group $G$, the notions of ***linear independence*** of subsets of $G$, and of a ***basis of*** $G$, have the same meanings as in Hewitt and Ross [1979], pp. 441-442, (A.10).

(b) If a basis for a group $H$ contains finitely many elements, say $n$, then the positive integer $n$ is called the ***dimension of*** the group $H$.

(c) Let $G = \mathbb{Z}^a$, and consider the elements

$$\mathbf{e}_\ell = (\delta_{\ell j})_{j=1}^{a}, \quad \ell = 1, \ldots, a$$

where $\delta_{\ell j}$ is ***Kronecker's delta function.*** We easily check that the set $\{\mathbf{e}_\ell, \ell = 1, \ldots, a\}$ forms a basis for $\mathbb{Z}^a$. We call this basis the ***standard basis for*** $\mathbb{Z}^a$. The same is true for $G = \mathbb{R}^a$. The set $\{\mathbf{e}_\ell, \ell = 1, \ldots, a\}$ is also called the ***standard basis for*** $\mathbb{R}^a$.

(d) If $A_1, A_2, \ldots$ is a sequence of subsets of $\mathbb{R}^a$, we say that ***the sets*** $A_n$ ***converge to*** $A \subset \mathbb{R}^a$, and we write $\lim_{n\to\infty} A_n = A$ if $\overline{\lim} A_n = A = \underline{\lim} A_n$, where

$$\overline{\lim} A_n = \bigcap_{n=1}^{\infty} \Big(\bigcup_{p=n}^{\infty} A_p\Big)$$

and

$$\underline{\lim} A_n = \bigcup_{n=1}^{\infty} \Big(\bigcap_{p=n}^{\infty} A_p\Big).$$

We now present some technical lemmas.

**(4.4) Lemma.** *Let $b$ be a positive integer. If $\mathbf{x}_1, \mathbf{x}_2, \ldots, \mathbf{x}_\nu$ are in $\mathbb{Z}^b \setminus \{0\}$, and if $r_1, r_2, \ldots, r_\nu$ are nonzero real numbers such that the $\mathbf{R}^b$-vector $\mathbf{x} = \sum_{j=1}^{\nu} r_j \mathbf{x}_j$ has positive components, then there are integers $y_1, y_2, \ldots, y_\nu$ such that the $\mathbf{R}^b$-vector $\mathbf{y} = \sum_{j=1}^{\nu} y_j \mathbf{x}_j$ has positive integer components.*

**Proof.** For $j = 1, \ldots, \nu$, write $\mathbf{x}_j = (x_{1j}, x_{2j}, \ldots, x_{bj})$ and $\mathbf{x} = (x_1, x_2, \ldots, x_b)$, so that $x_\ell > 0$ for $\ell = 1, 2, \ldots, b$. We have

$$x_\ell = r_1 x_{\ell 1} + r_2 x_{\ell 2} + \ldots + r_\nu x_{\ell \nu},$$

$\ell = 1, 2 \ldots, b$. Since $x_\ell > 0$ for $\ell = 1, 2, \ldots, b$, not all the $x_{\ell j}$, $\ell = 1, \ldots, b$, $j = 1, \ldots, \nu$ are zero. Define

$$\varepsilon = \min_{\substack{1 \le j \le \nu \\ 1 \le \ell \le b}} \left( \frac{x_\ell}{2\nu |x_{\ell j}|}, \quad \text{for} \quad x_{\ell j} \neq 0 \right); \tag{1}$$

$\varepsilon$ is a positive real number. For each integer $j = 1, \ldots, \nu$, choose a real number $\varepsilon_j$ such that $0 \le |\varepsilon_j| < \varepsilon$ and such that $q_j = r_j - \varepsilon_j$ is a nonzero rational number.

Consider the vector $\mathbf{c} = \sum_{j=1}^{\nu} q_j \mathbf{x}_j = (c_1, c_2, \ldots, c_b)$. For $\ell = 1, \ldots, b$ we have

$$c_\ell = \sum_{j=1}^{\nu} q_j x_{\ell j} = \sum_{j=1}^{\nu} (r_j - \varepsilon_j) x_{\ell j},$$

$$c_\ell = \sum_{j=1}^{\nu} r_j x_{\ell j} - \sum_{j=1}^{\nu} \varepsilon_j x_{\ell j} = x_\ell - \sum_{j=1}^{\nu} \varepsilon_j x_{\ell j}. \tag{2}$$

For $j = 1, \ldots, \nu$ and $\ell = 1, 2, \ldots, b$, either $x_{\ell j} = 0$, so that $\varepsilon_j x_{\ell j} = 0$, or $x_{\ell j} \neq 0$. Using (2), and the choice of $\varepsilon$; we get

$$|\varepsilon_j x_{\ell j}| < \frac{x_\ell}{2\nu},$$

$$\begin{aligned} c_\ell &\ge x_\ell - \sum_{j=1}^{\nu} |\varepsilon_j| \, |x_{\ell j}| \\ &\ge x_\ell - \sum_{j=1}^{\nu} \frac{x_\ell}{2\nu} \\ &= \tfrac{1}{2} x_\ell \\ &> 0. \end{aligned}$$

The relations (3) show that $c_\ell$ is a positive real number. Also, it is clear that, for each $\ell = 1, 2, \ldots, b$, $c_\ell$ is a rational number. Thus the vector $\mathbf{c} = (c_1, c_2, \ldots, c_b)$ has positive rational components. Multiplying the vector $\mathbf{c}$ by a suitable positive integer, we obtain the desired vector $\mathbf{y}$. □

**(4.5) Lemma.** *Let $A$ be an $m \times n$ matrix with integer entries. The set*

$$(i) \qquad K = \{\mathbf{x} \in \mathbf{R}^n : A \cdot \mathbf{x} = \mathbf{0} \text{ in } \mathbf{R}^m\}$$

*is a linear subspace of $\mathbf{R}^n$ spanned by vectors with integer coefficients.*

**Proof.** Apply Theorem 18 and Theorem 17 of Birkhoff and MacLane [1965, p. 220 to obtain an $m \times n$ matrix $B$, an $m \times m$ matrix $P$ and an $n \times n$ matrix $Q$ such that

$$(1) \qquad B = PAQ,$$

$$(2) \qquad B = \begin{pmatrix} \mathbf{I}_r & \mathbf{O}_{r,n-r} \\ \mathbf{O}_{m-r,r} & \mathbf{O}_{m-r.n-r} \end{pmatrix}$$

where $\mathbf{I}_r$ is the $r \times r$ identity matrix and $\mathbf{O}_{\ell,j}$ is the $\ell \times j$ matrix of zeros.

Theorem 13 and Corollary 1 on p. 216, and the results on p. 219 of Birkhoff and MacLane [1965] show that the matrices $P$ and $Q$ are nonsingular matrices that are products of elementary matrices. It is clear that the process of reducing the matrix $A$ to the matrix $B$ can be carried out by using only elementary matrices with rational entries. Hence the matrices $P$ and $Q$ may be taken with rational entries. Let $N = \{\mathbf{x} \in \mathbf{R}^n : B \cdot \mathbf{x} = \mathbf{0}\}$. The sets $N$ and $K$ are obviously linear subspaces of $\mathbf{R}^n$. A vector $\mathbf{x}$ in $\mathbf{R}^n$ is in $K$ if and only if $A \cdot \mathbf{x} = AQQ^{-1}\mathbf{x} = \mathbf{0}$. Since $P$ is nonsingular, $\mathbf{x}$ is in $K$ if and only if $PAQQ^{-1}\mathbf{x} = \mathbf{0}$. Hence $K = Q(N)$. Since $Q$ is nonsingular, the dimensions of $N$ and $K$ are the same, and the image under $Q$ of any basis for $N$ is a basis for $K$. For $\mathbf{x} = (x_1, x_2, \ldots, x_r, \ldots, x_n)$ in $\mathbf{R}^n$, (2) shows that the vectors $\mathbf{e}_\ell, \ell = r+1, \ldots, n$ span the subspace $N$. Hence the set $\{Q(\mathbf{e}_\ell) : \ell = r+1, \ldots, n\}$ is a basis for $K$. Since $Q$ has rational entries, we multiply each vector in this basis by a suitable positive integer to obtain the desired basis for $K$. □

**(4.6) Lemma.** *Let $P$ be an order in $\mathbb{Z}^a$. Let $\mathbf{x}_1, \mathbf{x}_2, \ldots, \mathbf{x}_n$ be vectors in $P \backslash \{0\}$, not necessarily distinct. Suppose that for some nonnegative real numbers $\alpha_1, \alpha_2, \ldots, \alpha_n$ we have*

$$\sum_{\ell=1}^{n} \alpha_\ell \mathbf{x}_\ell = \mathbf{0}.$$

*It then follows that $\alpha_1 = \alpha_2 = \ldots = \alpha_n = 0$.*

**Proof.** The proof is by contradiction. If $\alpha_1 = \alpha_2 = \ldots = \alpha_n = 0$, the proof is complete. If some $\alpha_\ell$ is nonzero, we may consider only the nonzero $\alpha_\ell$'s, and so lose no generality in assuming that no $\alpha_\ell$ vanishes.

For $\ell = 1, \ldots, n$ we write $\mathbf{x}_\ell = (x_{1\ell}, x_{2\ell}, \ldots, x_{a\ell})$. Let $A$ be the $a \times m$ matrix $(x_{j\ell})_{j=1}^{a}{}_{\ell=1}^{n}$. Our hypothesis states that the vector $\boldsymbol{\alpha} = (\alpha_1, \alpha_2, \ldots, \alpha_n)$ is a nonzero vector in the null-space of the matrix $A$. Since the matrix $A$ has integer entries, Lemma (4.5) shows that there are vectors $\mathbf{q}_1, \mathbf{q}_2, \ldots, \mathbf{q}_s$ $(1 \leq s \leq m-1)$ with integer components, and real numbers $\beta_1, \beta_2, \ldots, \beta_s$ such that $\boldsymbol{\alpha} = \sum_{\ell=1}^{s} \beta_\ell \mathbf{q}_\ell$. Since the vector $\boldsymbol{\alpha}$ has positive real components, Lemma (4.4) implies that there are integers $y_1, y_2, \ldots, y_s$ such that the vector $\mathbf{t} = \sum_{\ell=1}^{s} y_\ell \mathbf{q}_\ell = (t_1, t_2, \ldots, t_n)$ has positive integer components. Thus the vector $\mathbf{t} \in \mathbb{R}^m$ is a nonzero vector in the null-space of $A$. Equivalently, we have

$$A \cdot \mathbf{t} = \sum_{j=1}^{n} t_j \mathbf{x}_j = \mathbf{0}, \tag{1}$$

where $t_j$ are positive integers for $j = 1, \ldots, n$. Now apply Theorem (2.4) of Hewitt and Koshi [1983] to see that (1) is an impossibility. Therefore the assumption that some $\alpha_\ell$ is nonzero is false. □

We now consider $P$ and $\mathbb{Z}^a$ as subsets of $\mathbb{R}^a$.

**(4.7) Lemma.** *Let $P$ be an order in $\mathbb{Z}^a$. Let*

$$P^1 = \left\{ \sum_{j=1}^{n} \alpha_j \mathbf{x}_j : \mathbf{x}_j \in P,\ \alpha_j \in \mathbb{R}^+ \right\}$$

*(i) and*

$$P^2 = \left\{ \sum_{j=1}^{m} \beta_j \mathbf{y}_j : \mathbf{y}_j \in -P,\ \beta_j \in \mathbb{R}^+ \right\},$$

*where $n$ and $m$ are arbitrary positive integers. We then have*

$$P^1 \cap P^2 = \{\mathbf{0}\}. \tag{ii}$$

**Proof.** If for some $\alpha_1, \alpha_2, \ldots, \alpha_n$, $\beta_1, \beta_2, \ldots, \beta_m$ in $\mathbb{R}^+$, $\mathbf{x}_1, \mathbf{x}_2, \ldots, \mathbf{x}_n$ in $P \backslash \{\mathbf{0}\}$ and $\mathbf{y}_1, \mathbf{y}_2, \ldots, \mathbf{y}_m$ in $-(P \backslash \{\mathbf{0}\})$ we have $\sum_{j=1}^{n} \alpha_j \mathbf{x}_j = \sum_{j=1}^{m} \beta_j \mathbf{y}_j$, then $\sum_{j=1}^{n} \alpha_j \mathbf{x}_j + \sum_{j=1}^{m} \beta_j(-\mathbf{y}_j) = 0$. Since $\mathbf{x}_1, \mathbf{x}_2, \ldots, \mathbf{x}_n$, $-\mathbf{y}_2, \ldots, -\mathbf{y}_m$ are in $P \backslash \{\mathbf{0}\}$, Lemma (4.6) shows that $\alpha_1 = \alpha_2 = \ldots = \alpha_n = \beta_1 = \ldots = \beta_m = 0$. □

**(4.8) Lemma.** *Let $P$ be an order in $\mathbb{Z}^a$, and let $P^1$ be as in (4.7.i). Then*

*(i) $P^1$ has nonvoid interior;*

*(ii) $P^1 \backslash \{\mathbf{0}\}$ is a positively independent subset of $\mathbb{R}^a$.*

**Proof.** Let $\{\mathbf{e}_1, \mathbf{e}_2, \ldots, \mathbf{e}_a\}$ be the standard basis for $\mathbb{R}^a$. Since $P$ is an order on $\mathbb{Z}^a$, just one of $\mathbf{e}_\ell$ and $-\mathbf{e}_\ell$ is in $P$ for $\ell = 1, 2, \ldots, a$. Hence there is a sequence $(\varepsilon_1, \varepsilon_2, \ldots, \varepsilon_a)$ where $\varepsilon_\ell = \pm 1$ such that $\varepsilon_\ell \mathbf{e}_\ell$ is in $P$ for $\ell = 1, 2, \ldots, a$. The nonvoid set $U = \{\mathbf{x} = (x_1, x_2, \ldots, x_a) \in \mathbb{R}^a, \frac{1}{2} < \varepsilon_\ell x_\ell < \frac{3}{2}, \ell = 1, 2, \ldots, a\}$ is open and contained in $P^1$. This proves $(i)$. We prove $(ii)$ by contradiction. Let $\mathbf{x}_1, \mathbf{x}_2, \ldots, \mathbf{x}_\ell$ be in $P^1 \backslash \{0\}$, let $n_1, n_2, \ldots, n_\ell$ be nonnegative integers, and assume that

$$(1) \qquad \sum_{j=1}^{\ell} n_j \mathbf{x}_j = 0.$$

If $n_1 = n_2 = \ldots = n_\ell = 0$, the proof is complete. If some $n_j$ is nonzero, we may consider only the nonzero $n_j$'s, and so lose no generality in supposing that no $n_j$ vanishes. For $j = 1, 2, \ldots, \ell$ we write $\mathbf{x}_j = \sum_{k=1}^{m_j} \beta_{kj} \mathbf{x}_{kj}$, for some positive real numbers $\beta_{kj}$ and nonzero elements $\mathbf{x}_{kj}$ of $P$. Then (1) implies that

$$(2) \qquad \sum_{j=1}^{\ell} \Big( n_j \sum_{k=1}^{m_j} \beta_{kj} \mathbf{x}_{kj} \Big) = \sum_{j=1}^{\ell} \sum_{k=1}^{m_j} n_j \beta_{kj} \mathbf{x}_{kj} = 0.$$

The vectors $\mathbf{x}_{kj}$ $(j = 1, 2, \ldots, \ell,\ k = 1, 2, \ldots, m_j)$ are in $P \backslash \{0\}$ and each is multiplied by the positive real number $n_j \beta_{kj}$. By Lemma (4.6), every $n_j \beta_{kj}$ is zero. Since $\beta_{kj}$ is positive, each $n_j$ vanishes. This is a contradiction. □

We now prove a principal fact about orders in $\mathbb{Z}^a$.

**(4.9) Theorem.** *Let $P$ be any order in $\mathbb{Z}^a$. There exists a nonzero linear mapping $L_1$ from $\mathbb{Z}^a$ into $\mathbb{R}$ such that*

$$(i) \qquad L_1^{-1}(]0, \infty[) \subsetneqq P \subset L_1^{-1}([0, \infty[).$$

*Furthermore, the following are equivalent:*

*$(ii)$ $P$ is Archimedean;*

*$(iii)$ $L_1^{-1}(\{0\}) = \{0\}$;*

*$(iv)$ $P \backslash \{0\} = L_1^{-1}(]0, \infty[)$;*

*$(v)$ $P = L_1^{-1}([0, \infty[)$.*

**Proof.** Let $P^1$ be as in (4.7.i). Lemma (4.8.ii) shows that the set $P^1 \backslash \{0\}$ is a positively independent subset of $\mathbb{R}^a$. Apply Lemma (2.3) and Theorem (2.5) of Hewitt and Koshi [1983] to obtain an order $P^*$ on $\mathbb{R}^a$ such that

$$(1) \qquad P \subsetneqq P^1 \subset P^*.$$

Lemma (4.8.i) and (1) show that $P^*$ has nonvoid interior. A simple argument, which we omit, shows that $P^*$ is a nondense order in the additive group $\mathbb{R}^a$. From (1) we obtain $-P \subsetneqq P^2 \subset -P^*$. Since $P^*$ is an order on $\mathbb{R}^a$, we have $(-P^*) \cap P^* = \{0\}$, and so $(-P) \cap P^* = \{0\}$. We now cite Theorem (3.8) of Hewitt and Koshi [1983]: there is a linear mapping $L$ of $\mathbb{R}^a$ onto $\mathbb{R}$ such that

$$(2) \qquad L^{-1}(]0, \infty[) \subsetneqq P^* \subsetneqq L^{-1}([0, \infty[).$$

Thus we have $L^{-1}(]-\infty,0[) \subsetneqq -P^* \subsetneqq L^{-1}(]-\infty,0])$. Let $L_1$ be the restriction of $L$ to $\mathbb{Z}^a$. Plainly $L_1$ is a linear mapping of $\mathbb{Z}^a$ into $\mathbb{R}$. It is clear that $L(\mathbf{e}_\ell)$ is different from 0 for some $\ell$ in $\{1,2,\ldots,a\}$, since $L$ is linear and not the zero mapping. Hence $L_1$ is not the zero mapping on $\mathbb{Z}^a$. We have

$$L_1^{-1}(]0,\infty[) = L^{-1}(]0,\infty[) \cap \mathbb{Z}^a \subsetneqq P^* \cap \mathbb{Z}^a, \qquad \text{and}$$

$$L_1^{-1}(]-\infty,0[) = L^{-1}(]-\infty,0[) \cap \mathbb{Z}^a \subsetneqq (-P^*) \cap \mathbb{Z}^a.$$

Also, it is clear that $P^* \cap \mathbb{Z}^a = P$, and $(-P^*) \cap \mathbb{Z}^a = -P$. Putting this together we find that

$$L_1^{-1}(]0,\infty[) \subsetneqq P \quad \text{and} \quad L_1^{-1}(]-\infty,0[) \subsetneqq -P. \tag{3}$$

Using the second inclusions in (2), we find that

$$P = P^* \cap \mathbb{Z}^a \subset L^{-1}([0,\infty[) \cap \mathbb{Z}^a = L_1^{-1}([0,\infty[). \tag{4}$$

The relations (3) and (4) establish (*i*).

Suppose that $P$ is an Archimedean order on $\mathbb{Z}^a$. Assume that $L_1^{-1}(\{0\})$ contains a nonzero vector $\mathbf{u}$. Then $n\mathbf{u}$ is in $L_1^{-1}(\{0\})$ for every positive integer $n$. Given any vector $\mathbf{v}$ in $P \backslash L_1^{-1}(\{0\})$, we have

$$\begin{aligned} L_1(n\mathbf{u} - \mathbf{v}) &= L_1(n\mathbf{u}) - L_1(\mathbf{v}) \\ &= -L_1(\mathbf{v}) \\ &< 0, \end{aligned}$$

and so $n\mathbf{u} < \mathbf{v}$ for all positive integers $n$. This contradicts (4.1.b). Therefore (iii) holds. It is now a simple matter to show that (*ii*)–(*iv*) are equivalent. □

We offer yet another technicality.

**Lemma 4.10.** *Let $F = \{\mathbf{x}_\ell\}_{\ell=1}^s$ be a nonvoid finite subset of $\mathbb{Z}^a$. Suppose that there are a vector $\boldsymbol{\alpha}_1$ in $\mathbb{R}^a$ and a positive real number $\delta$ such that*

$$\boldsymbol{\alpha}_1 \cdot \mathbf{x}_\ell > \delta > 0 \tag{$i$}$$

*for $\ell = 1,2,\ldots,s$. Then there is a vector $\boldsymbol{\alpha}$ whose coordinates are a linearly independent set over $\mathbb{Q}$ and for which*

$$\boldsymbol{\alpha} \cdot \mathbf{x}_\ell \geq \delta/2 > 0 \tag{$ii$}$$

*for $\ell = 1,2,\ldots,s$.*

**Proof.** Write $\boldsymbol{\alpha}_1 = (\alpha_{11}, \alpha_{21}, \ldots, \alpha_{a1})$, $\mathbf{x}_\ell = x_{1\ell}, x_{2\ell}, \ldots, x_{a\ell}$ for $\ell = 1,2,\ldots,s$. Let $r$ be a positive irrational number that is not in the linear span over $\mathbb{Q}$ of $\{\alpha_{11}, \alpha_{21}, \ldots, \alpha_{a1}\}$. Since

$F$ cannot contain $\mathbf{0}$, for every $\ell$ in $\{1,2,\ldots,s\}$ there is at least one integer in $\{1,2,\ldots,a\}$ such that $x_{j\ell} \neq 0$. Choose a positive integer $n$ such that

$$0 < \frac{r}{n} < \min\left\{\frac{\delta}{2a}\frac{1}{|x_{j\ell}|}\right\}, \tag{1}$$

the minimum being taken over all $j$ in $\{1,2,\ldots,a\}$ and $\ell$ in $\{1,2,\ldots,s\}$ for which $x_{j\ell} \neq 0$. Now define

$$\alpha_j = \alpha_{j1} + \frac{r}{n} \tag{2}$$

for $j = 1,2,\ldots,a$. Since $r$ is not in the linear span over $\mathbb{Q}$ of $\{\alpha_{11},\alpha_{21},\ldots,\alpha_{a1}\}$, the definition (2) shows that the set $\{\alpha_1,\alpha_2,\ldots,\alpha_a\}$ is linearly independent over $\mathbb{Q}$. For $\ell = 1,2,\ldots,s$, (2) implies that

$$\sum_{j=1}^{a}\alpha_j x_{j\ell} = \sum_{j=1}^{a}\left(\alpha_{j1} + \frac{r}{n}\right)x_{j\ell} = \sum_{j=1}^{a}\alpha_{j1}x_{j\ell} + \sum_{j=1}^{a}\frac{r}{n}x_{j\ell}. \tag{3}$$

The inequalities (1) show that

$$\begin{aligned}\left|\sum_{j=1}^{a}\frac{r}{n}x_{j\ell}\right| &\leq \sum_{j=1}^{a}\frac{r}{n}|x_{j\ell}| \\ &\leq \sum_{j=1}^{a}\frac{\delta}{2a} \\ &= \frac{\delta}{2}.\end{aligned} \tag{4}$$

The relations $(i)$, (3) and (4) imply that $\sum_{j=1}^{a}\alpha_j x_{j\ell} \geq \frac{\delta}{2}$ for all $(x_{1\ell},x_{2\ell},\ldots,x_{a\ell})$ in $F$. This proves $(ii)$. □

The next two theorems are vital in our study of orders on $\mathbb{Z}^a$.

**Theorem 4.11.** *Let $S$ be a finite subset of $\mathbb{Z}^a\backslash\{0\}$. Let $P$ be an order on $\mathbb{Z}^a$. There is an Archimedean order $P_1$ in $\mathbb{Z}^a$ such that*

$$S \cap P = S \cap P_1 \tag{$i$}$$

*and*

$$S \cap (-P) = S \cap (-P_1). \tag{$ii$}$$

**Proof.** Clearly it suffices to prove $(i)$. We may suppose that $S = S \cup (-S) = -S$. We will find a vector $\boldsymbol{\alpha} = (\alpha_1,\alpha_2,\ldots,\alpha_a)$ in $\mathbb{R}^a$ such that the set $\{\alpha_1,\alpha_2,\ldots,\alpha_a\}$ is linearly independent over $\mathbb{Q}$ and such that

$$\boldsymbol{\alpha}\cdot\mathbf{x} > 0 \tag{1}$$

for every $\mathbf{x}$ in $S \cap P$, and

$$\boldsymbol{\alpha} \cdot \mathbf{y} < 0 \tag{2}$$

for every $\mathbf{y}$ in $S \cap (-P)$. We then define the order $P_1$ as in Theorem (4.2.i): $P_1 = \{\mathbf{x} \in \mathbb{Z}^a : \boldsymbol{\alpha} \cdot \mathbf{x} \geq 0\}$. It is clear that $(i)$ holds for this $P_1$.

We proceed to construct the vector $\boldsymbol{\alpha}$. Let $\operatorname{conv}(S \cap P)$ and $\operatorname{conv}(S \cap (-P))$ denote the convex hulls of $S \cap P$ and $S \cap (-P)$, respectively, in $\mathbb{R}^a$. We have:

$$S \cap (-P) = -(S \cap P); \tag{3}$$

$$\operatorname{conv}(S \cap P) = \Big\{\mathbf{x} \in \mathbb{R}^a : \mathbf{x} = \sum_{k=1}^{n} \lambda_k \mathbf{x}_k, \mathbf{x}_k \in S \cap P, \lambda_k \geq 0, \sum_{k=1}^{n} \lambda_k = 1\Big\}; \tag{4}$$

$$\operatorname{conv}(S \cap P) = -\operatorname{conv}(S \cap (-P)). \tag{5}$$

Since $S \cap P$ does not contain $\mathbf{0}$, Lemma (4.6) shows that $\operatorname{conv}(S \cap P)$ does not contain $\mathbf{0}$. From (5) we see that $\operatorname{conv}(S \cap (-P))$ does not contain $\mathbf{0}$. Since $\operatorname{conv}(S \cap P)$ and $\operatorname{conv}(S \cap (-P))$ are subsets of $P^1$ and $P^2$, respectively, not containing $\mathbf{0}$ ($P^1$ and $P^2$ being defined as in (4.7.i)), it follows from Lemma (4.7.ii) that $\operatorname{conv}(S \cap P)$ and $\operatorname{conv}(S \cap (-P))$ are disjoint. Plainly conv$(S \cap P)$ and $\operatorname{conv}(S \cap (-P))$ are compact. We apply Theorem (34.1) of Berberian [1b p. 134 (see also p. 122, (30.1)) to find real numbers $\alpha_{11}, \alpha_{21}, \ldots, \alpha_{a1}$, such that

$$\sum_{\ell=1}^{a} \alpha_{\ell 1} x_\ell > 1 \tag{6}$$

for all $(x_1, x_2, \ldots, x_a)$ in $\operatorname{conv}(S \cap P)$ and

$$\sum_{\ell=1}^{a} \alpha_{\ell 1} y_\ell < 1 \tag{7}$$

for all $(y_1, y_2, \ldots, y_a)$ in $\operatorname{conv}(S \cap (-P))$. The inequality (6) and the equality (5) show that

$$\sum_{\ell=1}^{a} \alpha_{\ell 1} y_\ell < -1 \tag{8}$$

for all $(y_1, y_2, \ldots, y_a)$ in $\operatorname{conv}(S \cap (-P))$. We apply Lemma (4.10) with $F = S \cap P$ and $\delta = 1$ to obtain a vector $\boldsymbol{\alpha} = (\alpha_1, \alpha_2, \ldots, \alpha_a)$ such that the set $\{\alpha_1, \alpha_2, \ldots, \alpha_a\}$ is linearly independent over $\mathbb{Q}$ and

$$\boldsymbol{\alpha} \cdot \mathbf{x} > 0$$

for every $\mathbf{x}$ in $S \cap P$. The relations (5) and (3) show that

$$\boldsymbol{\alpha} \cdot \mathbf{y} < 0$$

for every $\mathbf{y}$ in $S \cap (-P)$. This establishes (1) and (2), and completes the proof. □

**(4.12) Theorem.** *Let $P$ be an order in $\mathbb{Z}^a$. There is a sequence of Archimedean orders $P_1, P_2, \ldots$ such that*

$$(i) \qquad \overline{\lim} P_n = \underline{\lim} P_n = P.$$

**Proof.** We need only show that $\overline{\lim} P_n \subset P \subset \underline{\lim} P_n$. For $n \geq 1$, let $C_n$ denote the parallelepiped in $\mathbb{Z}^a$ defined by

$$(2) \qquad C_n = \{\mathbf{x} = (x_1, x_2, \ldots, x_a) \in \mathbb{Z}^a : |x_\ell| \leq n, \ell = 1, 2, \ldots, a\}.$$

Apply Theorem (4.11) with $S = C_n$ to obtain an Archimedean order $P_n$ such that

$$(3) \qquad P_n \cap C_n = P \cap C_n.$$

For positive integers $n$ and $N$ with $n \geq N$, use (3) to infer that $P_n \cap C_N = P_N \cap C_N$. For $\mathbf{x}$ in $P$, let $N$ be a positive integer such that $\mathbf{x}$ is in $C_N \cap P$. Then $\mathbf{x}$ is in $P_n$ for all $n \geq N$. Therefore, we have $P \subset \underline{\lim} P_n$. Replacing $P_n$ by $-P_n$ and $P$ by $-P$, we find that $-P \subset \underline{\lim}(-P_n)$. Suppose that $\mathbf{y}$ does not belong to $P$. Then $\mathbf{y}$ belongs to $-P$, and so $\mathbf{y}$ is in all but finitely many $-P_n$'s. Hence $\mathbf{y}$ is in only finitely many $P_n$'s. Therefore $\mathbf{y}$ is not in $\overline{\lim} P_n$ and so $\overline{\lim} P_n \subset P$. □

## §5. Orders on locally compact Abelian groups.

Like §4, this section is of an ancillary character. It is essential for our goal of extending M. Riesz's Theorem (1.6), as adumbrated in §3. We will use the results of §4 and of Hewitt and Koshi [1983] to give an analogue of Theorem (4.11) for arbitrary measurable orders on groups. Hewitt and Koshi [1983] have identified the measurable orders on certain locally compact Abelian groups. For the reader's convenience, we quote one of their results.

**(5.1) Theorem.** *Let $B$ be a torsion-free locally compact Abelian group that is the union of its compact open subgroups. Let $P$ be any nondense order in the group $\mathbb{R}^a \times B$, where $a$ is a positive integer. There are a nonzero real-valued linear function $L$ on $\mathbb{R}^a$ and an order $P_0$ in $L^{-1}(\{0\}) \times B$ such that*

$$(i) \qquad P = (L^{-1}(]0, \infty[) \times B) \cup P_0.$$

*The mapping $L_1$ defined on $\mathbb{R}^a \times B$ by*

$$L_1(\mathbf{x}, b) = L(\mathbf{x})$$

*is a continuous real-valued homomorphism such that*

$$L_1^{-1}(]0, \infty[) \subsetneqq P \subsetneqq L_1^{-1}([0, \infty[).$$

**Proof.** See Hewitt and Koshi [1983], Theorem (3.12). □

Throughout the present section, $H$ will denote a torsion-free locally compact Abelian group, $F$ will denote a torsion-free compact Abelian group, and $a$ will denote a positive integer. Unless otherwise stated, $H$ is noncompact and $F$ is infinite.

**(5.2) Definition.** A nondense order $P$ on the group $\mathbb{R}^a \times F \times H$ is called ***strongly nondense*** if $P$ is nondense in $\mathbb{R}^a \times F \times \{y\}$ for all $y$ in $H$.

The next theorem extends Theorem (3.7) of Hewitt and Koshi [1983].

**(5.3) Theorem.** *Suppose that $P$ is a strongly nondense order on $\mathbb{R}^a \times F \times H$. Then there is a linear function $L$ mapping $\mathbb{R}^a$ onto $\mathbb{R}$ such that*

$$(i) \qquad L^{-1}(]0, \infty[) \times F \times \{0\} \subsetneqq P \cap (\mathbb{R}^a \times F \times \{0\}) \subsetneqq L^{-1}([0, \infty[) \times F \times \{0\}.$$

*For every $y$ in $H$ there is a real number $\alpha(y)$ such that*

$$(ii) \qquad L^{-1}(]-\infty, \alpha(y)[) \times F \times \{y\} \subsetneqq -P$$

*and*

$$(iii) \qquad L^{-1}(]\alpha(y), \infty[) \times F \times \{y\} \subsetneqq P.$$

*(iv) The mapping defined by $y \mapsto \alpha(y)$ is a continuous real-valued homomorphism of $H$.*

**Proof.** The set $P \cap (\mathbb{R}^a \times F \times \{0\})$ is nondense in $\mathbb{R}^a \times F$. Theorem (5.1) yields a homomorphism $L$ of $\mathbb{R}^a$ onto $\mathbb{R}$ such that $(i)$ holds. For every $y$ in $H$, there are nonvoid open subsets $U$ of $\mathbb{R}^a$ and $V$ of $F$ such that $U \times V \times \{y\} \subset P$. Thus $P$ contains the set $U \times V \times \{y\} + (\{0\} \times F \times \{0\}) \cap P$. Since the set $(\{0\} \times F \times \{0\}) \cap P$ is dense in $\{0\} \times F \times \{0\}$ (Theorem (3.2)) of Hewitt and Koshi [1983]), it follows that $U \times F \times \{y\} \subset P$. The same argument, with $P$ replaced by $-P$, shows that for every $y$ in $H$ there is a nonvoid open subset $U'$ of $\mathbb{R}^a$ such that $U' \times F \times \{y\} \subset -P$. Now suppose that $(\mathbf{x}_1, f, y)$ is in $P$ and that $L(\mathbf{x}_2) > L(\mathbf{x}_1)$. From $(i)$ we see that $(\mathbf{x}_2 - \mathbf{x}_1, 0, 0)$ is in $P$, and so $(\mathbf{x}_1, f, y) + (\mathbf{x}_2 - \mathbf{x}_1, 0, 0) = (\mathbf{x}_2, f, y)$ is in $P$. Similarly, if $(\mathbf{x}_1, f, y)$ is in $-P$ and $L(\mathbf{x}_2) < L(\mathbf{x}_1)$ then $(\mathbf{x}_2, f, y)$ is in $-P$. From this we infer that the inequalities

$$-\infty < \sup\{L(\mathbf{x}) : \mathbf{x} \text{ in } \mathbb{R}^a, \{\mathbf{x}\} \times F \times \{y\} \subset -P\}$$

$$\leq \inf\{L(\mathbf{x}) : \mathbf{x} \text{ in } \mathbb{R}^a, \{\mathbf{x}\} \times F \times \{y\} \subset P\} < \infty \tag{1}$$

hold for every $y$ in $H$. We claim that the second inequality is in fact an equality. Assume the contrary: there is an interval $]a, b[$ with the property that, whenever $\mathbf{x}$ in $\mathbb{R}^a$ is such that $L(\mathbf{x})$ is in $]a, b[$, then $\{\mathbf{x}\} \times F \times \{y\} \cap P \neq \emptyset$ and $\{\mathbf{x}\} \times F \times \{y\} \cap (-P) \neq \emptyset$. It is clear then that if $\mathbf{x}$ in $\mathbb{R}^a$ is such that $L(\mathbf{x})$ is in $]-b, -a[$, then $\{\mathbf{x}\} \times F \times \{-y\} \cap P \neq \emptyset$ and $\{\mathbf{x}\} \times F \times \{-y\} \cap (-P) \neq \emptyset$. Consider the open nonvoid neighborhood $V$ of $\mathbf{0}$ in $\mathbb{R}^a$, defined by $V = L^{-1}(]a, b[) + L^{-1}(]-b, -a[) = L^{-1}(]a, b[) - L^{-1}(]a, b[)$. Let $\mathbf{x}$ be any element of $V$. Write $\mathbf{x} = \mathbf{x}_1 + \mathbf{x}_2$ where $L(\mathbf{x}_1)$ is in $]a, b[$ and $L(\mathbf{x}_2)$ is in $]-b, -a[$. Let $f_1$, $f_2$, $f_3$, and $f_4$ be elements of $F$ such that $(\mathbf{x}_1, f_1, y)$ and $(\mathbf{x}_2, f_2, -y)$ are in $P$ and $(\mathbf{x}_1, f_3, y)$ and $(\mathbf{x}_2, f_4, -y)$ are in $-P$. It follows that $(\mathbf{x}, f_1 + f_2, 0)$ is in $P$ and $(\mathbf{x}, f_3 + f_4, 0)$ is in $-P$. From $(i)$ it follows that $L(\mathbf{x}) = 0$. Plainly this is impossible since it implies that $L$ is identically zero. Thus the second inequality in (1) is an equality.

For every $y$ in $H$, let $\alpha(y)$ be the number defined by either the sup or the inf in (1). It is clear from the definition of $\alpha(y)$ that $(ii)$ and $(iii)$ hold. We now show that $\alpha$ is a continuous homomorphism. Let $y$ and $y'$ be in $H$ and let $\mathbf{x} \in \mathbb{R}^a$ be such that $L(\mathbf{x}) > \alpha(y) + \alpha(y')$. Let $\mathbf{x}'$ in $\mathbb{R}^a$ be such that

$$L(\mathbf{x}) - \alpha(y') > L(\mathbf{x}') > \alpha(y), \tag{2}$$

and let $\mathbf{z} = \mathbf{x} - \mathbf{x}'$. From (2) we see that $L(\mathbf{z}) > \alpha(y')$. Consequently we have $\{\mathbf{x}'\} \times F \times \{y\} \subset P$ and $\{\mathbf{z}\} \times F \times \{y'\} \subset P$. These inclusions imply that $\{\mathbf{x}' + \mathbf{z}\} \times F \times \{y + y'\} \subset P$ or $\{\mathbf{x}\} \times F \times \{y + y'\} \subset P$. We have

$$\begin{aligned} \alpha(y + y') &= \inf\{L(\mathbf{x}) : \{\mathbf{x}\} \times F \times \{y + y'\} \subset P\} \\ &\leq \inf\{L(\mathbf{x}) : L(\mathbf{x}) > \alpha(y) + \alpha(y')\} \\ &= \alpha(y) + \alpha(y'). \end{aligned}$$

Similarly, if $\mathbf{x}$ in $\mathbb{R}^a$ is such that $L(\mathbf{x}) < \alpha(y) + \alpha(y')$, write $\mathbf{x} = \mathbf{x}' + \mathbf{z}$, where $L(\mathbf{x}') < \alpha(y)$ and $L(\mathbf{z}) < \alpha(y')$. From $(ii)$ it follows that $\{\mathbf{x}'\} \times F \times \{y\} \subset -P$, and $\{\mathbf{z}\} \times F \times \{y\} \subset -P$, so that $\{\mathbf{x}\} \times F \times \{y + y'\} \subset -P$. We have

$$\begin{aligned}\alpha(y + y') &= \sup\{L(\mathbf{x}) : \{\mathbf{x}\} \times F \times \{y\} \subset -P\} \\ &\geq \sup\{L(\mathbf{x}) : L(\mathbf{x}) < \alpha(y) + \alpha(y')\} \\ &= \alpha(y) + \alpha(y').\end{aligned}$$

That is, $\alpha$ is a homomorphism of $H$ into $\mathbb{R}^a$. Finally, to establish the continuity of $\alpha$ it suffices to show that $\alpha$ is bounded on a neighborhood of 0 in $H$. Since $P$ is nondense, there are nonvoid open subsets $U, V$, and $W$ of $\mathbb{R}^a$, $F$ and $H$ respectively, such that $U \times F \times W$ is contained in $P$ and such that $U$ is bounded in $\mathbb{R}^a$. Let $u$ be a real number such that $L(\mathbf{x}) < u$ for all $\mathbf{x}$ in $U$. From the definition of $\alpha$ it follows that $\alpha(w) \leq u$ for all $w$ in $W$. Choose any $w_0$ in $W$. The homomorphism $\alpha$ is bounded above on the neighborhood $W - w_0$ of 0, and is bounded below on the neighborhood $-W + w_0$ of 0. Thus $\alpha$ is bounded on the neighborhood $(W - w_0) \cap (-W + w_0)$ of 0. □

We can now classify an important family of orders.

**(5.4) Theorem.** *Notation is as in (5.3). The mapping $\tau$ defined on $\mathbb{R}^a \times F \times H$ by*

$(i)$
$$(\mathbf{x}, f, y) \mapsto L(\mathbf{x}) - \alpha(y)$$

*is a continuous homomorphism of $\mathbb{R}^a \times F \times H$ onto $\mathbb{R}$ such that*

$(ii)$
$$\tau^{-1}(]0, \infty[) \subsetneqq P \subsetneqq \tau^{-1}([0, \infty[)$$

*and*

$(iii)$
$$\tau^{-1}(]-\infty, 0[) \subsetneqq -P \subsetneqq \tau^{-1}(]-\infty, 0]).$$

$(iv)$ *If $H$ is $\sigma$-compact, $\tau^{-1}(\{0\})$ has Haar measure zero and $P$ is Haar-measurable.*

**Proof.** Parts $(i)$ – $(iii)$ are immediate consequences of Theorem (5.3). To prove $(iv)$ we use the fact that every locally null subset of a $\sigma$-compact group has Haar measure zero. Assume that $\tau^{-1}(\{0\})$ is not locally null, and that $K \subset \tau^{-1}(\{0\})$ is compact with positive Haar measure. Then $K - K \subset \tau^{-1}(\{0\})$ contains an open neighborhood of 0. This is impossible since $\tau$ is not identically zero. □

The orders describes in Theorem (5.4) are analogues of the Archimedean orders on $\mathbb{Z}^a$, which are identified in Theorem (4.9). We now embark on the study of Haar-measurable orders.

**(5.5) Lemma.** *Suppose that $P$ is a Haar-measurable order on $\mathbb{R}^a \times F \times H$ where $H$ is a discrete torsion-free Abelian group. The set $P \cap (\mathbb{R}^a \times \{(0,0)\})$ is an order on $\mathbb{R}^a \times \{(0,0)\}$ that is measurable with respect to Haar measure on $\mathbb{R}^a$.*

**Proof.** By Theorem (3.8) of Hewitt and Koshi [1983] it is enough to show that $P \cap (\mathbb{R}^a \times \{(0,0)\})$ is nondense in $\mathbb{R}^a \times \{(0,0)\}$. Assume the contrary: $P \cap (\mathbb{R}^a \times (0,0)\})$ is dense in $\mathbb{R}^a \times \{(0,0)\}$. Since $F$ is compact, Theorem (3.2) of Hewitt and Koshi [1983] shows that the set $P \cap (\{0\} \times F \times \{0\})$ is dense in $\{0\} \times F \times \{0\}$. Hence the set $P \cap (\mathbb{R}^a \times F \times \{0\})$, which contains $P \cap (\mathbb{R}^a \times \{(0,0)\}) + P \cap (\{0\} \times F \times \{0\})$, is dense in $(\mathbb{R}^a \times F \times \{0\})$, and so is non Haar-measurable with respect to Haar measure on $\mathbb{R}^a \times F \times \{0\}$. Plainly this contradicts the fact that $P$ is a Haar-measurable subset of $\mathbb{R}^a \times F \times H$. □

**(5.6) Remark.** Lemma (5.5) need not hold for nondiscrete $H$. Consider the group $\mathbb{R}^2 = \mathbb{R} \times \mathbb{R}$. Let $P_0$ be any dense non-Lebesgue measurable order in $\mathbb{R}$ (Hewitt and Koshi [1983], Theorem (3.3) and Remarks (3.4,a,b)). The set $P = \{\mathbf{x} = (x_1, x_2) \text{ in } \mathbb{R}^2;\ x_1 > 0\} \cup \{\mathbf{x} = (0, x_2) \text{ in } \mathbb{R}^2,\ x_2 \text{ in } P_0\}$ is a Lebesgue-measurable order on $\mathbb{R}^2$.

The next lemma, while simple, will be very useful.

**(5.7) Lemma.** *Notation is as in (5.5). For every $y$ in $H$, exactly one of the following holds:*

(*i*) $\mathbb{R}^a \times F \times \{y\} \subset P$;

(*ii*) $(\mathbb{R}^a \times F \times \{y\}) \cap P = \emptyset$;

(*iii*) $P \cap (\mathbb{R}^a \times F \times \{y\})$ *is nondense in* $\mathbb{R}^a \times F \times \{y\}$.

**Proof.** Since $H$ is discrete, the set $P \cap (\mathbb{R}^a \times F \times \{0\})$ is measurable with respect to Haar measure on $\mathbb{R}^a \times F \times \{0\}$. It follows from Theorem (3.1) of Hewitt and Koshi [1983] that there are open nonvoid subsets $U$ and $V$ of $\mathbb{R}^a \times F$ such that $U \times \{0\} \subset P$ and $V \times \{0\} \subset -P$. Suppose that neither (*i*) nor (*ii*) holds, and that $(\mathbf{x}_1, f_1, y)$ is in $P$ and $(\mathbf{x}_2, f_2, y)$ is in $-P$. Then we have $(\mathbf{x}_1, f_1, y) + (U \times \{0\}) \subset P$ and $(\mathbf{x}_2, f_2, y) + (V \times \{0\}) \subset -P$. That is, the set $P$ is nondense in $\mathbb{R}^a \times F \times \{y\}$. □

We next construct a useful family of orders.

**(5.8) Theorem.** *Let $H$ be a countable discrete torsion-free Abelian group. Let $L$ be a nonzero continuous real-valued homomorphism on $\mathbb{R}^a$ and let $\alpha$ be a real-valued homomorphism on $H$ ($\alpha$ may be identically 0). Consider the mapping $\tau$ defined on $\mathbb{R}^a \times F \times H$ by*

$$(i) \qquad \tau(\mathbf{x}, f, y) = L(\mathbf{x}) - \alpha(y)$$

*and suppose that $F \neq \{0\}$ or that $a > 1$. Then $\tau$ is a continuous real-valued homomorphism, and there is a nondense order $P$ on $\mathbb{R}^a \times F \times H$ such that*

$$(ii) \qquad \tau^{-1}(]0, \infty[) \subsetneqq P \subsetneqq \tau^{-1}([0, \infty[).$$

*The set $\tau^{-1}(\{0\})$ has Haar measure zero, and the order $P$ is a strongly nondense Haar-measurable subset of $\mathbb{R}^a \times F \times H$.*

**Proof.** The first assertion is obvious. Next consider the set

$$\tau^{-1}(]0,\infty[) \cup (L^{-1}(]0,\infty[) \times \{(0,0)\} = \tau^{-1}(]0,\infty[). \tag{1}$$

The set (1) is nonvoid and positively independent. By Remark (2.6) of Hewitt and Koshi [1983], there is an order $P$ on $\mathbb{R}^a \times F \times H$ that contains (1). We have $\tau^{-1}(]0,\infty[) \subset P, \tau^{-1}(]-\infty,0[) \subset -P$, and so $P \subset \tau^{-1}([0,\infty[)$. It is easy to show that the inclusions in (*ii*) are strict. This proves (*ii*). That $\tau^{-1}(\{0\})$ is Haar-measurable follows from (5.4.*iv*). To prove the last assertion, we note that for every $y$ in $H$ there are $\mathbf{x}_1$ and $\mathbf{x}_2$ in $\mathbb{R}^a$ such that $\tau(\mathbf{x}_1, f, y) > 0$ and $\tau(\mathbf{x}_2, f, y) < 0$ for all $f$ in $F$. Thus the set $\mathbb{R}^a \times F \times \{y\}$ contains elements of $P$ and elements of $-P$. According to Lemma (5.7), the set $P \cap (\mathbb{R}^a \times F \times \{y\})$ is nondense in $\mathbb{R}^a \times F \times \{y\}$. That is, $P$ is strongly nondense. □

**(5.9) Remarks.** (a) One easily checks that the set $H'$ of all $y$ in $H$ such that $\mathbb{R}^a \times F \times \{y\}$ contains elements of $P$ and elements of $-P$ is a subgroup of $H$. It is also easy to see that if $ky$ is in $H'$ for some positive integer $k$, then $y$ in in $H'$.

(b) The case $H = \mathbb{Z}^b$, where $b$ is a positive integer, is of particular interest. In this case, (a) and Theorem (A.6) of Hewitt and Ross [1979] pp. 450–451, show that there is a basis $\mathbf{e}'_1, \mathbf{e}'_2, \ldots, \mathbf{e}'_b$ of $\mathbb{Z}^b$ such that $\mathbf{e}'_1, \mathbf{e}'_2, \ldots, \mathbf{e}'_h$ is a basis for $H', 1 \le h \le b$, whenever $H'$ is not $\{0\}$. Hence if $H' \neq \{0\}$, every element $\mathbf{y}$ in $\mathbb{Z}^b$ can be written uniquely

$$\mathbf{y} = \mathbf{y}_1 + \mathbf{y}_2$$

where $\mathbf{y}_1$ is a linear combination of $\mathbf{e}'_1, \mathbf{e}'_2, \ldots, \mathbf{e}'_h$ and hence is in $H'$.

In the remainder of this section, we take $H = \mathbb{Z}^b$, where $b$ is a positive integer. We continue with our study of Haar-measurable orders.

**(5.10) Lemma.** *Let $P$ be any order on $\mathbb{Z}^b$, and let $S$ be any finite subset of $\mathbb{Z}^b \backslash \{0\}$. Then there is a real-valued homomorphism $\alpha$ on $\mathbb{Z}^b$ such that $\alpha(\mathbf{y}) > 0$ for all $\mathbf{y}$ in $S \cap P$ and $\alpha(\mathbf{y}) < 0$ for all $\mathbf{y}$ in $S \cap (-P)$.* □

**Proof.** See the proof of Theorem (4.11). □

**(5.11) Lemma.** *Let $P$ be a Haar-measurable order on $\mathbb{R}^a \times F \times \mathbb{Z}^b$. Notation is as in (5.9). Let $J$ be a finite symmetric subset of $\mathbb{Z}^b \backslash H'$. There is a real-valued homomorphism $\alpha_1$ on $\mathbb{Z}^b$ such that $\alpha_1(\mathbf{y}_2) > 0$ for all $\mathbf{y}$ in $J$ for which $\mathbb{R}^a \times F \times \{\mathbf{y}\} \subset P$ and $\alpha_1(\mathbf{y}_2) < 0$ for all $\mathbf{y}$ in $J$ for which $\mathbb{R}^a \times F \times \{\mathbf{y}\} \subset -P$.*

**Proof.** If $J$ is void there is nothing to prove. If $H' = \{0\}$, we may appeal to Lemma (5.10). So suppose that $J$ is nonvoid and that $H' \neq \{0\}$. Denote by $P_1$ the order $P \cap (\{(0,0)\} \times \mathbb{Z}^b)$

on $\{(0,0)\}\times\mathbb{Z}^b$. Clearly we have $(\{(0,0)\}\times J)\cap P_1 = \{(0,0)\}\times\{\mathbf{y}\in J : \mathbb{R}^a\times F\times\{\mathbf{y}\}\subset P\}$, and $(\{(0,0)\}\times J)\cap(-P_1) = \{(0,0)\}\times\{\mathbf{y}\in J : \mathbb{R}^a\times F\times\{\mathbf{y}\}\subset -P\}$. Write each $\mathbf{y}$ in $J$ as in (5.9,b), and let

$$J_2 = \{\mathbf{y}_2 : \{(0,0)\}\times\{\mathbf{y}\}\subset P_1 \quad \text{and} \quad \mathbf{y}\in J\}.$$

We claim that $\{(0,0)\}\times J_2\subset P_1$. Assume the contrary. Then for some $(0,0,\mathbf{y})$ in $(\{(0,0)\}\times J)\cap P_1$, the corresponding $(0,0,\mathbf{y}_2)$ is not in $P_1$. Hence $(0,0,\mathbf{y}_2)$ must be in $-P_1$, and so $(0,0,\mathbf{y}_2)$ is in $-P$. Write $\mathbb{R}^a\times F\times\{\mathbf{y}\} = \mathbb{R}^a\times F\times\{\mathbf{y}_1\} + (0,0,\mathbf{y}_2)$. It follows that $\mathbb{R}^a\times F\times\{\mathbf{y}\}$ contains elements in $-P$. This is impossible since $\mathbf{y}$ is in $(\{(0,0)\}\times J)\cap P_1$. This establishes our claim. The lemma follows now by applying Lemma (5.10) to the set $S = J_2\cup(-J_2)$ and the order $P_1$ on $\mathbb{Z}^b$. □

We now give a set-theoretic technicality.

**(5.12) Lemma.** *Let $X$ be a locally compact torsion-free Abelian group. Suppose that $P$ and $P^*$ are Haar-measurable orders on $X$ and that $K$ is a symmetric subset of $X$. Denote by $\mu_X$ Haar measure on $X$. The following are equivalent:*

(*i*) $$\mu_X((K\cap P)\backslash(K\cap P^*)) = 0;$$

(*ii*) $$\mu_X((K\cap P^*)\backslash(K\cap P)) = 0;$$

(*iii*) $$\mu_X((K\cap(-P^*))\backslash(K\cap(-P))) = 0;$$

(*iv*) $$\mu_X((K\cap(-P))\backslash(K\cap(-P^*))) = 0.$$

**Proof.** It is easy to see we need only show that (*i*) implies (*ii*). Suppose that (*i*), and hence (*iv*) hold, and assume that (*ii*) fails. Then there is a subset $M$ of $X$ such that $M\subset K\cap P^*$, $M\cap K\cap P = \emptyset$, $\mu_X(M) > 0$. It follows that $M\cap P = \emptyset$, and so $M\subset -P$ and $M\cap(-P^*) = \emptyset$. Thus $M$ is contained in $(K\cap(-P))\backslash(K\cap(-P^*))$. This contradicts (*iv*), because $\mu_X(M) > 0$. □

The following theorem is fundamental. Its proof is long but not conceptually difficult.

**(5.13) Theorem.** *Let $P$ be a Haar-measurable order on the group $\mathbb{R}^a\times F\times\mathbb{Z}^b$, where $a$ and $b$ are positive integers and $F$ is a compact torsion-free Abelian group. Let $K$ be a compact subset of $\mathbb{R}^a\times F\times\mathbb{Z}^b$. There is a strongly nondense Haar-measurable order $P^*$ on $\mathbb{R}^a\times F\times\mathbb{Z}^b$ such that*

(*i*) $$\mu\Big((K\cap P^*)\backslash(K\cap P)\Big) = 0$$

*and*

$$(ii) \qquad \mu\Big((K \cap (-P^*))\backslash(K \cap (-P)\Big) = 0,$$

*where $\mu$ denotes a Haar measure on $\mathbb{R}^a \times F \times \mathbb{Z}^b$.*

**Proof.** Clearly if the conclusion of the theorem holds with some order $P^*$ and some set $K$, then it holds with the same order $P^*$ and all compact subsets of $K$. Thus we may suppose that $K$ is a compact symmetric subset of $\mathbb{R}^a \times F \times \mathbb{Z}^b$. We will construct the order $P^*$ by applying Theorem (5.8). Thus we need to define appropriate homomorphisms $L$ on $\mathbb{R}^a$ and $\alpha$ on $\mathbb{Z}^b$. By Lemma (5.5) the set $P \cap (\mathbb{R}^a \times (0, \mathbf{0}))$ is a Lebesgue-measurable order on $\mathbb{R}^a$. By Theorem (3.8) of Hewitt and Koshi [1983], there is a real-valued homomorphism on $\mathbb{R}^a$ such that

$$L^{-1}(]0, \infty[) \times \{(0, \mathbf{0})\} \subsetneqq P \cap (\mathbb{R}^a \times \{(0, \mathbf{0})\}) \subsetneqq L^{-1}([0, \infty[) \times \{(0, \mathbf{0})\}.$$

Write $K = (\bigcup_{j=1}^{c}(C_j \times \{\mathbf{y}_j\}) \cup (\bigcup_{\ell=1}^{d}(D_\ell \times \{\mathbf{z}_\ell\}))$, where $C_j$ and $D_\ell$ are compact subsets of $\mathbb{R}^a \times F$, the $\mathbf{y}_j$'s are in $H'$, and the $\mathbf{z}_\ell$'s are in $\mathbb{Z}^b \backslash H'$, where $H'$ is as in (5.9,b). We will use the decomposition $\mathbf{x} = \mathbf{x}_1 + \mathbf{x}_2$ for the element $\mathbf{x}$ in $\mathbb{Z}^b$, as we did in (5.9,b). By Lemma (5.7), the order $P \cap (\mathbb{R}^a \times F \times H')$ is strongly nondense in $\mathbb{R}^a \times F \times H'$. By Theorem (5.3) there is a real-valued homomorphism $\alpha_1$ on $H'$ such that $L^{-1}(]-\infty, \alpha_1(\mathbf{y})[) \times F \times \{\mathbf{y}\} \subsetneqq (-P) \cap (\mathbb{R}^a \times F \times H')$, and $L^{-1}(]\alpha_1(y), \infty[) \times F \times \{\mathbf{y}\} \subsetneqq P \cap (\mathbb{R}^a \times F \times H')$ for each $\mathbf{y} \in H'$. Also, from (5.4), the homomorphism $\tau_1$ defined on $\mathbb{R}^a \times F \times H'$ by $\tau_1(\mathbf{x}, f, \mathbf{y}) = L(\mathbf{x}) - \alpha_1(\mathbf{y})$ is continuous and has the following properties:

$$(1) \qquad \tau_1^{-1}(]0, \infty[) \subsetneqq P \cap (\mathbb{R}^a \times F \times H') \subsetneqq \tau_1^{-1}([0, \infty[);$$

and $\mu'(\tau_1^{-1}(\{0\})) = 0$, where $\mu'$ denotes Haar measure on $\mathbb{R}^a \times F \times H'$.

It is clear that this set also has $\mu$ measure zero.

By Lemma (5.11) there is a real-valued homomorphism $\alpha_2$ of $\mathbb{Z}^b$ such that for $\ell = 1, \ldots, d$, we have

$$(2) \qquad \alpha_2((\mathbf{z}_\ell)_2) > 0$$

if $\mathbb{R}^a \times F \times \{\mathbf{z}_\ell\}$ is contained in $P$, and

$$(3) \qquad \alpha_2((\mathbf{z}_\ell)_2) < 0$$

if $\mathbb{R}^a \times F \times \{\mathbf{z}_\ell\} \subset (-P)$.

Let

$$\kappa = \max\{|L(\mathbf{x})| + |\alpha_1(\mathbf{y}_1)| : (\mathbf{x}, f, \mathbf{y}) \in K\} + 1$$

and

$$\nu = \min\{|\alpha_2((\mathbf{z}_\ell)_2)|, \quad \ell = 1, \ldots, d\}.$$

We distinguish two cases.

**Case A.** $\nu = 0$. By (2) and (3), the set $K$ is contained in $\mathbb{R}^a \times F \times H'$. Extend the homomorphism $\alpha_1$ to a real-valued homomorphism $\alpha_1^*$ on all of $\mathbb{Z}^b$. Let $\tau_1^*$ be the homomorphism defined on $\mathbb{R}^a \times F \times \mathbb{Z}^b$ by

$$\tau_1^*(\mathbf{x}, f, \mathbf{y}) = L(\mathbf{x}) - \alpha_1^*(\mathbf{y}).$$

Notice that $\tau_1^*$ agrees with $\tau_1$ on the subgroup $\mathbb{R}^a \times F \times H'$. Thus we have

$$(4) \qquad \begin{aligned} (\mathbb{R}^a \times F \times H') \cap (\tau_1^*)^{-1}(]0,\infty[) &= (\mathbb{R}^a \times F \times H') \cap \tau_1^{-1}(]0,\infty[), \\ (\mathbb{R}^a \times F \times H') \cap (\tau_1^*)^{-1}([0,\infty[) &= (\mathbb{R}^a \times F \times H') \cap \tau_1^{-1}([0,\infty[), \end{aligned}$$

and

$$(\mathbb{R}^a \times F \times H') \cap (\tau_1^*)^{-1}(\{0\}) = (\mathbb{R}^a \times F \times H') \cap \tau_1^{-1}(\{0\}).$$

Also, the sets $(\tau_1^*)^{-1}(\{0\})$ and $\tau_1^{-1}(\{0\})$ have Haar measure zero. (See the proof of Theorem (5.4).) We now apply Theorem (5.8) to obtain a strongly nondense order $P^*$ on $\mathbb{R}^a \times F \times \mathbb{Z}^b$ such that

$$(5) \qquad (\tau_1^*)^{-1}(]0,\infty[) \subsetneqq P^* \subsetneqq (\tau_1^*)^{-1}([0,\infty[).$$

From (4) and (5) we infer that

$$(6) \qquad \begin{aligned} P^* \cap (\mathbb{R}^a \times F \times H') \cap (\tau_1^*)^{-1}(]0,\infty[) &= (\mathbb{R}^a \times F \times H') \cap (\tau_1^*)^{-1}(]0,\infty[) \\ &= (\mathbb{R}^a \times F \times H') \cap \tau_1^{-1}(]0,\infty[). \end{aligned}$$

From (1), we see that the set $(\mathbb{R}^a \times F \times H') \cap \tau_1^{-1}(]0,\infty[)$ differs from $(\mathbb{R}^a \times F \times H') \cap P$ by a set of $\mu$ measure zero. A simple calculation, which we omit, shows that

$$\mu((P \cap (\mathbb{R}^a \times F \times H')) \backslash (P^* \cap (\mathbb{R}^a \times F \times H'))) = 0.$$

Since $K$ is contained in $\mathbb{R}^a \times F \times H'$, we get

$$\mu((P \cap K) \backslash (P^* \cap K)) = 0.$$

Thus we have found the desired order $P^*$ in this case.

**Case B.** $\nu > 0$. Define the continuous real-valued homomorphism $\tau$ on $\mathbb{R}^a \times F \times \mathbb{Z}^b$ by

$$(7) \qquad \tau(\mathbf{x}, f, \mathbf{y}) = L(\mathbf{x}) - \alpha_1(\mathbf{y}_1) + \frac{\kappa}{\nu}\alpha_2(\mathbf{y}_2).$$

Define the homomorphism $\alpha$ on $\mathbb{Z}^b$ by

$$\alpha(\mathbf{y}) = \alpha_1(\mathbf{y}_1) - \frac{\kappa}{\nu}\alpha_2(\mathbf{y}_2),$$

so that

$$\tau(\mathbf{x}, f, \mathbf{y}) = L(\mathbf{x}) - \alpha(\mathbf{y}).$$

According to Theorem (5.8), there is a strongly nondense Haar measurable order $P^*$ on $\mathbb{R}^a \times F \times \mathbb{Z}^b$ with the property that

$$\tau^{-1}(]0,\infty[) \subsetneqq P^* \subsetneqq \tau^{-1}([0,\infty[). \tag{8}$$

It remains to show that $P^*$ satisfies the equalities $(i)$ and $(ii)$. Let $(\mathbf{x}, f, \mathbf{y})$ be in $K \cap P$. We distinguish three cases.

**Case B.I.** The element $\mathbf{y}$ is in $H'$ $(\mathbf{y}_2 = \mathbf{0})$ and $L(\mathbf{x}) = \alpha_1(\mathbf{y})$. Then $(\mathbf{x}, f, \mathbf{y})$ belongs to $\tau^{-1}(\{0\})$. This set has Haar measure zero.

**Case B.II.** The element $\mathbf{y}$ is in $H'$ and $L(\mathbf{x})$ is different from $\alpha_1(\mathbf{y})$. Clearly $(\mathbf{x}, f, \mathbf{y})$ belongs to $P \cap (\mathbb{R}^a \times F \times H')$. The relations (1) show that

$$L(\mathbf{x}) - \alpha_1(\mathbf{y}) > 0.$$

Since the corresponding element $\mathbf{y}_2$ to $\mathbf{y}$ is $\mathbf{0}$, it follows that $\alpha(\mathbf{y}) = \alpha_1(\mathbf{y})$ and so $\tau(\mathbf{x}, f, \mathbf{y}) > 0$. Thus $(\mathbf{x}, f, \mathbf{y})$ belongs to $\tau^{-1}(]0,\infty[)$. From (8) it follows that $(\mathbf{x}, f, \mathbf{y})$ belong to $P^*$.

**Case B.III.** The element $\mathbf{y}$ is not in $H'$. Then necessarily we have

$$\mathbb{R}^a \times F \times \{\mathbf{y}\} \subsetneqq P.$$

From (2) we get

$$\alpha_2(\mathbf{y}_2) > 0.$$

From the definition of $\nu$, it follows that

$$\alpha_2(\mathbf{y}_2) \geq \nu > 0$$

and so

$$\begin{aligned}\tau(\mathbf{x}, f, \mathbf{y}) &= L(\mathbf{x}) - \alpha_1(\mathbf{y}_1) + \frac{\kappa}{\mu}\alpha_2(\mathbf{y}_2)\\ &\geq L(\mathbf{x}) - \alpha_1(\mathbf{y}_1) + \kappa\\ &\geq 1.\end{aligned}$$

Hence $(\mathbf{x}, f, \mathbf{y})$ belongs to $\tau^{-1}(]0,\infty[)$, and so it belongs to $P^*$. From the above cases, we conclude that

$$\mu((K \cap P)\backslash(K \cap P^*)) = 0.$$

By Lemma (5.12) the proof is complete in this case.

We have now dealt with all cases. □

The next and last theorem of this section identifies a remarkable property of all Haar-measurable orders.

**(5.14) Theorem.** *Let $P$ be a Haar-measurable order on a locally compact torsion-free Abelian group $X$. Let $K$ be any compact subset of $X$. There are a continuous real-valued homomorphism $\psi$ on $X$ and a subset $N$ of $X$ of Haar measure zero such that:*

$$(i) \qquad \psi(x) > 0$$

*for all* $x \in (K \cap P) \backslash (N \cup \{0\})$;

$$(ii) \qquad \psi(x) < 0$$

*for all* $x \in (K \cap -P) \backslash (N \cup \{0\})$.

**Proof.** If $K$ is void there is nothing to prove. Also it is clear that if the conclusion of the theorem holds for some set $K$ and a homomorphism $\psi$, then it holds with $K$ replaced by any compact subset of $K$ and the same homomorphism $\psi$. So we may suppose that $K$ is a symmetric subset of $X$ containing a neighborhood of the identity. Let $B$ be the subgroup of $X$ generated by $K$. Then $B$ is compactly generated [for a discussion of compactly generated groups, see Hewitt and Ross [1979], p. 35, Definition (5.12)]. A structure theorem for locally compact Abelian groups (Hewitt and Ross [1979], p. 95, Theorem (9.14)) asserts that $B$ is topologically isomorphic with $\mathbb{R}^a \times F \times \mathbb{Z}^b$, where $a$ and $b$ are nonnegative integers and $F$ is a compact (obviously torsion-free) Abelian group. Since $X$ admits the measurable order $P$, either it is discrete in which case $B \cong \mathbb{Z}^b$, or $a$ is necessarily a positive integer. In the first case, apply Lemma (5.10). Suppose that we are in the second case. Apply Theorem (5.13) to obtain a strongly nondense Haar-measurable order $P^*$ on $B$ for which (5.13.*i*) holds. Apply Theorem (5.4) to see that $P^*$ differs from $\tau^{-1}(]0, \infty[)$ only by a subset of $\tau^{-1}(\{0\})$, which has Haar measure 0 in $B$. Combining these observations, we find a homomorphism $\tau$ on $B$ satisfying (*i*) and (*ii*). Extend $\tau$ in any way to a real-valued (plainly continuous) homomorphism $\psi$ on $X$. Since $B$ is open, $\psi^{-1}(\{0\})$ has Haar measure 0 on $X$, and evidently (*i*) and (*ii*) still hold. □

## §6. The Hilbert transform on locally compact Abelian groups.

We now begin our task of answering question (3.6).

**(6.1) Introduction.** Let $G$ be a locally compact Abelian group with character group $X$. *The group $X$ need not be ordered.* We suppose instead that there is a nonzero continuous homomorphism $\tau$ from $X$ into $\mathbb{R}$. Let $\varphi$ denote the adjoint homomorphism of $\tau$. The mapping $\varphi$ is also continuous and satisfies the identity

$$(i) \qquad \chi \circ \varphi(r) = \exp(i\tau(\chi)r)$$

for all $r$ in $\mathbb{R}$ and all $\chi$ in $X$ (Hewitt and Ross [1979], p. 392, (24.37)). With $\tau$ we associate the function $\mathrm{sgn}_\tau$ defined on $X$ by:

$$(ii) \qquad \mathrm{sgn}_\tau(\chi) = \begin{cases} 1 & \text{if } \tau(\chi) > 0; \\ 0 & \text{if } \tau(\chi) = 0; \\ -1 & \text{if } \tau(\chi) < 0. \end{cases}$$

For $f$ in $\mathcal{L}_1(G)$, Lemma (20.6) of Hewitt and Ross [1979], p. 286 shows that the function $(x,t) \mapsto f(x + \varphi(t))$ is $\mu \times \lambda$-measurable (here $\mu$ is Haar measure on $G$ and $\lambda$ is Lebesgue measure on $\mathbb{R}$). We adhere to this $G$ and the present notation through the present section.

Consider the one-parameter group of transformations $U^t$ acting on $G$ by translation by $\varphi(t)$. That is, $U^t(x) = x + \varphi(t)$, for all $x$ in $G$. We will apply the results of Calderón [1968] in this set-up, taking $M = G$ and $T_n$ to be the truncated Hilbert transform on $\mathbb{R}$. It is easy to check that the results of Calderón [1968] still hold when $M$ is replaced by any locally compact Abelian group $G$, not necessarily $\sigma$-compact (see Asmar [1986], section 2).

Let us recall some classical definitions and facts.

**(6.2) Definition.** Let $f$ be in $\mathcal{L}_1(\mathbb{R})$ and let $n$ be a positive integer. The *truncated Hilbert transform $H_n f$ of $f$* is defined by

$$H_n f(x) = \frac{1}{\pi}\int_{1/n \le |t| \le n} \frac{f(x-t)}{t}\,dt.$$

The *Hilbert transform $Hf$ of $f$* is defined by

$$Hf(x) = \lim_{n\to\infty} H_n f(x).$$

The *maximal Hilbert transform $MHf$ of $f$* is defined by

$$MHf(x) = \sup_n |H_n f(x)|.$$

**(6.3) Theorem.** *If $f$ is in $\mathcal{L}_p(\mathbb{R})$ $(1 < p < \infty)$, the Hilbert transform $Hf$ exists almost everywhere and satisfies the inequality*

$$(i) \qquad \|Hf\|_p \le M_p \|f\|_p$$

*where $M_p$ is as in (1.6.i).*

*(ii) If $f$ is in $\mathcal{L}_1(\mathbb{R})$, the Hilbert transform $Hf$ exists almost everywhere.*
*If $f$ is in $\mathcal{L}_p(\mathbb{R})$ $(1 < p < \infty)$, then the inequality*

$$(iii) \qquad ||H_n f||_p \leq M_p ||f||_p$$

*holds for every positive integer $n$, where $M_p$ is as in (i).*
*If $f$ is in $\mathcal{L}_p(\mathbb{R})(1 < p < \infty)$, there exists a constant $C_p$ depending only on $p$ such that*

$$(iv) \qquad ||MHf||_p \leq C_p ||f||_p.$$

*If $f$ is in $\mathcal{L}_1(\mathbb{R})$, there exist constants $A$ and $B$ such that for every positive real number $y$, the inequalities*

$$(v) \qquad \lambda(\{s \in \mathbb{R} : |H(f)(s)| > y\}) \leq A\frac{||f||_1}{y},$$

*and*

$$(vi) \qquad \lambda(\{s \in \mathbb{R} : |MHf(s)| > y\}) \leq \frac{B||f||_1}{y}$$

*hold.*

For (*i*) and (*iii*) see Zygmund [1959, Vol.II], Ch. XVI, §3, p. 256 Theorem (3.8). For (*ii*), (*iv*), (*v*) and (*vi*) see Garsia [1970], Sections 4.3, 4.4, pp. 112–128, 4.3.1, 4.3.9, 4.4.2, 4.4.3.

We recall that the best constant $M_p$ is $\tan\frac{\pi}{2p}$ if $1 < p \leq 2$ and $\cot\frac{\pi}{2p}$ if $2 \leq p < \infty$ (see Pichorides [1972]). For $p$ in $]1, \infty[$, this number will be called *M. Riesz's constant* and will be denoted by $M_p$.

**(6.4) Definition.** Let $p$ be a number in $[1, \infty[$, and let $f$ be in $\mathcal{L}_p(G)$. For every positive integer $n$, the function

$$(i) \qquad H_n^\tau f(x) = \frac{1}{\pi}\int_{1/n \leq |t| \leq n} f(x - \varphi(t))\frac{1}{t}dt$$

is defined for $\mu$-almost all $x$ in $G$[12]. The function $H_n^\tau f$ is called the $n^{\text{th}}$ ***truncated Hilbert transform of f on G***. The *maximal Hilbert transform of $f$ on $G$* is defined for $\mu$-almost all $x$ by

$$(ii) \qquad M^\tau f(x) = \sup_{1 \leq n < \infty} |H_n^\tau f(x)|.$$

Combining the results of Calderón [1968] and the properties of the classical Hilbert transform on $\mathbb{R}$, we obtain the basic properties of the Hilbert transform on $G$. We proceed to list them.

**(6.5) Theorem.** *Notation is as in (6.4). For every function $f$ in $\mathcal{L}_p(G)$ $(1 < p < \infty)$, the inequalities*

$$(i) \qquad \|H_n^\tau f\|_{p,G} \leq M_p\|f\|_{p,G}$$

*and*

$$(ii) \qquad \|M^\tau f\|_{p,G} \leq C_p\|f\|_{p,G}$$

*($C_p$ is as in (6.3.iv)) obtain. For $f$ in $\mathcal{L}_p(G)$ $(1 \leq p < \infty)$ and every positive real number $y$, we have*

$$(iii) \qquad \mu(\{x \in G : M^\tau f(x) > y\}) \leq \frac{B^p}{y^p}\|f\|_{p,G}$$

*for all $y > 0$, where $B_p = \max(A, C_p)$, $A$ is as in (6.3,v) and $C_p$ is as in (6.3,iv).*

*(iv) For every $f$ in $\mathcal{L}_p(G)$ $(1 \leq p < \infty)$, there is a function $H^\tau f$ such that $H_n^\tau f$ converges to $H^\tau f$, $\mu$-almost everywhere on $G$ and in the $\mathcal{L}_p(G)$ norm for $1 < p < \infty$. Moreover for $1 < p < \infty$ and $f \in \mathcal{L}_p(G)$, we have*

$$(v) \qquad \|H^\tau f\|_{p,G} \leq M_p\|f\|_{p,G}.$$

*For $1 \leq p < \infty$ and $f \in \mathcal{L}_p(G)$ we have*

$$(vi) \qquad \mu\Big(\{x : |H^\tau f(x)| > y\}\Big) \leq \frac{B_p^p}{y}\|f\|_{p,G}^p$$

*for all $y > 0$, where $B_p$ is as in (iii).*

The details of the proof of Theorem (6.5) are formidable. To keep this essay within reasonable bounds, we omit the proof. They are available in Asmar [1986]. A diligent reader can also construct them from Calderón [1968].

**(6.6) Definition.** For $f$ in $\mathcal{L}_p(G)$, $1 \leq p < \infty$, the function $H^\tau f$ given by (6.5,*iv*) is called the *Hilbert transform of $f$* (with respect to the homomorphism $\tau$).

**Theorem 6.7.** *Let $p$ be a number in $]1, 2]$, and let $f$ be in $\mathcal{L}_p(G)$. The Fourier transform of $H^\tau f$ satisfies the equality*

$$(i) \qquad \widehat{H^\tau f}(\chi) = -i \operatorname{sgn}_\tau(\chi)\hat{f}(\chi)$$

*for almost all $\chi$ in $X$ (with respect to Haar measure on $X$).*

**Proof.** It is enough to show that (i) holds for all $f$ in $\mathcal{L}_1(G) \cap \mathcal{L}_p(G)$. Because of (6.5,*iv*), (*i*) will be established for $f$ in $\mathcal{L}_1(G) \cap \mathcal{L}_p(G)$ if we succeed in showing, for example, that for almost all $\chi$ in $X$,

$$(1) \qquad \lim_{n\to\infty} \widehat{H_n^\tau f}(\chi) = -i \operatorname{sgn}_\tau(\chi)\hat{f}(\chi).$$

For every $\chi$ in $X$ we have

$$\begin{aligned}
\widehat{H_n^\tau f}(\chi) &= \frac{1}{\pi}\int_G \overline{\chi}(x)\int_{1/n\le|t|\le n}\frac{f(x-\varphi(t))}{t}\,dt\,d\mu(x)\\
&= \frac{1}{\pi}\int_{1/n\le|t|\le n}\frac{1}{t}\int_G \overline{\chi}(x)f(x-\varphi(t))\,d\mu(x)\,dt\\
&= \frac{1}{\pi}\int_{1/n\le|t|\le n}\frac{1}{t}\int_G \overline{\chi}(x+\varphi(t))f(x)\,d\mu(x)\,dt\\
&= \frac{1}{\pi}\int_{1/n\le|t|\le n}\frac{\overline{\chi}(\varphi(t))}{t}\,dt\int_G \overline{\chi}(x)f(x)\,d\mu(x)\\
&= \hat{f}(\chi)\frac{1}{\pi}\int_{1/n\le|t|\le n}\frac{\overline{\chi}(\varphi(t))}{t}\,dt.\\
&= \hat{f}(\chi)\frac{(-i)}{\pi}\int_{1/n\le|t|\le n}\frac{\sin(\tau(\chi)\tau)}{t}\,dt.
\end{aligned}$$

The equality (1) follows now from the classical identity

$$\lim_{n\to\infty}\frac{1}{\pi}\int_{1/n\le|t|\le n}\frac{\sin(\tau(\chi)t)}{t}\,dt = \operatorname{sgn}(\tau(\chi))$$

which can be found, for example, in Zygmund [1959, Vol. I], Ch. II., §7, p. 56, (7.4). □

**(6.8) Remark.** It is clear from the uniqueness of Fourier transforms, Theorem (6.7), and the definition in (2.7) of the conjugate function, that the function $H^\tau f$ is the conjugate function $\tilde{f}$ of $f$ with respect to a Haar-measurable order $P$ whenever the equality

$$(i) \qquad \operatorname{sgn}_\tau\chi = \operatorname{sgn}_P\chi$$

holds for almost all $\chi$ in $X$. In the following theorems, we list cases where all or some Haar-measurable orders admit $\tau$'s such that $(i)$ holds.

**(6.9) Theorem.** *Let $B$ be a torsion-free infinite locally compact Abelian group that is the union of its compact open subgroups, and let $a$ be a positive integer. Let $P$ be any Haar-measurable order in the group $\mathbb{R}^a \times B$ and let $p$ be a number in $[1,\infty[$. There is a continuous homomorphism $\tau$ from $\mathbb{R}^a \times B$ onto $\mathbb{R}$ such that (6.8,i) hold. For $f$ in $\mathcal{L}_p(G)$, where $G$ is the character group of $X$, the conjugate function $\tilde{f}$ of $f$ is obtained as the pointwise limit of the functions (6.4,i) and has the properties (6.5,i–vi).*

**Proof.** Apply Theorem (5.2) and Theorem (6.5).

**(6.10) Theorem.** *Let $H$ be a torsion-free countable discrete Abelian group and let $F$ be a compact torsion-free Abelian group. Let $P$ be any strongly nondense order in the group $\mathbb{R}^a \times H \times F$ and let $p$ be a number in $[1,\infty[$. There is a continuous homomorphism $\tau$ from $\mathbb{R}^a \times H \times F$ onto $\mathbb{R}$ such that (6.8,i) holds. For $f$ in $\mathcal{L}_p(G)$, where $G$ is the character* *group of $\mathbb{R}^a \times H \times F$, the conjugate function $\tilde{f}$ of $f$ is obtained as the pointwise limit of the function (6.4,i) and has the properties (6.5,i–vi).*

**Proof.** To obtain the homomorphism $\tau$ apply Theorem (5.4). □

**(6.11) Applications.** (a) Take $G = \mathbb{T}$, $X = \mathbb{Z}$, and $\tau$ the identity homomorphism of $\mathbb{Z}$ into $\mathbb{R}$. The adjoint homomorphism $\varphi$ of $\tau$ is the natural homomorphism of $\mathbb{R}$ onto $\mathbb{R}/\mathbb{Z}$. Theorem (6.5) yields the classical results about the conjugate function on $\mathbb{T}$. The summability method (6.5,*iv*) for $\tilde{f}$ is far from new; it can be found, for example, in Zygmund [1959, Vol. I], Ch. II, §7, pp. 56–57, formula (7.6).

(b) Take $G = \mathbb{R}^a$, where $a$ is an integer greater than one, so that $X = \mathbb{R}^a$. Let $P$ be a Haar measurable order on $\mathbb{R}^a$. Apply Hewitt and Koshi [1983], Theorem (3.8) to obtain a linear function $L$ mapping $\mathbb{R}^a$ onto $\mathbb{R}$ such that

$$L^{-1}(]0,\infty[) \subsetneqq P \subsetneqq L^{-1}([0,\infty[). \tag{1}$$

Let $\varphi$ be the (continuous) adjoint homomorphism of $L$, mapping $\mathbb{R}$ into $\mathbb{R}^a$. Form the functions

$$H_n f(x) = \frac{1}{\pi} \int_{1/n \le |t| \le n} \frac{f(x - \varphi(t))}{t} dt$$

for $n = 1, 2, \ldots$, and $f$ in $\mathcal{L}_p(\mathbb{R}^a)$ $(1 \le p < \infty)$. By Theorem (6.5), the functions $H_n f$ converge to a function $\tilde{f}$ with the properties given by this theorem.

(c) The conjugate function on $\Sigma_{\mathbf{a}}$. As we mentioned earlier, the conjugate function on $\Sigma_{\mathbf{a}}$ is studied in Hewitt and Ritter [1983]. We borrow from Hewitt and Ritter [1983] the notation, and use without proof, several facts about the structure of the $\mathbf{a}$-adic solenoid $\Sigma_{\mathbf{a}}$ and its character group $\mathbb{Q}_{\mathbf{a}}$. The homomorphism $\tau$ from $X$ into $\mathbb{R}$ is in this case the "identity" isomorphism, $\chi_{\ell/A_j} \mapsto \frac{\ell}{A_j}$ where $\chi_{\ell/A_j}$ is the character of $\Sigma_{\mathbf{a}}$ given by

$$\chi_{\ell/A_j}(t, \mathbf{x}) = \exp\left[2\pi i \frac{\ell}{A_j}\left(t + \sum_{\nu=0}^{\infty} x_\nu A_\nu\right)\right].$$

for all $(t, \mathbf{x})$ in $\Sigma_{\mathbf{a}}$. (See Hewitt and Ritter [1983], (1.2.4). The (continuous) adjoint homomorphism $\varphi$ of $\mathbb{R}$ into $\Sigma_{\mathbf{a}}$ is given by

$$\varphi(t) = \Big(t - [t + 1/2], [t + 1/2]\mathbf{u}\Big),$$

so that

$$\begin{aligned}\chi_{\ell/A_j}(\varphi(t)) &= \exp(2\pi i \tau(\chi_{\ell/A_j})t) \\ &= \exp(2\pi i \frac{\ell}{A_j} t).\end{aligned}$$

This is shown in Hewitt and Ritter [1983], following (3.2.4). The group $\mathbb{Q}_{\mathbf{a}}$ admits exactly one order under which 1 is in $P$. This is also the order for which (6.8,$i$) holds with $\tau$ the identity isomorphism. For every $f$ in $\mathcal{L}_1(\Sigma_{\mathbf{a}})$ and almost all $(t, \mathbf{x})$ in $\Sigma_{\mathbf{a}}$ we have from (6.4,$i$)

(1)
$$H_n f(s, \mathbf{x}) =$$
$$\frac{1}{\pi} \int_{1/n \le |t| \le n} f((s, \mathbf{x}) + (-t + [t + 1/2], -[t + 1/2]\mathbf{u})) \tfrac{1}{t} dt$$
$$\frac{1}{\pi} \int_{1/n \le |t| \le n} f((s - t + [t + 1/2] - [s - t + 1/2], [s - t + 1/2]\mathbf{u} + \mathbf{x} - [t + 1/2]\mathbf{u})) \tfrac{1}{t} dt.$$

As shown in (6.5) the function $H_n f$ given in (1) converges $\mu$-almost everywhere on $\Sigma_{\mathbf{a}}$ to the conjugate function $\tilde{f}$ of $f$. The function $\tilde{f}$ has all of the properties listed in (6.5). We improve on Hewitt and Ritter [1983], whose construction of $\tilde{f}$ succeeded only for $f \in \mathcal{L} \log^+ \mathcal{L}$.

(d) **The conjugate function on $\mathbb{T}^a$ ($a$ an integer > 1) with respect to Archimedean orders on $\mathbb{Z}^a$.** An order $P$ on $\mathbb{Z}^a$ is Archimedean if and only if there are real numbers $\alpha_1, \alpha_2, \ldots, \alpha_a$ that are independent over $\mathbb{Q}$ for which $P = \{(x_1, x_2, \ldots, x_a) \in \mathbb{Z}^a : \sum_{j=1}^{a} \alpha_j x_j \ge 0\}$ (Theorem (4.2)). The mapping $\tau$ defined on $\mathbb{Z}^a$ by

(1)
$$(x_1, x_2, \ldots, x_a) \mapsto \sum_{j=1}^{a} \alpha_j x_j$$

is a continuous isomorphism of $\mathbb{Z}^a$ onto a dense subgroup of $\mathbb{R}$. The adjoint homomorphism $\varphi$ of $\tau$ is the mapping of $\mathbb{R}$ into a dense subgroup of $\mathbb{T}^a$ such that

(2)
$$\chi \circ \varphi(t) = \exp(i\tau(\chi)t)$$

for every $\chi$ in $\mathbb{Z}^a$ and all $t$ in $\mathbb{R}$. Using (1) and (2) we get

(3)
$$\chi(\varphi(t)) = \exp\Big(i\Big(\sum_{j=1}^{a} \alpha_j x_j\Big)t\Big).$$

Write

$$\varphi(t) = (\varphi^{(1)}(t), \varphi^{(2)}(t), \ldots, \varphi^{(a)}(t)).$$

From (3) we get

$$\exp\Big(i\Big(\sum_{j=1}^{a} \varphi^{(j)}(t) x_j\Big)t\Big) = \exp\Big(i\Big(\sum_{j=1}^{a} \alpha_j x_j\Big)t\Big)$$

for all $t$ in $\mathbb{R}$. Therefore we must have

$$\varphi^j(t) = \alpha_j t \quad [\mathrm{mod}\ 2\pi]$$

for $j = 1, 2, \ldots, a$, and all $t$ in $\mathbb{R}$. We will simply write $\varphi^{(j)}(t) = \alpha_j t$. For $f$ in $\mathcal{L}_1(\mathbb{T}^a)$ we have

(4)
$$H_n f(\mathbf{x}) = \frac{1}{\pi} \int_{1/n \le |t| \le n} f(\mathbf{x} - \varphi(t)) \tfrac{1}{t} dt$$
$$= \frac{1}{\pi} \int_{1/n \le |t| \le n} f((x_1 - \alpha_1 t, x_2 - \alpha_2 t, \ldots, x_a - \alpha_a t)) 1/t dt$$

As shown in (6.5) the functions (4) converge $\mu$–almost everywhere to the conjugate function $\tilde{f}$ of $f$.

## §7. The conjugate function on locally compact Abelian groups.

In this section, we achieve the goal of this essay, *viz.*, a construction analogous to Privalov's theorem (1.2). Our construction yields the conjugate function $\tilde{f}$ *pointwise* $\mu$-almost everywhere. We begin with a generalization of the well-known fact that trigonometric polynomials are dense in $\mathcal{L}_p(G)$, $1 \le p < \infty$, where $G$ is a compact Abelian group.

**(7.1) Theorem.** *Let $G$ be a locally compact Abelian group with Haar measure $\mu$ and character group $X$. Let $p$ be a number in the interval $[1, \infty[$. The linear subspace of $\mathcal{L}_1(G) \cap \mathcal{L}_p(G)$ consisting of the functions with compactly supported Fourier transforms is dense in $\mathcal{L}_p(G)$.*

**Proof.** The case $p = 1$ is treated in Corollary (33.13) of Hewitt and Ross [1970], p. 301. Throughout the rest of the proof we suppose that $p$ is an arbitrary but fixed number in $]1, \infty[$. Given $\varepsilon > 0$ and a nonzero function $f$ in $\mathcal{L}_p(G)$, we want to find a function $g$ in $\mathcal{L}_p(G)$ such that $\hat{g}$ has compact support on $X$ and

$$\|g - f\|_p < \varepsilon. \tag{1}$$

Apply Theorem (20.4) of Hewitt and Ross [1979], p. 285 to obtain a neighborhood $U$ of 0 with compact closure in $G$ such that

$$\|f_{-y} - f\|_p < \frac{\varepsilon}{2} \tag{2}$$

for all $y$ in $U$. Define the function $h$ in $\mathcal{L}_1(G)$ to be $\frac{1}{\mu(U)} 1_U$. We claim that

$$\|f * h - f\|_p < \frac{\varepsilon}{2}. \tag{3}$$

For every function $F$ in $\mathcal{L}_{p'}(G)$, $(\frac{1}{p} + \frac{1}{p'} = 1)$, we use Fubini's Theorem, Hölder's inequality, and (2), and obtain

$$\begin{aligned}
&\left| \int_G (f * h(x) - f(x)) F(x) d\mu(x) \right| \\
&= \left| \int_G \Big( \int_G f(x - y) h(y) d\mu(y) - f(x) \Big) F(x) d\mu(x) \right| \\
&= \left| \int_G \int_G (f(x - y) - f(x)) h(y) d\mu(y) F(x) d\mu(x) \right| \\
&= \left| \int_G \int_G (f(x - y) - f(x)) F(x) d\mu(x) h(y) d\mu(y) \right| \\
&\le \int_G \|f_{-y} - f\|_p \|F\|_{p'} h(y) d\mu(y) \\
&= \|F\|_{p'} \frac{1}{\mu(U)} \int_U \|f_{-y} - f\|_p d\mu(y) \\
&< \|F\|_{p'} \frac{\varepsilon}{2}.
\end{aligned} \tag{4}$$

From (4) and Theorem (17.1) of Hewitt and Stromberg [1965], p. 223 it follows that (3) holds. Use Corollary (33.13) of Hewitt and Ross [1970], p. 301 to obtain a function $k$ in $\mathcal{L}_1(G)$ such that $\widehat{k}$ is in $\mathfrak{C}_{00}(X)$ and

$$(5) \qquad ||k - h||_1 < \frac{\varepsilon}{2||f||_p}.$$

Let $g$ be the function $f * k$. By Corollary (20.14.ii) of Hewitt and Ross [1979], p. 293, the function $g$ is in $\mathcal{L}_1(G) \cap \mathcal{L}_p(G)$. From the equality $\widehat{g} = \widehat{f}\widehat{k}$, it follows that $\widehat{g}$ has compact support. It remains to show that (1) holds. From (3), (5), and (20.14.ii) of Hewitt and Ross [1979], p. 293, we have

$$\begin{aligned} ||f - f * k||_p &\leq ||f - f * h||_p + ||f * h - f * k||_p \\ &< \frac{\varepsilon}{2} + ||f||_p ||h - k||_1 \\ &< \frac{\varepsilon}{2} + \frac{\varepsilon}{2} \\ &= \varepsilon. \end{aligned}$$

□

We now establish a pure existence theorem, our first step in proving the analogue of M. Riesz's theorem.

**(7.2) Theorem.** *Let $G$ be a locally compact Abelian group with character group $X$ such that $X$ admits a Haar-measurable order $P$. There is a linear operator $f \mapsto \tilde{f}$ defined on $\mathcal{L}_p(G)$ for $1 < p < \infty$ with the following properties:*

$$(i) \qquad ||\tilde{f}||_{p,G} \leq M_p ||f||_{p,G}$$

*for all $f$ in $\mathcal{L}_p(G)$ $(1 < p < \infty)$;*

$$(ii) \qquad \widehat{\tilde{f}}(\chi) = -i\, sgn_P(\chi)\widehat{f}(\chi)$$

*for almost all $\chi$ in $X$ and all $f$ in $\mathcal{L}_p(G)$ $(1 < p < \infty$ if $G$ is compact, $1 < p \leq 2$ if $G$ is not compact).*

**Proof.** We define the linear operator $f \mapsto \tilde{f}$ on a dense subset of $\mathcal{L}_p(G)$, show that it is continuous and that it satisfies $(i)$ and $(ii)$. We then extend the operator $f \mapsto \tilde{f}$ uniquely to all of $\mathcal{L}_p(G)$. It is clear then that the extended operator satisfies $(i)$ and $(ii)$. By Theorem (7.1), it is enough to consider functions with compactly supported Fourier transforms. We suppose that $f$ is in $\mathcal{L}_p(G)$ and that $\widehat{f}$ vanishes off of a compact subset $K$ of $X$. We denote by $\mu$ and $\mu_X$ Haar measures on $G$ and $X$, respectively. Apply Theorem (5.14) to obtain a continuous real-valued homomorphism $\psi$ on $X$ such that

$$\psi(\chi) > 0$$

for $\mu_X$-almost all $\chi$ in $K \cap P\backslash\{0\}$, and

$$\psi(\chi) < 0$$

for $\mu_X$-almost all $\chi$ in $K \cap (-P)\backslash\{0\}$. Equivalently, the equality

$$\operatorname{sgn}_\psi(\chi) = \operatorname{sgn}_P(\chi) \tag{1}$$

holds for $\mu_X$-almost all $\chi$ in $K$. Form the function (6.4,$i$) with $\tau$ replaced by $\psi$ and apply Theorem (6.5,$i, iv, v$) to obtain the Hilbert transform $H^\psi f$. Denote this function by $\tilde{f}$. Then ($i$) follows from (6.5,$v$). It remains to show that ($ii$) holds.

From (6.7,$i$) we have

$$\widehat{\tilde{f}}(\chi) = -i \operatorname{sgn}_\psi(\chi)\hat{f}(\chi). \tag{2}$$

Since $\hat{f} = 0$ off of $K$, (1) shows that the equality

$$\operatorname{sgn}_\psi(\chi)\hat{f}(\chi) = \operatorname{sgn}_P(\chi)\hat{f}(\chi)$$

holds for $\mu_X$-almost all $\chi$ in $X$. Putting this in (2) we see that ($ii$) holds. □

**(7.3) Remarks.** (a) For compact $G$, Theorem (7.2) is the Bochner-Helson theorem (2.6), with the unknown constant $A_p$ replaced by M. Riesz's constant $M_p$. Our proof is totally different from previously published proofs.

(b) We owe to Berkson and Gillespie [1985] the proof that the constant in (7.2.$i$) is actually $M_p$, in the case that $G$ is compact.

**(7.4) Construction.** In the remainder of this section, we present a summability method for $\tilde{f}$ that yields pointwise convergence for $f$ in $\mathcal{L}_p(G)$ $1 < p < \infty$. We will also obtain analogues of Kolmogorov's theorem (1.4) for $\tilde{f}$ giving weak (1, 1)-type inequalities for the conjugate function, and of Zygmund's theorem (1.11).

Let $X$ be a torsion-free locally compact Abelian group admitting a Haar-measurable order $P$, let $G$ be the dual group of $X$, and let $\mu$ and $\mu_X$ be Haar measures on $G$ and $X$ respectively. Let $\Gamma$ be any $\sigma$-compact open subgroup of $X$. Because $\Gamma \cap P$ is a Haar-measurable order on $\Gamma$, the subgroup $\Gamma$ has the form $\mathbb{R}^a \times H$, where either $a$ is a positive integer and $H$ is a $\sigma$-compact torsion-free Abelian group, or $a$ is zero and $H$ is countable and discrete. (See Hewitt and Koshi [1983], Remark (4.4.e), p. 453.) Write

$$\Gamma = \bigcup_{n=1}^{\infty} K_n$$

where each $K_n$ is compact with nonvoid interior, and $K_m \subset K_n$ if $m \leq n$. For each $n$, apply Theorem (5.14) and obtain a continuous real-valued homomorphism $\psi_n$ on $X$ and a null subset $N$ of $X$ (independent of $n$) such that

$$\psi_n(\chi) > 0$$

for all $\chi$ in $(P \cap K_n)\backslash(N \cup \{0\})$ and

$$\psi_n(\chi) < 0$$

for all $\chi$ in $((-P) \cap K_n)\backslash(N \cup \{0\})$. Let $G_0$ denote the annihilator of $\Gamma$ in $G$. According to (23.24.e) of Hewitt and Ross [1979], p. 365, the group $G_0$ is compact. Let $\mu_0$ be normalized Haar measure on $G_0$. We have

$$\widehat{\mu}_0 = 1_\Gamma$$

(see Hewitt and Ross [1970], p. 214, (31.7.$i$)). A function $f$ in $\mathcal{L}_p(G)$ $(1 \le p < \infty)$ is constant on the cosets of $G_0$ if and only if the equality

$$f = f * \mu_0$$

holds $\mu$-almost everywhere on $G$ (see Hewitt and Ross [1970], p. 105, (28.68)).

For a subset $S$ of $\mathcal{L}_p(G)$ $(1 \le p < \infty)$ we will write $S * \mu_0$ to denote the subset of $\mathcal{L}_1(G)$ consisting of functions of the form $f * \mu_0$ where $f$ is in $S$. Finally, we let $\check{\mathfrak{C}}_{00}(\Gamma)$ denote the subset of $\mathcal{L}_p(G) * \mu_0$ $(1 \le p \le 2)$ consisting of functions constant on cosets of $G_0$ with Fourier transforms vanishing off of compact subsets of $\Gamma$.

For $f$ in $\mathcal{L}_p(G)$ $(1 \le p \le 2)$, one can recognize the elements of $f$ in $\mathcal{L}_p(G) * \mu_0$ by looking at their Fourier transforms. Indeed, the equality $(i)$ and the uniqueness of the Fourier transform show that $f$ is in $\mathcal{L}_p(G) * \mu_0$ if and only if $\hat{f}$ vanishes off of $\Gamma$.

**(7.5) Lemma.** *Let $p$ be a number in $[1, \infty[$. The linear space $\mathcal{L}_p(G) \cap \mathcal{L}_1(G) \cap \check{\mathfrak{C}}_{00}(\Gamma)$ is dense in $\mathcal{L}_p(G) * \mu_0$.*

**Proof.** The set $\mathcal{L}_p(G) \cap \mathcal{L}_1(G) * \mu_0$ is plainly dense in $\mathcal{L}_p(G) * \mu_0$. Let $f$ be in $\mathcal{L}_p(G) * \mu_0$. Given $\varepsilon > 0$, use the Theorem (7.1) to choose $g$ in $\mathcal{L}_1(G) \cap \mathcal{L}_p(G)$ such that $\widehat{g}$ has compact support and

$$\|f - g\|_p < \varepsilon.$$

The function $g * \mu_0$ is in $\mathcal{L}_p(G) \cap \mathcal{L}_1(G) \cap \check{\mathfrak{C}}_{00}(\Gamma)$, and since $f = f * \mu_0$, we have

$$\begin{aligned} \|f - g * \mu_0\|_p &= \|f * \mu_0 - g * \mu_0\|_p \\ &\le \|f - g\|_p \|\mu_0\| \\ &< \varepsilon. \end{aligned}$$

□

**(7.6)** Notation is as in (7.3). For $f$ in $\mathcal{L}_p(G)$ $(1 \le p < \infty)$, let $H^n f$ denote the Hilbert transform of $f$ with respect to the homomorphism $\psi_n$. The function $H^n f$ is given by (6.5.$iv$) and has the following properties:

$(i)$
$$\|H^n f\|_{p,G} \le M_p \|f\|_{p,G}$$

for all $f$ in $\mathcal{L}_p(G)$, $1 < p < \infty$;

$(ii)$
$$\mu(\{x : x \in G : |H^n f(x)| > y\}) \le \frac{B_p^p}{y^p} \|f\|_{p,G}^p,$$

where $B_p$ is as in (6.7.$iii$) and $f$ is in $\mathcal{L}_p(G)$ $(1 \le p < \infty)$.

**(7.7) Theorem.** *Notation is as in (7.4)–(7.6). Let $f$ be in $\mathcal{L}_p(G) * \mu_0$ $(1 \le p < \infty)$. There is a function $Hf$ defined $\mu$-almost everywhere on $G$ with the following properties:*

*(i) The function $H^n f$ converges in measure to $Hf$;*

*(ii) for all $y > 0$ and all $f$ in $\mathcal{L}_p(G)$, $1 \le p < \infty$, the inequality*

$$\mu(\{x : x \in G, |Hf(x)| > y\}) \le \frac{B_p^p}{y^p}||f||_{p,G}^p$$

*holds.*

**Proof.** Note that $f * \mu_0 = f$. Suppose that $g$ is in $\mathcal{L}_1(G) \cap \mathcal{L}_p(G)$ and that $\widehat{g}$ vanishes off of a compact subset $K$ of $\Gamma$. Let $n_0$ be a positive integer such that $K \subset K_n$ for all $n \ge n_0$. It is easy to check that the equality

$$H^n g = H^m g \tag{1}$$

holds $\mu$-almost everywhere for all $m, n \ge n_0$. (Compute the Fourier transforms of both sides of (1) and use (7.2.1).) We define $Hg$ to be $H^{n_0} g$. For $g$, property $(i)$ is trivial and $(ii)$ is $(7.6.ii)$.

Given $f$ in $\mathcal{L}_p(G)$, $1 \le p < \infty$, and an arbitrary $\varepsilon > 0$, choose a function $g$ as above and such that

$$||f - g||_p < 2^{-1-1/p}\varepsilon^{1+1/p}B_p$$

(use Lemma (7.5)). Choose a positive integer $n_0$ so that (1) holds for $g$. For $m, n \ge n_0$ we have

$$\begin{aligned}
&\mu(\{x : |H^n f(x) - H^m f(x)| > \varepsilon\}) \\
&= \mu(\{x : |H^n f(x) - H^n g(x) + H^m g(x) - H^m f(x)| > \varepsilon\}) \\
&\le \mu(\{x : |H^n(f-g)(x)| > \tfrac{\varepsilon}{2}\}) + \mu(\{x : |H^m(f-g)| > \tfrac{\varepsilon}{2}\}) \\
&\le 2B_p^p\left(\tfrac{2}{\varepsilon}\right)^p ||f - g||_p^p \\
&< \varepsilon.
\end{aligned}$$

The last inequality but one follows from $(7.6.ii)$. Thus the sequence of functions $(H^n f)_{n=1}^{\infty}$ is Cauchy in measure. Hence there is a function $Hf$ such that $H^n f$ converges in measure to $Hf$. That is, $(i)$ holds. The inequality $(ii)$ holds automatically for $Hf$. □

We next show that $Hf$ is the function $\tilde{f}$ of Theorem (7.2).

**(7.8) Theorem.** *Notation is borrowed from (7.7). For $f$ in $\mathcal{L}_p(G) * \mu_0$, $1 < p < \infty$, the functions $H^n f$ converge in the $\mathcal{L}_p(G)$-norm to $Hf$. We have*

$$||Hf||_p \le M_p||f||_p. \tag{i}$$

*If $G$ is compact, the equality*

$$\widehat{Hf}(\chi) = -i\ sgn_P(\chi)\hat{f}(\chi) \tag{ii}$$

holds everywhere on $X$ for $1 < p < \infty$. If $G$ is noncompact, $(ii)$ holds $\mu_X$-almost everywhere on $X$ for $1 < p \leq 2$. This is to say, $Hf$ is the conjugate function of $f$ with respect to the order $P$.

**Proof.** Suppose that $g \in \mathcal{L}_1(G) \cap \mathcal{L}_p(G) \cap \check{\mathfrak{C}}_{00}(\Gamma)$. The inequality $(i)$ for $g$ follows from (7.7.1) and $(7.6.i)$. The identity $(ii)$ for $g$ follows from $(6.7.i)$. Because of $(7.7.i)$ it is enough to show that the sequence $(H^n f)_{n=1}^{\infty}$ is Cauchy in $\mathcal{L}_p(G)$ to conclude that it converges in the $\mathcal{L}_p(G)$-norm to $Hf$.

Given $f$ in $\mathcal{L}_p(G) * \mu_0$, $(1 < p < \infty)$, and $\varepsilon > 0$, choose $g$ in $\mathcal{L}_1(G) \cap \mathcal{L}_p(G) \cap \check{\mathfrak{C}}_{00}(\Gamma)$ such that

$$||f - g||_p \leq \frac{1}{2M_p}\varepsilon.$$

Let $n_0$ be a positive integer such that (7.7.1) holds. Then, for all $m, n \geq n_0$ we have

$$\begin{aligned} ||H^n f - H^m f||_p &= ||H^n f - H^n g + H^m g - H^n f||_p \\ &\leq ||H^n (f - g)||_p + ||H^m (f - g)||_p \\ &\leq 2M_p ||f - g||_p \\ &< \varepsilon. \end{aligned} \tag{1}$$

The last equality but one follows from $(7.6.i)$. Since $(i)$ and $(ii)$ hold for all $f$ in a dense subset of $\mathcal{L}_p(G) * \mu_0$, they hold for all $f$ in $\mathcal{L}_p(G) * \mu_0$. □

**(7.9) Pointwise convergence for the conjugate function.** Suppose that we have a sequence $(F_n)_{n=1}^{\infty}$ of functions in $\mathcal{L}_1(G)$ with the following properties:

$(i)$ $\widehat{F}_n \in \check{\mathfrak{C}}_{00}(\Gamma)$ for all $n$;

$(ii)$ $f * F_n$ converges to $f$ pointwise for all $f$ in $\mathcal{L}_p(G) * \mu_0$, $1 < p < \infty$;

$(iii)$ $f * F_n$ converges in the $\mathcal{L}_p(G)$-norm to $f$, for all $f$ in $\mathcal{L}_p(G) * \mu_0$, $1 < p < \infty$.

Let $K_n$ denote the support of $\widehat{F}_n$. We lose no generality in supposing that $K_n \subset K_m$ for all $m \geq n$ and that

$$\Gamma = \bigcup_{n=1}^{\infty} K_n.$$

Let $\psi_n$, $H^n$, and $H$ have the same meanings as in (7.4)–(7.7). Then we have

$$H^n F_n = HF_n \tag{iv}$$

almost everywhere on $G$ for all $n$. The identity $(iv)$ follows from (7.2.1) and the fact that $\widehat{F}_n$ vanishes off of $K_n$, when we take the Fourier transforms of the two sides of $(iv)$.

**(7.10) Theorem.** *Notation is as in (7.9).*

$(i)$ *If $(7.9.ii)$ holds then the functions $(H^n f) * F_n$ converge pointwise $\mu$-almost everywhere to $Hf$.*

*(ii) If (7.9.iii) holds then the functions $(H^n f) * F_n$ converge in the $\mathcal{L}_p(G)$-norm to $Hf$.*

**Proof.** For almost all $\chi$ in $X$, we have

$$\begin{aligned}((H^n f) * F_n)^\wedge(\chi) &= \widehat{H^n f}(\chi)\widehat{F}_n(\chi)\\ &= -i\ \mathrm{sgn}_{\psi_n}(\chi)F_n(\chi)\widehat{f}(\chi)\\ &= (H^n F_n)^\wedge(\chi)\widehat{f}(\chi)\\ &= (HF_n)^\wedge(\chi)\widehat{f}(\chi)\\ &= -i\ \mathrm{sgn}_P(\chi)\widehat{f}(\chi)\widehat{F}_n(\chi)\\ &= ((Hf) * F_n)^\wedge(\chi).\end{aligned}$$

Since Fourier transforms are unique, we get

$$(H^n f) * F_n = (Hf) * F_n$$

$\mu$-almost everywhere on $G$. The conclusions (*i*) and (*ii*) now follow from (7.9.*ii*) and (7.9.*iii*). □

**(7.11) Remarks.** (a) The existence of sequences $(F_n)_{n=1}^\infty$, as described by (7.9), has been established by Edwards and Hewitt [1965] (see also Hewitt and Ross [1970], section **44**, pp. 631 679). Edwards and Hewitt were looking for an analogue on locally compact groups of the Fejér-Lebesgue theorem for functions in $\mathcal{L}_1(\mathbb{T})$. They found sequences $(F_n)_{n=1}^\infty$ for a restricted class of groups. For general groups, an *iterated* sequence $(F_{m,n})_{m=1}^\infty{}_{n=1}^\infty$ is needed. All of (7.10) holds for these iterated sequences. The reader may refer to Hewitt and Ross, *loc. cit.*, for details.

(b) Our construction of $Hf$ is of course restricted to functions in $\mathcal{L}_p(G) * \mu_0$ $(1 \le p < \infty)$. If $X$ is not $\sigma$-compact, we will need a great gallimaufry of $\mu_0$'s to accommodate all of $\mathcal{L}_p(G)$. Naturally every $f$ in $\mathcal{L}_p(G)$ belongs to some $\mathcal{L}_p(G) * \mu_0$, with $\hat{\mu}_0$ vanishing off of a $\sigma$-compact subgroup of $X$.

We complete this section with Kolmogorov's and Zygmund's theorems in our present context.

**(7.12) Theorem.** *Let $f$ be in $\mathcal{L}_1(G)$, and let $\mu_0$ be such that $f = f * \mu_0$. For $0 < p < 1$, we have*

(*i*) $$\int_G |Hf|^p d\mu \le \frac{1}{1-p} B_1^p \|f\|_1^p.$$

**Proof.** We showed in (7.7.*ii*) that $H$ if of weak type $(1,1)$. A standard argument now yields (*i*). See, for example, Hewitt and Stromberg [1965], p. 422, Corollary (21.72). □

**(7.13) Theorem.** *Suppose that $f$ is in $\mathcal{L}_1 \log^+ \mathcal{L}_1$. The function $Hf$ is in $\mathcal{L}_1(G)$, and*

(*i*) $$\|f\|_{1,G} \le B + C \int_G |f \log^+ f| d\mu,$$

*for certain constants $B$ and $C$.*

**Proof.** The constant $M_p = \tan\left(\frac{\pi}{2p}\right)$ is $o\left(\frac{1}{p-1}\right)$. Then a theorem of Yano can be applied ***verbatim*** to give $(i)$. See Zygmund [1959, Vol. II], Ch. XII, §4, Theorem (4.4.1), pp. 119–120. □

## §8. The conjugate function on $\mathbb{T}^a$.

Specific examples are the lifeblood of mathematics. A general theory is a mere wraith until it is fleshed out with particular cases. Therefore we will carry out our construction in detail for $\mathbb{T}^a$.

**(8.1) Introduction.** Throughout this section, $a$ denotes an integer greater than 1. The symbol $P$ denotes an arbitrary but fixed order on $\mathbb{Z}^a$. For a function $f$ in $\mathcal{L}_p(\mathbb{T}^a)$ $(1 \le p < \infty)$, the conjugate function of $f$ with respect to the order $P$ is denoted by $\tilde{f}$. We will describe a method of recapturing in a concrete way the function $\tilde{f}$ from the function $f$. Our method uses the Fejér kernel on $\mathbb{T}^a$. It enables us to obtain $\tilde{f}$ as a pointwise limit of certain trigonometric polynomials obtained from $f$. Also, this method yields the desired bound on the norm of the conjugation operator. Namely, we obtain the inequality

$$(1) \qquad \|\tilde{f}\|_{p,\mathbb{T}^a} \le M_p \|f\|_{p,\mathbb{T}^a}$$

for all functions $f$ in $\mathcal{L}_p(\mathbb{T}^a)(1 < p < \infty)$.

**(8.2)** Let $f$ be a function in $\mathcal{L}_p(\mathbb{T}^a)$. Apply Theorem (4.12) to obtain a sequence of Archimedean orders $P_n$, $n = 1, 2, \ldots$, satisfying (4.12.$i$) and (4.12.3). Since each $P_n$ is Archimedean, the construction of Example (6.10.$iv$) applies and yields the conjugate function of $f$ with respect to $P_n$. We denote this function by $\tilde{f}^n$. For each order $P_n$, Theorem (4.9) provides an order-preserving isomorphism $\tau_n$ of $\mathbb{Z}^a$ into $\mathbb{R}$ for which the equality

$$\mathrm{sgn}_{P_n} \chi = \mathrm{sgn}_{\tau_n} \chi$$

obtains for every $\chi$ in $\mathbb{Z}^a$. The adjoint homomorphism of $\tau_n$ will be denoted by $\varphi_n$.

For a positive integer $n$, the $n^{\text{th}}$ *Fejér kernel on* $\mathbb{T}$ is given by

$$(i) \qquad K_n(t) = \sum_{k=-n}^{n} \left(1 - \frac{|k|}{n+1}\right) \exp(ikt) = \begin{cases} \frac{1}{n}\left[\frac{\sin(\frac{1}{2}(n+1)t)}{\sin((\frac{1}{2})t)}\right]^2 & \text{if } \sin(\frac{1}{2}t) \ne 0; \\ n+1 & \text{otherwise}. \end{cases}$$

For the above definitions and other properties of the Fejér kernel on $\mathbb{T}$, see Hewitt and Stromberg [1965], p. 292.

It is obvious from $(i)$ that

$$(ii) \qquad \widehat{K}_n(j) = \begin{cases} 1 - \frac{|j|}{n+1} & \text{if } |j| \le n; \\ 0 & \text{otherwise}. \end{cases}$$

For a positive integer $n$, let $\mathbf{n}$ denote the $a$-tuple $(n, n, \ldots, n)$. The $n^{\text{th}}$ *Fejér kernel on* $\mathbb{T}^a$ is defined by

$$(iii) \qquad K_{\mathbf{n}}(\mathbf{t}) = \prod_{j=1}^{a} K_n(t_j)$$

where $K_n$ is given by (1) and $\mathbf{t} = (t_1, t_2, \ldots, t_a)$ is in $\mathbb{T}^a$. For $\boldsymbol{\ell} = (\ell_1, \ell_2, \ldots, \ell_a)$ in $\mathbb{Z}^a$, $(iii)$ yields

$$(iv)\qquad \begin{aligned} \widehat{K}_n(\boldsymbol{\ell}) &= \int_{\mathbb{T}^a} \left(\prod_{j=1}^{a} K_n(t_j)\right)\left(\prod_{j=1}^{a} \exp(-i\ell_j t_j)\right) d\mu(\mathbf{t}) \\ &= \prod_{j=1}^{a} \widehat{K}_n(\ell_j). \end{aligned}$$

It follows from $(ii)$ and $(iv)$ that

$$(v)\qquad \widehat{K}_{\mathbf{n}}(\boldsymbol{\ell}) = 0$$

if $\ell_j > n$ for some $j$ in $\{1, 2, \ldots, a\}$.

The Fejér kernel on $\mathbb{T}^a$ has the following properties:

$(vi)$ the sequence of functions $(f * K_{\mathbf{n}})_{n=1}^{\infty}$ converges to $f$, $\mu$-almost everywhere on $\mathbb{T}^a$, for every function $f$ in $\mathcal{L}_1(\mathbb{T}^a)$;

$(vii)$ the sequence of functions $(f * K_{\mathbf{n}})_{n=1}^{\infty}$ converges to $f$ in the $\mathcal{L}_p(\mathbb{T}^a)$-norm for every function $f$ in $\mathcal{L}_p(\mathbb{T}^a)$ $(1 \leq p < \infty)$.

For a proof of $(vi)$, see Zygmund [1959, Vol. II], Ch. XVII, §3, Theorem (3.1), p. 309. For $(vii)$, see Zygmund [1959, Vol. II], Ch. XVII, §1, Theorem (1.23), p. 304.

The Fejér kernel on $\mathbb{T}^a$ is a positive summability kernel: for we have

$$(viii)\qquad K_{\mathbf{n}}(\mathbf{t}) > 0$$

for all $\mathbf{t}$ in $\mathbb{T}^a$,

$$(ix)\qquad \int_{\mathbb{T}^a} K_{\mathbf{n}}(\mathbf{t})\, d\mu(\mathbf{t}) = 1,$$

$$(x)\qquad \lim_{n\to\infty} \int_{|\mathbf{t}|>\delta} K_{\mathbf{n}}(\mathbf{t})\, d\mu(\mathbf{t}) = 0$$

for every fixed positive real number $\delta$, where $|\mathbf{t}| = \sqrt{t_1^2 + t_2^2 + \ldots + t_a^2}$.

Properties $(viii)$–$(x)$ are easily obtained from the properties of the Fejér kernel on $\mathbb{T}$ listed on p. 292 of Hewitt and Stromberg [1965]. We omit the proofs.

**Theorem 8.3.** *Let $p$ be a real number greater than 1 and let $f$ be in $\mathcal{L}_p(\mathbb{T}^a)$. Let $\tilde{f}^n$, $P$, $P_n$, $K_{\mathbf{n}}$ be as in (8.2).*

*$(i)$ The functions $\tilde{f}^n$, $n = 1, 2, \ldots$, are in $\mathcal{L}_p(\mathbb{T}^a)$ and converge in the $\mathcal{L}_p(\mathbb{T}^a)$-norm to a function $\tilde{f}$.*

(*ii*) *The function* $\tilde{f}$ *is the conjugate function of* $f$ *with respect to the order* $P$.

(*iii*) *The polynomials* $\tilde{f}^n * K_{\mathbf{n}}$, $1, 2, \ldots$, *converge in the* $\mathcal{L}_p(\mathbb{T}^a)$*-norm to the function* $\tilde{f}$.

(*iv*) *The polynomials* $\tilde{f}^n * K_{\mathbf{n}}$, $n = 1, 2, \ldots$ *converge* $\mu$*-almost everywhere on* $\mathbb{T}^a$ *to the function* $\tilde{f}$.

(*v*) *The function* $\tilde{f}$ *satisfies the inequality*

$$\|\tilde{f}\|_p \leq M_p \|f\|_p.$$

**Proof.** Let $g$ be a trigonometric polynomial on $\mathbb{T}^a$. We write $g$ as

$$g = \sum_{\chi \in S} a_\chi \chi$$

where $S$ is a finite subset of $\mathbb{Z}^a$ and $a_\chi$ are nonzero complex numbers. In the notation of (4.12.2), choose a positive integer $n(g)$ such that $C_n$ contains $S$ for all $n \geq n(g)$. For $n \geq n(g)$, (4.12.3) yields

$$\begin{aligned} P_n \cap S &= P_n \cap (C_n \cap S) \\ &= (P_n \cap C_n) \cap S \\ &= (P \cap C_n) \cap S \\ &= P \cap S, \end{aligned} \tag{1}$$

and so

$$(-P_n) \cap S = (-P) \cap S \tag{2}$$

for all $n \geq n(g)$. The identities (1) and (2) imply that

$$P_n \cap S = P_{n(g)} \cap S$$

and

$$(-P_n) \cap S = (-P_{n(g)}) \cap S \tag{3}$$

for all $n \geq n(g)$. The definitions of $\tilde{g}^n, \tilde{g}^{n(g)}$ and $\tilde{g}$, (1), (2), and (3) show that

$$\tilde{g}^n = \tilde{g}^{n(g)} = \tilde{g} \tag{4}$$

for all $n \geq n(g)$. We return to our $f$ in $\mathcal{L}_p(\mathbb{T}^a)$. Given a positive real number $\varepsilon$, choose a trigonometric polynomial $g$ such that

$$\|f - g\|_p \leq \frac{\varepsilon}{2M_p}. \tag{5}$$

For every positive integer $n$, the conjugate function of $f - g$ with respect to $P_n$ is $\tilde{f}^n - \tilde{g}^n$. Since the order $P_n$ is Archimedean, we may apply (6.10.4) and obtain from (5) the inequality

$$\begin{aligned} \|\tilde{f}^n - \tilde{g}^n\|_p &\leq M_p\|f - g\|_p \\ &\leq M_p\frac{\varepsilon}{2M_p} \\ &= \frac{\varepsilon}{2} \end{aligned} \tag{6}$$

for every positive integer $n$. Combining (4) and (6) we see that

$$\|\tilde{f}^n - \tilde{g}\|_p < \frac{\varepsilon}{2} \tag{7}$$

for all $n \geq n(g)$. Now, for any positive integers $m$ and $n$ greater than or equal to $n(g)$, (7) shows that

$$\begin{aligned} \|\tilde{f}^n - \tilde{f}^m\|_p &\leq \|\tilde{f}^n - \tilde{g}\|_p + \|\tilde{f}^m - \tilde{g}\|_p \\ &< \frac{\varepsilon}{2} + \frac{\varepsilon}{2} \\ &= \varepsilon. \end{aligned} \tag{8}$$

It follows from (8) that the sequence of functions $(\tilde{f}^n)_{n=1}^{\infty}$ is Cauchy in $\mathcal{L}_p(\mathbb{T}^a)$. By the completeness of $\mathcal{L}_p(\mathbb{T}^a)$, there is a function $f'$ in $\mathcal{L}_p(\mathbb{T}^a)$ such that

$$\tilde{f}^n \to f' \tag{9}$$

in the $\mathcal{L}_p(\mathbb{T}^a)$ norm. From (9) we infer that

$$\widehat{\tilde{f}^n}(\chi) \to \widehat{f'}(\chi) \tag{10}$$

for every $\chi$ in $\mathbb{Z}^a$. If $\chi$ is any character of $\mathbb{T}^a$, we see from (1) and (2) with $S = \{\chi\}$ that there is a positive integer $n(\chi)$ such that

$$P_n \cap \{\chi\} = P \cap \{\chi\}$$

and

$$(-P_n) \cap \{\chi\} = (-P) \cap \{\chi\} \tag{11}$$

for every $n \geq n(\chi)$. Plainly, the equalities (11) imply that

$$\text{sgn}_n \chi = \text{sgn}_P\, \chi \tag{12}$$

for every $n \geq n(\chi)$. The equality (12) and the defining property of the conjugate function of $f$ with respect to $P$ yield the equalities

$$\begin{aligned} \widehat{\tilde{f}^n}(\chi) &= -i\ \text{sgn}_n(\chi)\hat{f}(\chi) \\ &= -i\ \text{sgn}\ (\chi)\hat{f}(\chi) \\ &= \hat{\tilde{f}}\ (\chi). \end{aligned} \tag{13}$$

Uniqueness of the Fourier transform, (10), and (13) show that

$$f' = \tilde{f} \tag{14}$$

$\mu$-almost everywhere on $\mathbb{T}^a$.

For every positive integer $n$, we have

$$\|\tilde{f}^n\|_p \leq M_p \|f\|_p. \tag{15}$$

The inequality ($v$) follows on combining (9), (14), and (15).

For $f$ in $\mathcal{L}_p(\mathbb{T}^a)$ $(1 < p < \infty)$, the function $\tilde{f}^n$ is in $\mathcal{L}_p(\mathbb{T}^a)$ for every positive integer $n$. Thus the convolution $K_{\mathbf{n}} * \tilde{f}^n$ is a well-defined element of $\mathcal{L}_p(\mathbb{T}^a)$. Let us compute the Fourier transforms of the functions $K_{\mathbf{n}} * \tilde{f}$, $K_{\mathbf{n}} * \tilde{f}^n$ and $\widetilde{K_n * f}$. If $\chi$ is in $C_n$, we see from (4.12.3) that

$$\operatorname{sgn}_P \chi = \operatorname{sgn}_{P_n} \chi. \tag{16}$$

Hence, for every $\chi$ in $C_n$, (16) shows that

$$\begin{aligned} \hat{\tilde{f}}(\chi) &= -i \operatorname{sgn}_P \chi \hat{f}(\chi) \\ &= -i \operatorname{sgn}_{P_n} \chi \hat{f}(\chi) \\ &= \widehat{\tilde{f}^n}(\chi). \end{aligned} \tag{17}$$

Combining (17) and (8.2.$v$), we obtain

$$\widehat{\tilde{f} * K_{\mathbf{n}}}(\chi) = \hat{\tilde{f}}(\chi)\hat{K}_{\mathbf{n}}(\chi) = \begin{cases} 0 & \text{if } \chi \notin C_n, \\ \widehat{\tilde{f}^n}(\chi)\hat{K}_{\mathbf{n}}(\chi) & \text{if } \chi \in C_n. \end{cases} \tag{18}$$

Also from (8.2.$v$), we have

$$\widehat{\tilde{f}^n * K_{\mathbf{n}}}(\chi) = \begin{cases} 0 & \text{if } \chi \notin C_n, \\ \widehat{\tilde{f}^n}(\chi)\hat{K}_{\mathbf{n}}(\chi) & \text{if } \chi \in C_n. \end{cases} \tag{19}$$

For every $\chi$ in $\mathbb{Z}^a$, we have

$$\begin{aligned} \widehat{f \widetilde{*} K_{\mathbf{n}}}(\chi) &= -i \operatorname{sgn}_P(\chi)\hat{f}(\chi)\hat{K}_{\mathbf{n}}(\chi) \\ &= \hat{\tilde{f}}(\chi)\hat{K}_{\mathbf{n}}(\chi) \\ &= \widehat{\tilde{f} * K_{\mathbf{n}}}(\chi). \end{aligned} \tag{20}$$

Uniqueness of the Fourier transform, (18), (19), and (20) show that for every positive integer $n$, the equalities

$$f \widetilde{*} K_{\mathbf{n}} = \tilde{f} * K_{\mathbf{n}} = \tilde{f}^n * K_{\mathbf{n}} \tag{21}$$

hold $\mu$-almost everywhere on $\mathbb{T}^a$. The second equality in (21) shows that the functions $\tilde{f}^n * K_n$ converges to the same limit function as $\tilde{f} * K_{\mathbf{n}}$ $\mu$-almost everywhere on $\mathbb{T}^a$, and in the $\mathcal{L}_p(\mathbb{T}^a)$-norm. The properties (8.2.*ii*) and (8.2.*vii*) of the Fejér kernel show that this limit function is $\tilde{f}$. This proves (*iii*) and (*iv*). □

**(8.4) Remark.** Single sequences of functions that converge to $\tilde{f}$ are described in Asmar [1986], section 6, for $G = \mathbb{T}^a$ and $f$ in $\mathcal{L}_p(\mathbb{T}^a)$, $1 \leq p < \infty$, where $a$ is a positive integer. Also single sequences of functions that yield pointwise convergence to $\tilde{f}$ are described by Theorem (6.9) and Theorem (6.8). We were unable to determine whether the iterated sequences used for pointwise convergence to $\tilde{f}$ in the general case treated in Theorem (7.10) are actually necessary. We conjecture that they are.

**Footnotes.**

[1] This essay is based on the first-named author's doctoral dissertation (Asmar [1986]) and on three lectures given by the second-named author at the Analysis Conference Singapore 1986. The authors are grateful to the editor, Professor S. T. L. Choy, for the opportunity he has graciously given us to present this extended essay for publication in the Proceedings of the Conference. They are also grateful to Professors Gerald B. Folland and Paul Pascal for valuable counsel, and to Ms. Laura Plaut for typesetting our MS.

[2] There is a bibliography at the end of the essay. Authors are listed alphabetically by surname and then by year of publication with superscripts where needed.

[3] This is the Austrian mathematician Alfred Tauber who published the first *Tauberian theorem* (Tauber [1897]). He lived from 1866 to about 1942. On 28 June 1942 he was reported as being in the Konzentrationslager Theresienstadt and was heard of no more.

[4] The spoken form for the symbol $\tilde{f}$ varies widely from language to language. Here is a small sample: eff wiggle (U. S. A.); eff twiddle (Great Britain); eff tilde (French); eff tilde (Spanish); eff Schlange (German); eff slang (Dutch); eff mato (Finnish); eff hullám (Hungarian); eff fala (Polish); eff s vol'noi (Russian); eff aakfah (Arabic); eff nami (Japanese); eff bolang (Mandarin).

[5] Hardy and Littlewood are referring to Privalov [$1916^{(1)}$] and Privalov [1917].

[6] This view was held by Riesz himself. See the footnote to page 218 of Riesz [1928], in which Riesz's irritation is partly concealed. In 1954, Riesz expressed himself vigorously to the second-named author about Titchmarsh and in particular about his methods of proof.

[7] It was proved only recently (Pichorides [1972]) that the smallest value of $M_p$ is $\tan\left(\frac{\pi}{2p}\right)$ for $1 < p \leq 2$ and $\operatorname{ctn}\left(\frac{\pi}{2p}\right)$ for $2 \leq p < \infty$.

[8] So christened in Hardy [1925], where historical reasons for the name are given.

[9] For a positive real number $t$, $\log^+ t = \max(\log t, 0)$.

[10] See also Bochner [1959]. Bochner retained a life-long interest in M. Riesz's Theorem (1.6) and its congeners. In the mid 1960's he told the second-named author of his admiration for this theorem. The mapping $f \mapsto \tilde{f}$ is a linear mapping of the linear space $\bigcup_{p>1} \mathfrak{L}_p(-\pi, \pi)$ into itself. Its norms on the subspaces $\mathfrak{L}_p(\pi, \pi)$ go to $\infty$ as $p \downarrow 1$ and as $p \uparrow \infty$. This phenomenon Bochner found particularly charming.

[11] Plainly the definition of the conjugate polynomial $\tilde{f}$ depends on our choice of an order $P$ in $X$. It would seem pedantic, however, to use a term like "$P$-conjugate polynomial."

[12] It is far from obvious that ($i$) exists as a Lebesque integral on $\mathbb{R}$ $\mu$-almost everywhere. Details of an almost identical construction may be found in Hewitt and Ritter [1983], pp. 823–824.

**Bibliography.**

**Asmar, Nakhlé.** The conjugate function on locally compact Abelian groups. Doctoral Dissertation, University of Washington, Seattle, Washington (1986).

**Berberian, Sterling.** *Lectures in Functional Analysis and Operator Theory*, Graduate Text in Mathematics, **15**. Berlin, Heidelberg, New York: Springer, 1973.

**Berkson, Earl, and T.A. Gillespie.** The Generalized M. Riesz Theorem and Transference. *Pacific J. Math.* **120** (**2**), Dec. 1985, 279–288.

**Birkhoff, Garrett, and Saunders Mac Lane.** *A Survey of Modern Algebra*, Third edition. New York: The Macmillan Company, 1965.

**Bochner, S..** Additive set functions on groups, *Ann. of Math.* (**2**) **40** (1939), 769–799.

**Bochner, S..** Generalized conjugate and analytic functions without expansion. *Proc. Nat. Acad. Sci. U.S.A.* **45** (1959), 855–857.

**Calderón, Alberto P..** Ergodic theory and translation-invariant operators. *Proc. Nat. Acad. Sci.* U.S.A., **59** (**2**) (1968), 349–353.

**Edwards, R.E., and Edwin Hewitt.** Pointwise limits for sequences of convolution operators. *Acta Math.* **113** (1965), 181–218.

**Fatou, Pierre.** Séries trigonométriques et séries de Taylor. *Acta Math.* **30** (1906), 335–400.

**Fejér, Leopold.** Über die Bestimmung des Sprunges der Funktion aus ihrer Fourierreihe. *J. für die reine u. angewandte Math.* **142** (1913), 165–188. Also in *Gesammelte Arbeiten*, Vol. I, 718–742. Budapest: Akadémiai Kiadó, 1970.

**Gårding, Lars.** Marcel Riesz in memoriam. *Acta Math.* **124** (1970), I–XI.

**Garsia, Adriano.** *Topics in almost everywhere convergence.* Chicago: Markham Publishing Company, 1970.

**Hardy, G. H.** Notes on some points in the integral calculus. LVIII. On Hilbert transforms. *Messenger of Math.* **54** (1925), 20–27. Also in *Collected Papers of G. H. Hardy*, Vol. III, 121–128. Oxford, England: The Clarendon Press, 1979.

**Hardy, G. H. and J. E. Littlewood.** The allied series of a Fourier series. *Proc. London Math. Soc.* (**2**) **22** (1923), *xliii–xlv*. Also in *Collected Papers of G. H. Hardy*, Vol. III, 167–168. Oxford, England: The Clarendon Press, 1979.

**Hardy, G. H., and J. E. Littlewood.** The allied series of a Fourier series. *Proc. London Math. Soc.* (**2**) **24** (1925), 211–246. Also in *Collected Papers of G. H. Hardy*, Vol. III, 171–246. Oxford, England: The Clarendon Press, 1979.

**Helson, Henry.** Conjugate series and a theorem of Paley. *Pacific J. Math.* **8** (1958), 437–446.

**Helson, Henry.** Conjugate series in several variables, *Pacific J. Math.* **9** (1959), 513–523.

**Hewitt, Edwin, and Shozo Koshi.** Orderings in locally compact Abelian groups and the theorem of F. and M. Riesz, *Math. Proc. Camb. Phil. Soc.* **93** (1983), 441–457.

**Hewitt, Edwin, and Gunter Ritter.** Fourier series on certain solenoids. *Math. Ann.* **257** (1981), 61–83.

**Hewitt, Edwin, and Gunter Ritter.** Conjugate Fourier series on certain solenoids. *Trans. Amer. Math. Soc.* **276** (1983), 817–840.

**Hewitt, Edwin, and Kenneth Ross.** Abstract harmonic analysis I. *Grundlehren der mathematischen Wissenschaften*, **115**, Second edition. Berlin, Heidelberg, New York: Springer-Verlag 1979.

**Hewitt, Edwin, and Kenneth Ross.** Abstract harmonic analysis II. *Grundlehren der mathematischen Wissenschaften* **152**. Berlin, Heidelberg, New York: Springer-Verlag, 1970.

**Hewitt, Edwin, and Karl Stromberg.** *Real and Abstract Analysis.* Graduate Texts in Mathematics. Berlin, Heidelberg, New York: Springer-Verlag, 1965.

**Hilb, E. and Marcel Riesz.** Neuere Untersuchungen über trigonometrische Reihen. *Enzyklopädie der Math. Wiss.*, **Band II C 10**, 1189–1228. Leipzig: B. G. Teubner, 1924.

**Hobson, E. W.** *The theory of functions of a real variable and the theory of Fourier's series.* 2nd edition. Cambridge, England: Cambridge University Press, 1926.

**Kolmogorov, A.N..** Sur les fonctions harmoniques conjuguées et les séries de Fourier. *Fund. Math.* **7** (1925), 24–29.

**Lichtenstein, L.** Review of I. I. Privalov's "Das Cauchysche Integral." *Jahrbuch über die Fortschritte der Math.* **47** (Jahrgang 1919–1920, publ. 1924/1926), 296–298.

**Lukácz, Franz.** Über die Bestimmung des Sprunges einer Funktion aus ihrer Fourierreihe. *J. für die reine u. angewandte Math.* **150** (1920), 107–112.

**Pichorides, S.K..** On the best values of the constants in the theorems of M. Riesz, Zygmumd and Kolmogorov *Stud. Mat.* **44** (1972), 165–179.

**Plessner, A.** Zur Theorie der konjugierten trigonometrisch Reihen. *Mitteilungen Math.* Seminarium Universität Giessen **10** (1923), 1–36.

**Pringsheim, A.** Ueber das Verhalten von Potenzreihen auf dem Convergenzkreise. *Münch.* Sitzungsber. **30** (1900), 37–100.

**Privalov, I. I.** Sur la convergence des séries trigonométriques conjuguées. *C. R. Acad. Sci.* Paris **162** (1916), 123–126.

**Privalov, I. I.** Sur la convergence des séries trigonométriques conjuguées. *C. R. Acad. Sci.* Paris **165** (1917), 96–99.

**Privalov, I. I.** Sur les fonctions conjuguées. *Bull. Soc. Math . France* **44** (1916), 100–103.

**Privalov, I. I.** [Привалов, И. И.]. Cauchy's integral [Интеграл Cauchy]. Comm. of the physico-mathematical Faculty of the University of Saratov. Saratov, USSR: 1919.

**Privalov, I. I.** Sur les séries trigonométriques conjuguées.
[К теории сопряженных тригонометрических рядов]. *Mat. Sbornik* **31** (1923), 224–228.

**Privalov. I. I.** On the convergence of conjugate trigonometric series
[О сходимости сопряженных тригонометрических рядов]. *Mat. Sbornik* **32** (1925), 357–363.

**Privalov. I. I.** Boundary values of analytic functions
[Граничные своиства аналитических функции]. Moskva: Izdatel'stvo Universiteta, 1941. 2nd edition, greatly revised and augmented. Editor A. I. Markuševič. Moskva Leningrad: Gostehizdat, 1950.

**Riesz, Marcel.** Les fonctions conjuguées et les séries de Fourier. *C. R. Acad. Sci.* Paris **178** (1924), 1464–1467.

**Riesz, Marcel.** Sur les fonctions conjuguées, *Math. Z.* **27** (1928), 218–244.

**Rudin, Walter.** Fourier Analysis on Groups. Interscience tracts in pure and applied mathematics. New York: Interscience Publishers, 1962.

**Smirnov, V. I.** Sur les valeurs limites des fonctions analytiques. *C. R. Acad. Sci.* Paris **188** (1929), 131–133.

**Tauber, A.** Über den Zusammenhang des reellen und imaginären Theiles einer Potenzreihe. *Monatshefte für Math. u. Phys.* **2** (1891), 79–118.

**Tauber, A.** Ein Satz aus der Theorie der unendlichen Reihen. *Monatschefte für Math. u. Phys.* **8** (1897), 273–277.

**Titchmarsh, E. C.** A contribution to the theory of Fourier transforms. *Proc. London Math. Soc.* **(2) 23** (1924), 279–289.

**Titchmarsh, E. C.** Reciprocal formulas involving series and integrals. *Math. Z.* **25** (1926), 321–347.

**Titchmarsh, E. C.** On conjugate functions. *Proc. London Math. Soc.* **(2) 29** (1929), 49–80.

**Titchmarsh, E. C.** *Introduction to the theory of Fourier integrals.* Oxford, England: The Clarendon Press, 1937.

**Young, W. H.** Konvergenzbedingungen für die verwandte Reihe einer Fourierschen Reihe. *Münch.* Sitzungsber. **41** (1911), 361–371.

**Young, W. H.** On the convergence of a Fourier series and of its allied series. *Proc. London Math. Soc.* **(2) 10** (1912), 254–272.

**Young, W. H., and G. C. Young.** On the theorem of Riesz-Fisher. *Quarterly J. Math.* **44** (1913), 49–88.

**Zygmund, A.** Sur la sommation des séries trigonométriques conjuguées. *Bull. Acad. Polonaise* 1924, A, 251–258.

**Zygmund, A.** Sur les fonctions conjuguées. *Fund. Math.* **13** (1929), 284–303. *Corrigenda*, ibid., **18** (1932), 312.

**Zygmund, A.** Trigonometric series. 2nd edition, Vols. I & II. Cambridge, England: Cambridge University Press, 1959, repr. with corrections and additions 1969.

California State University, Long Beach
Long Beach, California 90840

and

University of Washington
Seattle, Washington 98195

Proceedings of the Analysis Conference, Singapore 1986
S.T.L. Choy, J.P. Jesudason, P.Y. Lee (Editors)



# ON NONLINEAR INTEGRALS

Chew Tuan Seng

Department of Mathematics
National University of Singapore
Singapore

Nonlinear integrals of Denjoy type and Perron type are given in this note. It is shown that these two nonlinear integrals, as in the linear case, are equivalent to nonlinear integrals of Henstock-Kurzweil type. In this note, it is also shown that a nonlinear measure is nothing but a linear integral of a kernel function.

It is known that orthogonally additive functionals, which are not necessarily linear, on certain function spaces can be characterized in terms of linear integrals with kernel functions [2; 4; 12; 13]. This type of functionals can also be characterized in terms of nonlinear integrals [1; 3; 5; 8; 11]. In [1; 5; 11], those nonlinear integrals are defined in abstract spaces. When realized on the real line, they correspond to the Lebesgue theory whereas in [3; 8], nonlinear integrals are of Henstock-Kurzweil type, which correspond to the theory of Henstock. It is more general. In this note, we shall define nonlinear integrals of Denjoy type and Perron type in section one. Furthermore, we shall point out that, as in the linear case, these three nonlinear integrals are equivalent. In section two, we shall show that a nonlinear measure is nothing but a linear integral of a kernel function.

## 1. EQUIVALENCE OF THREE TYPES OF NONLINEAR INTEGRALS

Let $\phi = \phi(s,I)$ be defined for s being real and I a subinterval of a compact interval $[a,b]$. Moreover, we assume that $\phi$ satisfies the following conditions:

(N1) $\phi(0,I) = 0$;

(N2) $\phi(\cdot,I)$ is continuous;

(N3) $\phi(s, I_1 \cup I_2) = \phi(s,I_1) + \phi(s,I_2)$ whenever $I_1$ and $I_2$ are disjoint;

(N4) Given $M > 0$, for every $\varepsilon > 0$ there exists $\eta > 0$ such that

$$\left| \sum_{i=1}^{n} \phi(s_i,I_i) - \sum_{i=1}^{n} \phi(t_i,I_i) \right| < \varepsilon$$

whenever $|s_i - t_i| < \eta$, $|s_i| \leq M$ and $|t_i| \leq M$ for $i = 1,2,\ldots,n$, and $\{I_1,I_2,\ldots,I_n\}$ is a partial division of $[a,b]$.

(N5) Given $M > 0$, for every $\varepsilon > 0$, there exists $\eta > 0$ such that

$$\left| \sum_{i=1}^{n} \phi(s_i,I_i) \right| < \varepsilon$$

whenever $I_i$, for $i = 1,2,\ldots,n$, are pairwise disjoint intervals with the total length less than $\eta$ and $|s_i| \leq M$, for $i = 1,2,\ldots,n$.

Obviously, (N4) implies (N2). For simplicity, we shall further assume that

$$\phi(s,[u,v]) = \phi(s,(u,v)) = \phi(s,[u,v)) = \phi(s,(u,v]).$$

We may say that $\phi(s,I)$ represents the measure of the single-step function having values s on I and zero elsewhere. Since $\phi(s,I)$ is not necessarily linear in s, the measure is a nonlinear measure.

DEFINITION 1 [3;8]. A function f defined on [a,b] is said to be Henstock-Kurzweil integrable with respect to $\phi$ on [a,b] if there is a number A such that for every $\varepsilon > 0$ there exists a function $\delta(\xi) > 0$ defined on [a,b] such that for every division D given by

$$a = x_0 < x_1 < \ldots < x_n = b \text{ with } \xi_1, \xi_2, \ldots, \xi_n,$$

and satisfying $\xi_i - \delta(\xi_i) < x_{i-1} \leq \xi_i \leq x_i < \xi_i + \delta(\xi_i)$ for each i we have

$$\left| \sum_{i=1}^{n} \phi(f(\xi_i), I_i) - A \right| < \varepsilon$$

where $I_i = [t_{i-1}, t_i]$ for $i = 1,2,\ldots, n$. The value A is defined to be the Henstock-Kurzweil integral of f over [a,b].

DEFINITION 2. Let F be a function defined on [a,b]. Write $F(I) = F(v) - F(u)$ if u and v are the endpoints of an interval I. Let $x \in [a,b]$ and $I_x$ be an interval of the form $(x, x+h)$ or $(x-h, x)$, where $h > 0$. Then F is said to have derivative $F'_\phi(x)$ at x with respect to $\phi$ if

$$\lim_{h \to 0} \frac{F(I_x) - \phi(F'_\phi(x), I_x)}{h} = 0.$$

A function f is said to be Denjoy integrable with respect to $\phi$ on [a,b] if there exists a function F which is continuous on [a,b] and $ACG_*$ such that its derivative $F'_\phi(x)$ with respect to $\phi$ is equal to f(x) almost everywhere on [a,b]. The value F(a,b) is said to be the Denjoy integral of f over [a,b].

DEFINITION 3. A function H is said to be a Perron major function of a function f in [a,b] with respect to $\phi$ if

$$-\infty \neq \underline{D}H(x) \geq \overline{\lim_{|I| \to 0}} \frac{\phi(f(x), I)}{|I|} \quad \text{for every } x,$$

where $\underline{D}$ denotes the lower derivative, I an interval containing x and $|I|$ the length of I. A function G is said to be a Perror minor function of f in [a,b] with respect to $\phi$ if $-G$ is a Perror major function of f in [a,b] with respect to $-\phi$ , i.e.,

$$\underline{\lim_{|I| \to 0}} \frac{\phi(f(x),I)}{|I|} \geq \overline{D}G(x) \neq +\infty \quad \text{for every } x.$$

A function f is said to be Perron integrable on [a,b] with respect to $\phi$ if

$$\inf\{H(b) - H(a)\} = \sup\{G(b) - G(a)\} \neq \pm\infty$$

where the infimum is over all Perron major functions H of f in [a,b] with respect to $\phi$, similarly for the supremum. The common value is defined to be the Perron integral of f on [a,b].

**THEOREM 1.** The above three nonlinear integrals are equivalent.

The proof follows the same argument as in the proof of the linear case [6, pp. 123-126; 9; 14, pp. 250-251]. We shall not elaborate here.

We remark that the technique of the proof employed in the generalized dominated convergence theorem [10] can be modified to give a similar convergence theorem for a nonlinear Perron integral.

## 2. CHARACTERIZATION OF A NONLINEAR MEASURE

**DEFINITION 4.** Let R be the real line. A function $k : [a,b] \times R \to R$ is called a Caratheodory function if $k(x,\cdot)$ is continuous, for almost all $x \in [a,b]$ and $k(\cdot,s)$ is measurable for every $s \in R$.

In what follows, $L_\infty$ denotes the space of all essentially bounded measurable functions on [a,b].

**DEFINITION 5.** Let $\{f_n\} \subset L_\infty$, then $\{f_n\}$ is said to be boundedly convergent to f, if $\{f_n\}$ converges to f pointwise almost everywhere and $\{f_n\}$ is uniformly bounded almost everywhere by some constant.

**DEFINITION 6.** A functional F defined on $L_\infty$ is said to be boundedly continuous if $F(f_n) \to F(f)$ as $n \to \infty$ whenever $\{f_n\}$ is boundedly convergent to f.

**DEFINITION 7.** Let $h \in L_\infty$ and $N_h = \{g;\ |g| \leq h,\ g \in L_\infty\}$. A functional F defined on $L_\infty$ is said to be uniformly $\|\ \|_\infty$-continuous on $N_h$ if for every $\varepsilon > 0$, there is $\delta > 0$ such that

$$|F(f) - F(g)| < \varepsilon$$

whenever $\|f-g\|_\infty < \delta$ with $f, g \in N_h$, where $\|\ \|_\infty$ is the usual norm of $L_\infty$.

**DEFINITION 8.** A functional F defined on $L_\infty$ is said to be orthogonally additive if

$$F(f+g) = F(f) + F(g)$$

whenever $f,g \in L_\infty$ with $f(x)g(x) = 0$ for almost all $x \in [a,b]$.

**LEMMA 1.** [4, Theorem 2.6] If F is a boundedly continuous and orthogonally additive functional on $L_\infty$, then F is uniformly $\|\ \|_\infty$-continuous on $N_h$, for every $h \in L_\infty$.

**THEOREM 2.** A functional F is boundedly continuous and orthogonally additive on $L_\infty$ iff

$$F(f) = \int_a^b k(x, f(x))dx \quad \text{for} \quad f \in L_\infty$$

where k is a Caratheodory function from $[a,b] \times R$ to R with $k(x,0) = 0$ for almost all x and satisfies the following property :

(*) for every natural number m, there exists a Lebesgue integrable function $g_m$ such that

$$|k(x,s)| \leq g_m(x)$$

for all $|s| \leq m$ and almost all x.

This theorem is a special case of the representation theorem in [4, Theorem 3.2]. We remark that the condition that a function $k(x,f(x))$ is Lebesgue integrable for every $f \in L_\infty$ is equivalent to the condition (*). This can be seen easily by observing the following fact : for every m, there exists $u(X) \in L_\infty$ such that

$$|k(x,s)| \leq |k(x, u(x))|$$

for all $|s| \leq m$ and almost all x. It is so since $k(x,\cdot)$ is continuous, for almost all x.

We remark that the above representation theorem was also proved by A. D. Martin and V. J. Mizel [12], but with a stronger assumption.

**THEOREM 3.** A function $\phi$ is a nonlinear measure defined as in section one iff

$$\phi(s,I) = \int_I k(x,s)dx \quad \text{for all s and all I}$$

where k is a Caratheodory function from $[a,b] \times R$ to R with $k(x,0) = 0$ for almost all x and satisfies the condition (*).

The sufficiency of the theorem follows immediately from Theorem 2 and Lemma 1. To prove the necessity, we need the following theorem and Lemmas.

**THEOREM 4.** Let $\{f_n\}$ be a sequence of measurable and Henstock-Kurzweil integrable functions with respect to $\phi$ on $[a,b]$. If (i) $f_n(x) \to f(x)$ almost everywhere in $[a,b]$, as $n \to \infty$ and (ii) the primitives $F_{\phi,n}$ with respect to $\phi$ of $f_n$ are absolutely continuous uniformly in n, then f is Henstock-Kurzweil integrable with respect to $\phi$ over $[a,b]$ and we have

$$\int_a^b f_n d\phi \to \int_a^b f d\phi \quad \text{as} \quad n \to \infty.$$

The proof is standard [7; pp. 86-87], and therefore omitted here.

**LEMMA 2.** Let $\{f_n\}$ be a sequence of functions which are Henstock-Kurzweil integrable with respect to $\phi$. If all functions $f_n$ are bounded by some constant M, then the functions $F_{\phi,n}$ defined by

$$F_{\phi,n}(x) = \int_a^x f_n d\phi$$

are absolutely continuous uniformly in n.

PROOF. First, by (N5), for every $\varepsilon > 0$, there exists $\eta > 0$ such that

$$\left| \sum_{i=1}^{n} \phi(s_i, I_i) \right| < \varepsilon$$

whenever $I_i$, for $i = 1,2,\ldots,n$, are pairwise disjoint intervals with the total length less than $\eta$ and $|s_i| \leq M$, for $i = 1,2,\ldots,n$. Since each $f_n$ is Henstock-Kurzweil integrable with respect to $\phi$, therefore, there exists $\delta_n(\xi) > 0$ such that for any $\delta_n$-fine division of $[a,b]$, we have

$$\left| \int_a^b f_n d\phi - \sum \phi(f_n(\xi), [u,v]) \right| \leq \varepsilon.$$

Let $J_j$, $j = 1,2,\ldots,m$ be pairwise disjoint intervals with the total length less than $\eta$, then, by Henstock's Lemma [7, Theorem 5],

$$\sum_{j=1}^{m} \left| \int_{J_j} f_n d\phi - \sum \phi(f_n(\xi), [u,v]) \right| \leq 4\varepsilon.$$

Note that, by (N5),

$$\sum_{j=1}^{m} \left| \sum \phi(f_n(\xi), [u,v]) \right| \leq 2\varepsilon.$$

Thus

$$\sum_{j=1}^{m} \left| \int_{J_j} f_n d\phi \right| \leq 6\varepsilon.$$

Consequently, the functions $F_{\phi,n}$ are absolutely continuous uniformly in n.

LEMMA 3. Every function in $L_\infty$ is Henstock-Kurzweil integrable with respect to $\phi$.

PROOF. First, it is easy to see that every step function is Henstock-Kurzweil integrable with respect to $\phi$ on $[a,b]$. Then, by Theorem 4 and Lemma 2, we can show that every simple function and, therefore, every function in $L_\infty$ are Henstock-Kurzweil integrable with respect to $\phi$ on $[a,b]$.

LEMMA 4. A functional G defined by

$$G(f) = \int_a^b f d\phi \quad \text{for all} \quad f \in L_\infty$$

is orthogonally additive and boundedly continuous on $L_\infty$.

PROOF. It follows from Lemma 3, Lemma 2 and Theorem 4.

PROOF OF NECESSITY. It follows from Lemma 4 that a functional G defined by

$$G(f) = \int_a^b f d\phi \quad \text{for all } f \in L_\infty$$

is boundedly continuous on $L_\infty$. Hence, by Theorem 2, there exists a Caratheodory function k satisfying the required conditions such that

$$G(f) = \int_a^b k(x,f(x))dx \quad \text{for all } f \in L_\infty.$$

Let I be a subinterval of [a,b] and s be a real number. Then

$$G(s\chi_I) = \int_I k(x,s)dx = \int_I s d\phi = \phi(s,I)$$

where $\chi_I$ is the characteristic function of I.

The proof of the necessity is complete.

## REFERENCES

1. J. Batt, Nonlinear integral operators on C(S,E), Studia Math. 48(1973), 145-177.
2. R. V. Chacon and N. Friedman, Additive functionals, Arch. Rational Mech. Anal. 18(1965), 230-240.
3. T. S. Chew and P. Y. Lee, Nonlinear Henstock-Kurzweil integrals and representation theorems, submitted.
4. L. Drewnonwski and W. Orlicz, Continuity and representation of orthogonally additive functionals, Bull. Acad. Polon, Sci., Ser. Math. 17(1969), 647-653.
5. N. Friedman and A. E. Tong, On additive operators, Canad. J. Math. 23(1971), 468-480.
6. R. Henstock, Theory of integration, Betterworths, London, 1963.
7. R. Henstock, A Riemann-type integral of Lebesgue power, Canadian J. Math. 20(1968), 79-87.
8. P. Y. Lee, Riesz representation theorems, SEA Bull. Math. 10(1986), no.2.
9. P. Y. Lee and Wittaya Naak-in, A direct proof that Henstock and Denjoy integrals are equivalent, Bull. Malaysian Math. Soc. (2) 5(1982), 43-47.
10. P. Y. Lee and T. S. Chew, On convergence theorems for nonabsolute integrals, Bull. Austral. Math. Soc., vol. 34(1986), 133-140.
11. M. Marcus and V. J. Mizel, A Radon-Nikodym theorem for functionals, J. Functional Analysis 23(1976), 285-309.
12. A. D. Martin and V. J. Mizel, A representation theorem for certain nonlinear functionals, Arch. Rational Mech. Anal. 15(1964), 353-367.
13. V. J. Mizel and K. Sundaresan, Representation of additive and biadditive functions, Arch. Rational Mech. Anal 30(1968), 102-126.
14. S. Saks, Theory of the integral, 2nd edition, Warsaw, 1937.

Proceedings of the Analysis Conference, Singapore 1986
S.T.L. Choy, J.P. Jesudason, P.Y. Lee (Editors)




# THE CONTROLLED CONVERGENCE THEOREM FOR THE APPROXIMATELY CONTINUOUS INTEGRAL OF BURKILL

Soeparna Darmawijaya

Universitas Gajah Mada, Indonesia

Lee Peng Yee

National University of Singapore, Singapore

We prove the controlled convergence theorem for the approximately continuous integral of Burkill and deduce other convergence theorems as corollaries.

## 1. Preliminaries

We know that the Henstock integral is included in the approximately continuous integral of Burkill [1, 2]. For the former, there is a theorem which is called "the controlled convergence theorem" [4,5] and which implies the other convergence theorems, for example, the dominated convergence theorem, the monotone convergence theorem and the uniform convergence theorem.

The aim of this paper is to show that the controlled convergence theorem can be extended to the approximately continuous integral of Burkill. Consequently, it also implies the other corresponding convergence theorems.

We recall the following notions [1].

Let $F : [a,b] \to R$, the real line, and $E$ be a closed subset of $[a,b]$. We say that $F$ is $AC^*_{ap}$ on $E$ or, simply, $F \in AC^*_{ap}(E)$ if (i) $F \in AC(E)$, and (ii) for every $\lambda \in (0,1)$ and every closed contiguous interval $[a_n, b_n]$ of $E$ there exists a set $E_n^\lambda \subset [a_n, b_n]$ with $a_n, b_n \in E_n^\lambda$ and $\mu(E_n^\lambda) > (1-\lambda)(b_n - a_n)$ such that

$$\sum_{n \in N} \omega(F; E_n^\lambda) < \infty$$

where $\omega$ denotes the oscillation of $F$ on $E_n^\lambda$ and $N$ the collection of all $n$ for which $[a_n, b_n]$ is a closed contiguous interval of $E$. If there exists a sequence of closed sets $\{X_i\}$ such that $[a,b] = \bigcup_{i=1}^\infty X_i$ and $F \in AC^*_{ap}(X_i)$ for every $i$, we say that $F$ is $ACG^*_{ap}$ on $[a,b]$ or, simply, $F \in ACG^*_{ap}[a,b]$.

Given $x \in [a,b]$, we associate a set $D_x$ of density 1 at $x$ and $x \in D_x$. The collection $\Delta$ of all closed sub-intervals $[u,v]$ of $[a,b]$ is called an approximately full cover (AFC) of $[a,b]$ if $u,v \in D_x$, $u < x < v$ and $x \in [a,b]$. If $\Delta$ is an AFC of $[a,b]$, a $\Delta$-partition is a partition $\{a = a_0, a_1, a_2, \dots, a_n = b;$ $x_1, x_2, \dots, x_n\}$ with $a_{i-1}, a_i \in D_{x_i}$ and $a_{i-1} \leq x_i \leq a_i$, $i = 1, 2, \dots, n$, or alternatively, $\{[u,v]; x\}$ with $u, v \in D_x$ and $u \leq x \leq v$.

A function $f : [a,b] \to R$ is said to be $R^*_{ap}$-integrable on $[a,b]$ or, simply, $f \in R^*_{ap}[a,b]$ if there exists $A$ such that for every $\varepsilon > 0$ there is an AFC, $\Delta$, of $[a,b]$ such that for every $\Delta$-partition $\{a = a_0, a_1, a_2, \ldots, a_n = b; x_1, x_2, \ldots, x_n\}$ or $\{[u,v]; x\}$ we have

$$\left|A - \sum_{i=1}^{n} f(x_i)(a_i - a_{i-1})\right| < \varepsilon$$

or, simply,

$$\left|A - \sum f(x)(v-u)\right| < \varepsilon .$$

The value $A$ is called an $R^*_{ap}$-integral of $f$ and we write

$$(R^*_{ap}) \int_a^b f(t)\, dt = A .$$

**THEOREM 1.1** [1]. Let $f \in R^*_{ap}[a,b]$. Then there exists a function $F : [a,b] \to R$ such that (i) $F \in C_{ap}[a,b]$, (ii) $F \in ACG^*_{ap}[a,b]$ and (iii) $F'_{ap}(x) = f(x)$ a.e. on $[a,b]$.

Note that $C_{ap}[a,b]$ is the set of all approximately continuous functions on $[a,b]$, and $F$ is the $R^*_{ap}$-primitive of $f$ on $[a,b]$.

## 2. The controlled convergence theorem

First of all, we need to generalize the uniformly convergent sequence of functions. Given functions $F_n$, $F : [a,b] \to R$, a sequence of functions $\{F_n\}$ is said to be locally approximately convergent to $F$ on $[a,b]$ if for every $x \in [a,b]$ and $\varepsilon > 0$ there exist a positive integer $n_0$ and a set $D_x$ of density 1 at $x$ with $x \in D_x$ such that

$$|F_n(u) - F(u)| < \varepsilon$$

for every $u \in D_x$ and $n > n_0$.

It is easy to prove the following theorem.

**THEOREM 2.2.** If a sequence of functions $\{F_n\}$ is locally approximately convergent to $F$ on $[a,b]$ and $F_n \in C_{ap}[a,b]$ for every $n$, then $F \in C_{ap}[a,b]$.

If for every $n$ we have $f_n \in R^*_{ap}[a,b]$ with $F_n$ being its $R^*_{ap}$-primitive, then the sequence of functions $\{f_n\}$ is said to be control-convergent to $f$ on $[a,b]$ if it satisfies the following conditions :

(i) $\{f_n\}$ converges to $f$ a.e. on $[a,b]$.

(ii) $\{F_n\}$ is $ACG^*_{ap}$ on $[a,b]$ uniformly in $n$, i.e., there exists a sequence of closed sets $\{X_i\}$ such that $[a,b] = \bigcup_{i=1}^{\infty} X_i$ and $F_n \in AC^*_{ap}(X_i)$ uniformly in $n$.

(iii) $\{F_n\}$ is locally approximately convergent to some $F$ on $[a,b]$.

**THEOREM 2.3.** (The controlled convergence theorem) If $\{f_n\}$ is control-convergent to f on [a,b], then $f \in R^*_{ap}[a,b]$ and

$$\lim_{n\to\infty} (R^*_{ap}) \int_a^b f_n(t)dt = (R^*_{ap}) \int_a^b f(t)dt .$$

To prove the above theorem, we need the following lemmas.

For convenience, we write in what follows $F_n(u,v) = F_n(v) - F_n(u)$.

**LEMMA 2.4** Let $f_n$ be Lebesgue integrable on [a,b] with primitive $F_n$ for every n. If $\{f_n\}$ satisfies conditions (i) and

(iv) $\{F_n\}$ is absolutely continuous on [a,b] uniformly in n, then for every $\varepsilon > 0$ there exists a positive integer $n_0$ such that for every partial division of [a,b] :

$$a \leq a_1 \leq b_1 \leq a_2 \leq b_2 \dots < b_p \leq b$$

we have

$$\left|\sum_{i=1}^{n} \{F_m(a_i, b_i) - F_n(a_i, b_i)\}\right| < \varepsilon$$

for every $m,n > n_0$.

**LEMMA 2.5** Let $F_n$ be the $R^*_{ap}$-primitive of $f_n \in R^*_{ap}[a,b]$ for every n. If $\{f_n\}$ satisfies conditions (i), (iii) and

(v) $\{F_n\}$ is $AC^*_{ap}$ on a closed set X uniformly in n, then the consequence of Lemma 2.4 holds with $a_i, b_i \in X$ for $i = 1, 2, \dots, p$.

**Proof** We define functions $H_n$ such that $H_n(x) = F_n(x)$ for every $x \in X$, and $H_n$ are piecewise linear and continuous on $X^c = [a,b] - X$ and $\{H_n\}$ converges, say to H, on [a,b]. In view of conditions (iii) and (v), the functions $H_n$ are indeed absolutely continuous uniformly in n on [a,b]. Again, we define functions $h_n(x) = H_n'(x)$ a.e. on [a,b], $h(x) = f(x)$ when $x \in X$ and $h(x) = H'(x)$ when $x \in X^c$. Thus, $\{h_n\}$ converges to h a.e. on [a,b]. By Lemma 2.4, for every $\varepsilon > 0$ there exists a positive integer $n_0$ such that for every $n > n_0$ and $a \leq a_1 < b_1 < a_2 < \dots < b_p \leq b$ we have

$$\left|\sum_{i=1}^{p} \{H_m(a_i, b_i) - H_n(a_i, b_i)\}\right| < \varepsilon .$$

Expecially, for $a_i, b_i \in X$ with $i = 1, 2, \dots, p$,

$$\left|\sum_{i=1}^{p} \{F_m(a_i, b_i) - F_n(a_i, b_i)\}\right| < \varepsilon$$

and the proof is complete.

**Proof of Theorem 2.3.** We may assume that $\{f_n\}$ converges to f everywhere on [a,b]. Also we have $f_n \in R^*_{ap}[a,b]$, that is, for every $\varepsilon > 0$ there exists an AFC, $\Delta_n$, of [a,b] such that for any $\Delta_n$-partition $\mathcal{D}_n = \{[u,v]; x\}$ we have

$$|F_n(a,b) - \sum f_n(x)(v-u)| < \varepsilon.2^{-n-1} . \quad (1)$$

Let $D_x^n$ be the set of density 1 at x and $x \in D_x^n$ from which we construct $\Delta_n$, that is, for every $u,v \in D_x^n$ with $u \leq x \leq v$ we have $[u,v] \in \Delta_n$. For every $x \in [a,b]$ there exists a positive integer m such that $x \in D_x^m$ and

$$|f_m(x) - f(x)| < \frac{\varepsilon}{b-a} . \quad (2)$$

The collection $\Delta = \{[u,v];\ u,v \in D_x^m,\ u \leq x \leq v \text{ and } x \in [a,b]\}$ is an AFC of [a,b]. By Theorem 2.2. and condition (iii), we have $F \in C_{ap}[a,b]$. For every $\Delta$-partition $\mathcal{D} = \{[u,v];\ x\}$ we have

$$|F(a,b) - \sum f(x)(v-u)| \leq |F(a,b] - \sum F_m(u,v)|$$
$$+ |\sum F_m(u,v) - \sum f_m(x)(v-u)| + |\sum f_m(x)(v-u) - \sum f(x)(v-u)|.$$

By (2), the third term is less than $\varepsilon(b-a)$. By (1) and Henstock's lemma on partial sums [3, Theorem 5], the second term of the above inequality is less than $2 \sum_{n=1}^{\infty} \varepsilon.2^{-n-1}$. Therefore, it remains to show that the first term of the above inequality is small.

By condition (ii), there exists a sequence of closed sets $\{X_i\}$ such that $[a,b] = \cup X_i$ and $F_n \in AC_{ap}^*(X_i)$ uniformly in n. For a fixed $X_i$, there exists a subsequence $\{F_{n(i,j)}\}$ such that for any partial devision $a_1 < b_1 \leq a_2 \ldots < b_s$ with $a_k, b_k \in X_i$, $k = 1, 2, \ldots, s$, we have

$$\left|\sum_{k=1}^{s} \{F_{n(i,j)}(a_k,b_k) - F(a_k,b_k)\}\right| < \varepsilon.2^{-i-j}. \quad (3)$$

We may assume that for each i, $\{F_{n(i,j)}\}$ is a subsequence of $\{F_{n(i-1,j)}\}$. Now we consider $F_{n(j)} = F_{n(j,j)}$ in place of $F_n$. If $D_x$ is the set of density 1 at x from which we construct the AFC, $\Delta$, above, we modify $D_x$ as follows. We put $D_x = D_x^m$ such that $m = n(j) \geq n(i)$ whenever $x \in Y_i = X_i - (X_1 \cup X_2 \cup \ldots \cup X_{i-1})$ and $|f_m(x) - f(x)| < \varepsilon/(b-a)$.

If in our new AFC, $\Delta$, the endpoints of a typical interval [u,v] in a $\Delta$-partition always lie in $X_i$ for some i, then we have

$$|F(a,b) - \sum F_m(u,v)| = |\sum \{F(u,v) - F_m(u,v)\}|$$
$$\leq \sum_{i=1}^{\infty} \sum_{j=1}^{\infty} \varepsilon.2^{-i-j} = \varepsilon$$

and the proof is complete. But, however in general u, v may not lie in the same $X_i$. To overcome this we proceed as follows.

Since $F_n \in AC_{ap}^*(X_i)$ uniformly in n, then for every j there exists $\delta_{ij} > 0$ such that for every sequence of non-overlapping intervals $[\alpha_k, \beta_k]$ with $\alpha_k, \beta_k \in X_i$ and $\sum_{k=1}^{\infty}(\beta_k - \alpha_k) < \delta_{ij}$ and for every $\lambda \in (0,1)$ there exists sets

$E^{\lambda}_{k} \subset [\alpha_k, \beta_k]$ with $\alpha_k, \beta_k \in E^{\lambda}_{k}$ and $\mu(E^{\lambda}_{k}) > (1-\lambda)(\beta_k - \alpha_k)$ such that

$$\sum_{k=1}^{\infty} \omega(F_{n(j)}; E^{\lambda}_{k}) < \varepsilon.2^{-i-j} .$$

Note that $(a,b) - X_i$ is open and is therefore the union of a finite or countable number of open intervals. Let $E_{ij}$ be the union of all but finite number of intervals in $[a,b] - X_i$ such that the total length of $E_{ij}$ is less than $\delta_{ij}$. So, if $E_{ij} = \bigcup_k [u_k, v_k]$ then for every $\lambda \in (0,1)$ there exists $H^{\lambda}_{k} \subset [u_k, v_k]$ such that $u_k, v_k \in H^{\lambda}_{k}$ and $\mu(H^{\lambda}_{k}) > (1-\lambda)(v_k - u_k)$. Let $E^{\lambda}_{ij} = \bigcup_k H^{\lambda}_{k}$ and $E^{*}_{ij} = \bigcup_n E^{1/n}_{ij}$. If $x \in Y_i \cap (X_i \cup E_{ij})^{o}$ and $m = n(j)$ we replace $D_x$ by $D_x \cap (X_i \cup E^{*}_{ij})$. If $x$ is a boundary point of $X_i \cup E_{ij}$, the number of such points is countable, say, $x_1, x_2, \ldots$ . Since $F_{n(j)}$ and $F$ are approximately continuous at each $x_k$, then there is a positive integer $n_k$ and a set $D_k$ of density 1 at $x_k$ and $x_k \in D_k$ such that for any $u, u \in D_k$ we have

$$|F_{n(j)}(u,v)| < \varepsilon.2^{-k} \quad \text{and} \quad |F(u,v)| < \varepsilon.2^{-k}.$$

Again, in this case when $x = x_k$ we replace $D_x$ by $D^{m}_{x} \cap D_k$ with $m \geq n_k$.

After some modifications above, we have a new AFC, $\Delta$, of $[a,b]$. Let $\mathcal{D} = \{[u,v]; x\}$ be a $\Delta$-partition. Note that $x \in Y_i$ for some $i$. If $x = x_k$, then the cooresponding sum, say $\sum_1$, in

$$|F(a,b) - \sum F_m(u,v)| = |\sum F(u,v) - \sum F_m(u,v)|$$

is small and in fact

$$|\textstyle\sum_1 F(u,v) - \sum_1 F_m(u,v)| \leq \sum_1 |F(u,v)| + \sum_1 |F_m(u,v)|$$

$$\leq \sum_{k=1}^{\infty} (\varepsilon.2^{-k} + \varepsilon.2^{-k}) = 2.\varepsilon .$$

Now suppose $x \neq x_k$ for every $k$. In this case we write $[u,v] = [u,x] \cup [x,v]$. If $u \in X_i$ we apply (3). Otherwise $u \in E^{*}_{ij}$ and $u \in E^{\lambda}_{ij}$ for some $\lambda$; then write $[u,x] = [u,w] \cup [w,x]$ where $w \in X_i$ and $[u,w] \subset E_{ij}$. Apply the inequality (4) to $[w,x]$ if $w \neq x$. Consequently, we have

$$|F(a,b) - \sum F_m(u,v)| < 5\varepsilon$$

and the proof is complete.

## 3. Some corollaries of the controlled convergence theorem

The following theorems may be considered as corollaries of the controlled convergence theorem. The proofs are similar to those in [4] and therefore omitted.

**THEOREM 3.1.** The converse of Theorem 1.1 holds.

**THEOREM 3.2.** The controlled convergence theorem holds true if conditions (ii) and (iii) are replaced by condition (iv).

**THEOREM 3.3.** The controlled convergence theorem holds true if conditions (ii) and (iii) are replaced by condition :

(vi) There exist functions G and H such that G, H $\in C_{ap}[a,b]$, G, H $\in ACG^*_{ap}[a,b]$, and $G(u,v) \leqslant F_n(u,v) \leqslant H(u,v)$ for every u, v $\in [a,b]$, $u < v$ and for every n.

**THEOREM 3.4.** (The dominated convergence theorem) The controlled convergence theorem holds true if conditions (ii) and (iii) are replaced by condition :

(vii) $g(x) \leqslant f_n(x) \leqslant h(x)$ a.e. on $[a,b]$, where g, h $\in R^*_{ap}[a,b]$.

**THEOREM 3.5.** (The monotone convergence theorem) The controlled convergence theorem holds true if conditions (ii) and (iii) are replaced by contitions :

(viii) $f_1(x) \leqslant f_2(x) \leqslant f_3(x)$ ... a.e. on $[a,b]$, and

(ix) The sequence $\{F_n(a,b)\}$ converges as $n \to \infty$.

**THEOREM 3.6.** (The locally approximate convergence theorem) If $f_n \in R^*_{ap}[a,b]$, for every n, and $\{f_n\}$ is locally approximate convergent to f on $[a,b]$, then the consequence of the controlled convergence theorem holds.

**References :**


1. Bullen, P. S., "The Burkill approximately continuous integral", J. Austral. Math. Soc. (Series A) 35(1983), 236-253.
2. Burkill, J. C., "The approximately continuous Perron integral", Math. Z. 34(1931), 270-278.
3. Henstock, R., "A Riemann-type integral of Lebesgue power", Canad. J. Math. 20(1968), 79-87.
4. Lee Peng Yee and Chew Tuan Seng, "A better convergence theorem for Henstock integral", Bull. London Math. Soc. 17(1985), 557-564.
5. Lee Peng Yee and Chew Tuan Seng, "A short proof of the controlled convergence theorem for Henstock integrals", Bull. London Math. Soc. 19(1987).

Proceedings of the Analysis Conference, Singapore 1986
S.T.L. Choy, J.P. Jesudason, P.Y. Lee (Editors)



# HARMONIC ANALYSIS ON CLASSICAL GROUPS

Gong Sheng

The Chinese University of Science and Technology

Li Shi Xiong and Zheng Xue An

Department of Mathematics, Anhui University, Hefei, People's Republic of China

The purpose of this article is to introduce briefly the principal results in harmonic analysis on classical groups and its extension on compact Lie groups in China, and also to introduce briefly some important results in this direction abroad.

Prof. Hua Luo Geng having accomplished his famous work "Harmonic Analysis on Classical Domains in the Theory of Functions of Several Complex Variables", applied his theory to the harmonic analysis on unitary groups, deepened the well-known Peter-Weyl Theorem and initiated the research on harmonic analysis on classical groups in China [1].

In the late 1950's, based on Hua's work a systematic research on harmonic analysis on unitary groups was carried on well [2]-[6]. A series of concepts, definitions and methods were set up, which subsequently influenced the research for harmonic analysis on classical groups and compact Lie groups. It is a pity that the research was suspended for a long time. The reason is now well-known.

In the late 1970's, the research regained its strength in China. The basic idea is that unitary groups are the characteristic manifolds for the complex classical domains of the first class and both rotation groups and unitary symplectic groups are characteristic manifolds for real classical domains of the first class and for the classical domains of the quaternions respectively. From this point of view, Wang Shi Kun, Dong Dao Zhen, He Zu Qi, Chen Guang Xiao and others systematically studied the harmonic analysis on rotation groups and unitary symplectic groups. Later Li Shi Xiong, Zheng Xue An, Fan Da Shan, Chen Shun Fu continued this research and extended it to compact Lie groups.

At present the research in this direction is carried on well.

Since the late 1960's, on the other hand, many researchers abroad have studied the harmonic analysis on compact Lie groups such as E. M. Stein, R. Coifman and G. Weiss, J. L. Clerc, R. J. Stanton and P. A. Tomas, R. S. Strichartz, N. J. Weiss, D. L. Ragozin, B. Dreseler, R. A. Mayer, M. E. Taylor, M. Sugiura, S. Giulin, P. M. Soard, G. Travaglin, B. George, H. Johnen and others.

The purpose of this article is to introduce briefly the principal results in harmonic analysis on classical groups and its extension on compact Lie groups in China, and also to introduce briefly some important results in this direction abroad. The related proofs are omitted.

## 1. Poisson Kernels and Abel Summation

The research in harmonic analysis on unitary groups was initiated from the Poisson kernels on the classical domains of several complex variables defined by Hua Luo Geng.

Let $R_I$ be the classical domain consisting of all complex matrices of order n such that

$$I - Z\bar{Z}' > 0.$$

It is well-known that the characteristic manifolds of $R_I$ are the unitary groups $U_n$ of order n, and the element in the analytic automorphism group can be represented by

$$W = (AZ + B)(CZ + D)^{-1} \tag{1.1}$$

where $W, Z \in R_I$ and $2n\times 2n$ matrix

$$F = \begin{pmatrix} A & B \\ C & D \end{pmatrix}$$

satisfies the following three conditions :

$$\bar{F}\begin{pmatrix} I & 0 \\ 0 & I \end{pmatrix}\bar{F}' = \begin{pmatrix} I & 0 \\ 0 & -I \end{pmatrix}, \tag{1.2}$$

$$F'\begin{pmatrix} I & 0 \\ 0 & I \end{pmatrix}\bar{F} = \begin{pmatrix} I & 0 \\ 0 & -I \end{pmatrix}, \tag{1.3}$$

$$\det F = 1.$$

On $U_n$, (1.1) changes into

$$V = (AU + B)(CU + D)^{-1} \tag{1.4}$$

which transforms a unitary matrix U into another unitary matrix V.

Let $\dot{U}$ and $\dot{V}$ denote the respective volume elements of U and V, then

$$\dot{V} = |\det(CU + D)|^{-2n}\dot{U}. \tag{1.5}$$

Let a point Z of $R_I$ become 0 under (1.1) and, accordingly, a point U of $U_n$ become V. Then Hua Luo Geng starting from the theory of harmonic functions in several complex variables, defined the Poisson Kernel as follows :

$$P(Z,U) = \frac{\det(I - Z\bar{Z}')^n}{|\det(Z - U)|^{2n}} \tag{1.6}$$

and proved the following

**THEOREM 1.1.** Let $\phi(U)$ be a continuous function on $U_n$, then

$$\phi(U) = \lim_{r\to 1} \frac{1}{\omega_n} \int_{U_n} P(rU,V)\phi(V)\,\dot{V}. \tag{1.7}$$

The Poisson kernel $P(rI,U)$ on unitary groups in (1.7) has the following expansion

$$P(rI,U) = \sum_f \rho_f(r)N(f)\chi_f(U),$$

where $N(f) = (f_1, f_2, \ldots, f_n)$ is the order of the single-valued irreducible unitary representation $A_f(U)$ of $U_n$ which takes $f = (f_1, f_2, \ldots, f_n)$ as its labels ($f_1 \geqslant f_2 \geqslant \ldots \geqslant f_n$ all are integers), $\chi_f(U)$ are the corresponding characters, and

$$\rho_f(r) = \frac{1}{N(f)\omega_n} \int_{U_n} P(rI,U)\chi_f(U)\dot{U}. \tag{1.8}$$

If $u(U)$ is an integrable function on $U_n$ and its Fourier series is

$$u(U) \sim \sum_f N(f)\mathrm{tr}(C_f A_f(U)), \tag{1.9}$$

where

$$C_f = \frac{1}{\omega_n} \int_{U_n} u(U)A_f(U')\dot{U}.$$

Then the Abel sum of (1.9) is

$$\sum_f \rho_f(r)N(f)\mathrm{tr}(C_f A_f(U)). \tag{1.10}$$

The concrete formula for $\rho_f(r)$ is included in the following theorem.

**THEOREM 1.2.** [2] If $\ell_1 = f_1+n-1, \ldots, \ell_k = f_k+n-k, \ldots, \ell_n = f_n$, when $\ell_1 > \ell_2 > \ldots > \ell_s \geqslant 0 > \ell_{s+1} > \ldots > \ell_n$ $(n \geqslant s \geqslant 0)$, we have

$$\rho_f(r) = r^{f_1+\ldots+f_s-f_{s+1}-\ldots-f_n} \times \sum_{s\geqslant g_{s+1}\geqslant\ldots\geqslant g_n\geqslant 0} \frac{N(f,g)N(g,f)}{N(f)N(g)} r^{2(g_{s+1}+\ldots+g_n)}, \tag{1.11}$$

where $N_s(a,b) = N(a_1,\ldots,a_s,b_{s+1},\ldots,b_n)$, $g_1+n-1$ and $g_2+n-2,\ldots,g_n$ in $g = (g_1,\ldots,g_n)$ are a permutation of $0,1,2,\ldots,n-1$ and $0\geqslant g_1\geqslant g_2\geqslant\ldots\geqslant g_s, s-n$. (1.11) can also be written as

$$\rho_f(r) = r^{|\ell_1|+\ldots+|\ell_n|-n(n-1)/2} \times \sum_{n-1\geqslant v_{s+1}>\ldots>v_n\geqslant 0} \prod_{j=1}^{s}\prod_{k=1}^{n} \frac{(\ell_j-v_k)(v_j-\ell_k)}{(v_j-v_k)(\ell_j-\ell_k)} r^{2(v_{s+1}+\ldots+v_n)}.$$

The proof of Theorem 1.2 is complicated and needs highly skillful calculation. Here a sketch is given and readers are referred to [2] for details.

We introduce the following notations

$$\binom{p}{q,t} = \frac{1}{2\pi}\int_0^{2\pi} \frac{e^{ip\theta}d\theta}{(1-re^{i\theta})^q(1-re^{-i\theta})^t} \qquad (0 < r < 1) \tag{1.12}$$

where $q,t$ are non-negative integers, and $p$ is an integer.

If $q = 0$ and $p < 0$, then we have

$$\binom{p}{0,t} = 0, \qquad \text{for } p < 0, \tag{1.13}$$

since $e^{ip\theta}(1-re^{-i\theta})^{-t}$ is a Fourier series whose terms have only negative powers. Similarly,

$$\binom{p}{q,0} = 0, \qquad \text{for } p > 0. \tag{1.14}$$

In virtue of

$$\frac{e^{ip\theta}}{(1-re^{i\theta})^q(1-re^{-i\theta})^t} - \frac{e^{ip\theta}}{(1-re^{i\theta})^{q-1}(1-re^{-i\theta})^t} = \frac{re^{i(p+1)\theta}}{(1-re^{i\theta})^q(1-re^{-i\theta})^t},$$

we have

$$\binom{p}{q,t} = r\binom{p+1}{q,t} + \binom{p}{q-1,t}. \tag{1.15}$$

In the same way, we have

$$\binom{p}{q,t} = r\binom{p-1}{q,t} + \binom{p}{q,t-1}. \tag{1.16}$$

Again, from

$$\frac{1-r^2}{(1-re^{i\theta})(1-re^{-i\theta})} = \frac{1}{(1-re^{i\theta})} + \frac{1}{(1-re^{-i\theta})} - 1$$

it is deduced that

$$(1-r^2)\binom{p}{q,t} = \binom{p}{q,t-1} + \binom{p}{q-1,t} - \binom{p}{q-1,t-1} \tag{1.17}$$

(1.15), (1.16) and (1.17) are three basic rules of calculation. Repeated application of (1.15) leads to

$$\begin{aligned}\binom{p}{q,t} &= \binom{p}{q-1,t} + r\binom{p+1}{q,t} \\ &= \binom{p}{q-2,t} + r\binom{p+1}{q-1,t} + \binom{p+1}{q,t} \\ &= \cdots\cdots \\ &= r\sum_{j=0}^{q-1}\binom{p+1}{q-j,t} + \binom{p}{0,t}.\end{aligned}$$

If $p < 0$, then it is easy to see from (1.13) that

$$\binom{p}{q,t} = r\sum_{j=0}^{q-1}\binom{p+1}{q-j,t} \tag{1.18}$$

By using (1.18) repeatedly, we can obtain

$$\binom{p}{q,t} = r^2 \sum_{j=0}^{q-1} \sum_{k=0}^{q-j-1} \binom{p+2}{q-j-k,t} = r^2 \sum_{s=0}^{q-1} (s+1)\binom{p+2}{q-s,t}.$$

If $p < 0$, then there exists

$$\binom{p}{q,t} = r^{-p} \sum_{k=0}^{q-1} \binom{k-p-1}{k}\binom{0}{q-k,t}. \tag{1.19}$$

Similarly, it can be deduced from (1.16) that if $p > 0$, we have

$$\binom{p}{q,t} = r^{p} \sum_{k=0}^{t-1} \binom{k+p-1}{k}\binom{0}{q,t-k}. \tag{1.20}$$

In virtue of (1.19) and (1.20), the calculation of (1.12) is reduced to that of $\binom{0}{p,q}$. Furthermore, by (1.17), the calculation can be reduced to that of $\binom{0}{0,q}$ and $\binom{0}{p,0}$. On the other hand, it is easy to see that

$$\binom{0}{p,0} = \binom{0}{0,q} = 1. \tag{1.21}$$

The formulae mentioned above being applied to (1.8), a complicated and highly skillful calculation can yields that

$$\begin{aligned}
\rho_f(r) &= \frac{1}{N(f)\omega_n} \int_0^{2\pi} \cdots \int_0^{2\pi} \frac{(1-r^2)^{n^2}}{|\det(I-rV)|^{2n}} \chi_f(V)\, \dot{V} \\
&= \frac{1}{(2\pi)^n} \int_0^{2\pi} \cdots \int_0^{2\pi} \frac{(1-r^2)^{n^2}}{\prod_{j=1}^{n} |1-re^{i\theta_j}|^{2n}} \times \\
&\quad \times e^{i(k_1\theta_1+\dots+k_n\theta_n)} D(e^{-i\theta_1}, \dots, e^{-i\theta_n}) d\theta_1 \dots d\theta_n \\
&= (-1)^{n(n-1)/2}(1-r^2)^{n^2} \begin{vmatrix} \binom{k_1-0}{n,n} & \binom{k_2-0}{n,n} & \cdots & \binom{k_n-0}{n,n} \\ \binom{k_1-1}{n,n} & \binom{k_2-1}{n,n} & \cdots & \binom{k_n-1}{n,n} \\ \cdots & \cdots & \cdots & \cdots \\ \binom{k_1-(n-1)}{n,n} & \binom{k_2-(n-1)}{n,n} & \cdots & \binom{k_n-(n-1)}{n,n} \end{vmatrix}
\end{aligned} \tag{1.22}$$

By using (1.15), (1.16) and (1.17) repeatedly, the value of the preceding determinant can be calculated out. Thus the conclusion of the theorem follows.

In virtue of Theorem 1.2, the Fourier series (1.10) of $\int_{U_n} u(V)P(rU,V)\,\dot{V}$ is absolutely convergent, therefore

$$\int_{U_n} u(V)P(rU,V)V = \sum_f \rho_f(r)N(f)\mathrm{tr}(C_f A_f(U)). \tag{1.23}$$

From Theorem 1.1, it follows that

$$u(U) = \lim_{r\to 1} \int_{U_n} u(V)P(rU,V)\dot{V}$$

$$= \lim_{r\to 1} \sum_f \rho_f(r)N(f)tr(C_f A_f(U)).$$

Thus the Fourier series of $u(U)$ is Abel-summable to itself.

Evidently, as far as application is concerned a concrete theorem on convergence is superior to an abstract existence theorem on approximation. Thus Theorem 1.1 sharpens the famous Peter-Weyl Theorem.

Assuming that $u(U)$ has sufficient smoothness, we can deduce the difference between

$$S_N = \sum_{N\geq f_1\geq f_2\geq \ldots \geq f_n \geq -N} \rho_f(r)N(f)tr(C_f A_f(U))$$

and $u(U)$. In addition, this proves, in the meantime, that the function system $\{a_{ij}^f(U)\}$ consisting of all elements of the matrices of the single-valued irreducible unitary representations $A_f(U) = (a_{ij}^f(U))$ for a unitary group is complete. As a corollary, we can immediately deduce the approximation theorem for any compact group and any compact homogeneous space.

Let us consider the real classical domain $R_n$ consisting of all real matrices of order n such that

$$I - XX' > 0,$$

the characteristic manifold of which is orthogonal group $O(n) = \{\Gamma, \Gamma\Gamma' = I\}$. The analytic automorphism

$$W = (AX + B)(CX+D)^{-1} \tag{1.24}$$

of $R_n$ maps onto itself $R_n$, $O(n)$ and $SO(n)$ onto $R_n$, $O(n)$ and $SO(n)$ respectively.

Lu Qi Keng with the help of the theory of harmonic functions in several real variables, defined Poisson kernels on $R_n$ as follows:

$$P(X,\Gamma) = \frac{\det(I - XX')^{(n-1)/2}}{\det(I - X\Gamma')^{n-1}} \tag{1.25}$$

where $X \in R_n$, $\Gamma \in SO(n)$.

As in Theorem 1.1, we can prove the following.

**THEOREM 1.3.** [1] If $u(\Gamma)$ is a continuous function on rotation group $SO(n)$, then

$$u(\Gamma_0) = \lim_{r\to 1} \frac{1}{c} \int_{SO(n)} P(r\Gamma_0, \Gamma)u(\Gamma)\dot{\Gamma}.$$

In the early 1960's, Zhong Jia Qing, using the method of generating function, proved the expansion of Poisson kernels on rotation groups.

**THEOREM 1.4.** [7] The Poisson kernel $P(r\Gamma,\Gamma)$ of rotation groups has the following expansion

$$P(rI,\Gamma) = \sum_m \rho_m(r)\sigma_m(\Gamma), \tag{1.26}$$

where

$$\rho_m(r) = \sum_{(q_1,\ldots,q_{n-k-1})\in\dot{D} \text{ or } \dot{E}} \varepsilon(q_1,\ldots,q_{n-k-1}) \times$$
$$\times N(m_1,\ldots,m_k,q_1,\ldots,q_{n-k-1})r^{\sum_1^k m_i + \sum_1^{n-k-1} q_j} \tag{1.27}$$

or, equivalent to

$$\rho_m(r) = \frac{(-1)^{(n-1)(n-2)/2}}{1!2!\ldots(n-2)!}\begin{vmatrix} \xi_1(r) & \xi_2(r) & \ldots & \xi_{n-1}(r) \\ \xi_1'(r) & \xi_2'(r) & \ldots & \xi_{n-1}'(r) \\ \cdots & \cdots & \cdots & \cdots \\ \xi_1^{(n-2)}(r) & \xi_2^{(n-2)}(r) & \ldots & \xi_{n-1}^{(n-2)}(r) \end{vmatrix}.$$

In (1.27), when $n = 2k+1$, we take $(q_1,q_2, \ldots, q_{n-k-1})$ from $\dot{D}$ which satisfies

1) $n-k-1 \geqslant q_1 \geqslant \ldots \geqslant q_{n-k-1} \geqslant 0$,

2) $q_i + q_j \neq i+j-1$, for all $i \neq j$,

and

$$\varepsilon(q_1,\ldots,q_{n-k-1}) = \begin{cases} 1, & \text{if } \sum_1^{n-k-1} q_i = 0, \\ 1, & \text{if } \sum_1^{n-k-1} q_i = 1, 2 \pmod 4, \\ -1, & \text{if } \sum_1^{n-k-1} q_i = 3, 4 \pmod 4; \end{cases}$$

when $n = 2k$, it is taken from $\dot{E}$ which satisfies

1) $n-k \geqslant q_1 \geqslant \ldots \geqslant q_{n-k-1} \geqslant 0$,

2) $q_i \neq i$ for all $i$,

3) $q_i + q_j \neq i+j$ for any $i \neq j$

and
$$\varepsilon(q_1, \ldots, q_{n-k-1}) = (-1)^{(q_1+\ldots+q_{n-k-1})/2} \tag{1.29}$$

Moreover, $N(m_1,\ldots,m_k, q_1,\ldots,q_{n-k-1})$ is the order of the irreducible unitary representation of a unitary group or order $n-1$ which takes $(m_1,\ldots,m_k,q_1,\ldots, q_{n-k-1})$ as its labels, $m_1 \geqslant m_2 \geqslant \ldots \geqslant m_k \geqslant 0$. If $n = 2k+1$ or $n = 2k$ and $m_1 \geqslant m_2 \geqslant \ldots \geqslant m_k = 0$, then $\sigma_m(\Gamma)$ is the character of the irreducible representation of $SO(n)$ which takes $m = (m_1,\ldots,m_k)$ as its labels. If $n = 2k$ and $m_1 \geqslant \ldots \geqslant m_k > 0$, then $\sigma_m(\Gamma)$ is the sum of two characters of the irreducible representations of $SO(n)$ which takes $(m_1,\ldots,\pm m_k)$ as its labels.

In (1.27), there exist $\xi_1(r) = r^{m_1+n-2}$, $\xi_2(r) = r^{m_2+n-3}$, ..., $\xi_k(r) = r^{m_k+n-k-1}$,

$$\xi_{k+1}(r) = \begin{cases} r^{n-k-2} + r^{n-k-1} & \text{for } n = 2k+1, \\ r^{n-k-2} - r^{n-k} & \text{for } n = 2k, \end{cases}$$

$$\xi_{k+2}(r) = \begin{cases} r^{n-k-3} + r^{n-k} & \text{for } n = 2k+1, \\ r^{n-k-3} - r^{n-k-1} & \text{for } n = 2k, \end{cases} \tag{1.30}$$

$$\cdots\cdots\cdots\cdots$$

$$\xi_{n-1}(r) = \begin{cases} 1 + r^{2n-2k-3} & \text{for } n = 2k+1, \\ 1 - r^{2n-2k-4} & \text{for } n = 2k. \end{cases}$$

The proof of Theorem 1.4 is complicated and needs highly skillful calculation. For details, see [7].

Let us consider a domain consisting of quaternion matrices X of order n such that $I - X\overline{X}' > 0$, the characteristic manifold of which is the unitary symplectic group USP(2n). As mentioned above, considering the analytic automorphism group on the domain $I - X\overline{X}' > 0$, we can obtain the corresponding Poisson kernels

$$P(rI,U) = \frac{(1-r^2)^{n(2n+1)}}{\det(I-rU)^{2n+1}}, \tag{1.31}$$

where $0 < r < 1$ and $U \in USP(2n)$, by using the theory of harmonic functions on the quaternion domain.

Employing the generating function methods used by Zhong Jia Qing in the proof of Theorem 1.4 leads to the following.

**THEOREM 1.5.** (He Zu Qi and Chen Guang Xiao, see [1]). In the expansion of Poisson kernels on unitary symplectic groups

$$P(rI,U) = \sum_f \rho_f(r)N(f)\chi_f(U),$$

the coefficients have the expression

$$\rho_f(r) = \frac{(-1)^n}{N(f)1!2!\dots(2n)!}\begin{vmatrix} \xi_1(r), & \dots, & \xi_{2n+1}(r) \\ \xi'_1(r), & \dots, & \xi'_{2n+1}(r) \\ \cdots & \cdots & \cdots \\ \xi_1^{(2n)}(r), & \dots, & \xi_{2n+1}^{(2n)}(r) \end{vmatrix},$$

where $\xi_1(r) = r^{f_1+2n}$, ..., $\xi_n(r) = r^{f_n+n+1}$, $\xi_{n+1}(r) = r^n$, $\xi_{n+2}(r) = r^{n-1} + r^{n+1}$, ..., $\xi_{2n+1}(r) = 1 + r^{2n}$.

As in the case of unitary groups, we are able to study the Abel summation of Fourier series on rotation or unitary symplectic groups, and on the basis of Theorems 1.4 and 1.5 we obtain the following corresponding result:

The Fourier series of any continuous function on the latter two classical groups is always Abel-summable to itself.

## 2. The Cessaro Summation

The series of methods established in the research for the harmonic analysis on unitary groups are widely applied to the research for the harmonic analysis on classical groups and on compact Lie groups. For example, the methods "from sums to kernels" and "from kernels to sums" to be introduced in this section just come from the ideas used in the research for the harmonic analysis on unitary groups. By applying these two methods, those results obtained by unitary groups in this section and the subsequent ones can be established on classical groups and on compact Lie groups. In fact, as far as we know, these two methods are almost applicable to various types of summation, central operators and central multipliers established on classical groups and on compact Lie groups at home and abroad.

The summation coefficients of those summations and central multipliers established by the method "from kernels to sums", such as Abel- and Cesaro-summation in this article and the class of central multipliers established through the Fourier transformation for Lie algebras by R. S. Strichartz [14], are usually very complicated. Here only those coefficients in Theorems 1.2, 1.4, 1.5, 2.3, 2.7 and 2.9 are concretely given and their determination depends on the complicated calculation and skillful methods mentioned above.

For studying the properties of Fourier series, such as the convergence of Cesaro summations in this section, the following method is established on unitary groups. Let $u(U) \in L(U_n)$, and

$$\psi_U(V) = c^{-1} \int_{U_n} u(UWVW^{-1})\dot{W}.$$

The method is to study Fourier series of $u(U)$ through the class of functions $\{\psi_U(V),\ U \in U_n\}$. As $\psi_U(V)$ is a class function, we only need study multiple Fourier series of a class of functions $\{\psi_U(e^{i\theta_1}, \ldots, e^{i\theta_n}),\ U \in U_n\}$ on torus, where $\psi_U(e^{i\theta_1}, \ldots, e^{i\theta_n})$ are the values of $\psi_U(V)$ at the maximum torus, i.e. at diagonal unitary matrices. Later on, this method was also used in the research for the harmonic analysis on classical groups and on compact Lie groups. The related examples can be found in [10] excluding those conducted at home.

Inspired by the Abel-summation on unitary groups we have defined Cesaro means of Fourier series (1.9) on unitary groups in [2]. Not only can the kernels be explicitly represented by matrices, but both the summation coefficients and the related integral constants can be calculated out explicitly. For understanding of the general Cesaro means, Fejer means, which is one of the most typical and most important example of Cesaso means, was studied carefully. This example indicated that the other coefficients and constants related to the general Cesaro means can be obtained in the same way.

Let $u(U)$ be an integrable function on $U_n$ and the Cesaro $(C,\alpha)$ means of its Fourier series (1.9) be

$$\sum_n^{\alpha}(U) = \sum_{nN \geq f_1 \geq \ldots \geq f_n \geq -nN} B_f^{\alpha}(N)N(f)\mathrm{tr}(C_f A_f(U)), \tag{2.1}$$

where

$$B_f^{\alpha}(N) = \frac{1}{cN(f)} \int_{U_n} \chi_f(\overline{V})K_N^{\alpha}(V)\dot{V} \tag{2.2}$$

and $K_N^{\alpha}(V)$ in (2.2) is Cesaro $(C,\alpha)$ kernel which is equal to

$$\frac{\det^n[\sum_{k=0}^{N} A_{N-k}^{\alpha-1} V^K(I - \overline{V}'^{2k+1})]}{B_N^{\alpha}(2A_N^{\alpha})^{n^2}\det^n(I - \overline{V}')}, \tag{2.3}$$

where

$$B_N^{\alpha} = \frac{1}{c}\int_{U_n} \frac{\det^n \sum_{k=0}^{N} A_{N-k}^{\alpha-1} V^k(I - \overline{V}'^{2k+1})}{(2A_N^{\alpha})^{n^2}\det^n(I - \overline{V}')}\dot{V}, \tag{2.4}$$

and

$$A_N^{\alpha} = \frac{(\alpha+N)\ \ldots\ (\alpha+1)}{N!},$$

moreover, $\alpha > -1$ is needed.

For Cesaro means on unitary groups we have

**THEOREM 2.1** [3]. Cesaro $(C,\alpha)$ means (2.1) of Fourier series (1.9) of any integrable function $u(U)$ on unitary groups can be expressed as

$$\frac{1}{c}\int_{U_n} u(\overline{V}'U)K_N^{\alpha}(V)\dot{V}. \tag{2.5}$$

**PROOF.** From (2.2) and (2.3), we have

$$K_N^{\alpha}(V) = \sum_{nN \geq f_1 \geq \ldots \geq f_n \geq -nN} B_f^{\alpha}(N)N(f)\chi_f(V). \tag{2.6}$$

Therefore (2.5) can be written as

$$\frac{1}{c}\int_{U_n} u(\overline{V}',U) \sum_{nN \geq f_1 \geq \ldots \geq f_n \geq -nN} B_f^{\alpha}(N)N(f)\chi_f(V)\dot{V},$$

which is just the (2.1).

It can be immediately seen that the relation between $(C,\alpha)$ kernels of Cesaro $(C,\alpha)$ means defined as above and Poisson kernels of Abel means defined by Hua is the same as in the case of Fourier series, i.e, $(C,\alpha)$ kernels become Poisson kernels if $\alpha$ tends to infinity.

**THEOREM 2.2.** [3] Let $u(U)$ be continuous on $U_n$. Then, when $\alpha > (n-1)/n$, Fourier series (1.9) of $u(U)$ is $(C,\alpha)$ summable to itself, and when $u(U)$ belongs to class Lip $P$ $(0 < P \leq 1)$, $\sum_N^\alpha(U)$ (see (2.1)) satisfies

1) $\quad |u(U) - \sum_N^\alpha(U)| \leq A_1 N^{-P}$, if $\alpha n-n+1 > P$;

2) $\quad |u(U) - \sum_N^\alpha(U)| \leq A_2 N^{-P}\log N$, if $\alpha n-n+1 = P$;

3) $\quad |u(U) - \sum_N^\alpha(U)| \leq A_3 N^{-\alpha n+n-1}$, if $\alpha n-n+1 < P$;

where $A_1$, $A_2$ and $A_3$ are absolute constants.

**PROOF.** Take $\eta = \max\{s-(\alpha+1)(n-1), 0\}$. From an estimate of one-dimensional Cesaro kernels $K_N^\alpha(\theta)$, we can obtain

$$|K_N^\alpha(\theta)|^{(n-1)}|\theta|^s \leq BN^{n-1-s}|N\theta|^\eta.$$

Take $s = 2(n-1)$ in the definition of $\eta$, and

$$I = |u(U) - \textstyle\sum_N^\alpha(U)| = \frac{1}{c}\left|\int_{U_n}(u(\overline{V}'U) - u(U))K_N^\alpha(V)\dot{V}\right|,$$

then

$$I \leq A\int_0^{2\pi}\cdots\int_0^{2\pi}|\psi_U(e^{i\theta_1},\ldots,e^{i\theta_n}) - u(U)||K_N^\alpha(\theta_1)\ldots K_N^\alpha(\theta_n)|$$

$$\times\ |N\theta_1|^\eta \ldots |N\theta_n|^\eta d\theta_1 \ldots d\theta_n. \tag{2.7}$$

The direct estimate of (2.7) leads to the conclusion.

Among Cesaro kernels, the kernel of Fejer means has the simplest expression

$$\frac{1}{B_N(N+1)^{n^2}}\left|\frac{\det(I - V^{N+1})}{\det(I - V)}\right|^{2n}, \tag{2.8}$$

where $B_N$ is a number such that the integral of (2.8) on $U_n$ is equal to 1 and

$$B_f(N) = B_f^1(N) = \frac{1}{N(f)c\,B_N(N+1)^{n^2}}\int_{U_n}\chi_f(V)\left|\frac{\det(I - V^{N+1})}{\det(I - V)}\right|^{2n}\dot{V}. \tag{2.9}$$

Fejer means of Fourier series (1.9) of $u(U)$ reads

$$\sum_{nN\geq f_1\geq\ldots\geq f_n\geq -nN} B_f(N)N(f)\mathrm{tr}(C_fA_f(U)) = \frac{1}{B_N(N+1)^{n^2}c}\int_{U_n}u(VU)\left|\frac{\det(I-V^{N+1})}{\det(I-V)}\right|^{2n}\dot{V}.$$

The following theorem gives the Fejer means coefficients and the integral constants of Fourier series on unitary groups. In the proof of the theorem, a complicated and ingenious skill for matrix integration is used. The same method can also be applied to the calculation of the coefficients and the integral constants for general (C.α) means.

**THEOREM 2.3.** [3] The Fejer means coefficients of Fourier series on unitary groups are

$$B_f(N) = \frac{1!2! \ldots (2n)!}{N(f)(2n!)^n n!(N+1)^{n^2}} (-1)^{n(n-1)/2}(1+O(1/N))$$

$$\times \sum_{\substack{s_1=0 \\ k_1 \geq 0}}^{2n} \cdots \sum_{\substack{s_n=0 \\ k_n \geq 0}}^{2n} C_{s_1}^{2n} C_n^{n+k_1} \cdots C_{s_n}^{2n} C_n^{n+k_n} N((N+1)s_1-f_1,\ldots,(N+1)s_n-f_n), \quad (2.10)$$

where $k_j = f_j+n-j+nN-(N+1)s_j$, $j = 1,2,\ldots,n$, and the integral constants $B_N$ are equal to

$$\frac{(n!(2n)!)^{n-1}}{((2n-1)! \ldots (n+1)!)^2} (1 + O(1/N)). \quad (2.11)$$

From the definition of N(f) and Theorem 2.3, it can be seen that

$$B_f = 1 + O(1/N).$$

Here we sketch the proof. For details, see [3].

Let

$$a_p^q = \frac{1}{2\pi} \int_0^{2\pi} \frac{e^{iq\theta}(1 - e^{-i(N+1)\theta})^{2n}}{(1 - e^{-i\theta})^p} d\theta. \quad (2.12)$$

Then if $p \leq 2n$, we have

$$a_p^q = \frac{(2n)!}{(p-1)!} \sum_{\substack{k=0 \\ k+(N+1)s=q}}^{\infty} \sum_{s=0}^{2n} \frac{(p+k-1)!}{k!s!(2n-s)!}$$

$$= \frac{(2n)!}{(p-1)!} \sum_{\substack{s=0 \\ q-(N+1)s \geq 0}}^{2n} \frac{(p+q-(N+1)s-1)!}{(q-(N+1)s)!s!(2n-s)!} \quad (2.13)$$

and

$$B_f(N)N(f) = (N+1)^{-n^2}(-1)^{n(n-1)/2} B_N^{-1} \begin{vmatrix} a_{2n}^{\ell_1+nN} & a_{2n}^{\ell_2+nN} & \ldots, & a_{2n}^{\ell_n+nN} \\ a_{2n-1}^{\ell_1+nN} & a_{2n-1}^{\ell_2+nN} & \ldots, & a_{2n-1}^{\ell_n+nN} \\ \cdot & \cdot & \cdot & \cdot \\ a_{n+1}^{\ell_1+nN} & a_{n+1}^{\ell_2+nN} & \ldots, & a_{n+1}^{\ell_n+nN} \end{vmatrix} \quad (2.14)$$

$$= \frac{B_N^{-1}((2n)!)^n(-1)^{n(n-1)/2}}{(2n-1)!(2n-2)!\dots n!(N+1)n^2} \begin{vmatrix} \sum\limits_{\substack{s_1=0\\k_1\geq 0}}^{2n} \frac{(2n+k_1-1)!}{k_1!s_1!(2n-s_1)!}, \dots, \sum\limits_{\substack{s_n=0\\k_n\geq 0}}^{2n} \frac{(2n+k_n-1)!}{k_n!s_n!(2n-s_n)!} \\ \sum\limits_{\substack{s_1=0\\k_1\geq 0}}^{2n} \frac{(2n-1+k_1)!}{k_1!s_1!(2n-s_1)!}, \dots, \sum\limits_{\substack{s_n=0\\k_n\geq 0}}^{2n} \frac{(2n-1+k_n-1)!}{k_n!s_n!(2n-s_n)!} \\ \cdots\cdots\cdots \\ \sum\limits_{\substack{s_1=0\\k_1\geq 0}}^{2n} \frac{(n+1+k_1-1)!}{k_1!s_1!(2n-s_1)!}, \dots, \sum\limits_{\substack{s_n=0\\k_n\geq 0}}^{2n} \frac{(n+1+k_n-1)!}{k_n!s_n!(2n-s_n)!} \end{vmatrix} \tag{2.15}$$

where $k_1 = \ell_1+nN-(N+1)s_1, \dots, k_n = \ell_n+nN-(N+1)s_n$; $\ell_1 = f_1+n-1, \dots, \ell_k = f_k+n-k, \dots, \ell_n = f_n$. Simplifying (2.15) further and applying ingenious calculation easily lead to (2.10). Especially, we have $B_f(N) = 1$ if $f = (0,0,\dots,0)$. Thus

$$B_n = \frac{(-1)^{n(n-1)/2}(n!)^n(n-1)!\dots 2!1!}{(N+1)^{n^2}(2n-1)!\dots n!}$$

$$\times \sum_{\substack{s_1=0\\k_1\geq 0}}^{2n} \dots \sum_{\substack{s_n=0\\k_n\geq 0}}^{2n} (-1)^{s_1+\dots+s_n} C_{s_1}^{2n} C_n^{n+k_1} \dots C_{s_n}^{2n} C_n^{n+k_n} N((N+1)s_1,\dots,(N+1)s_n) \tag{2.16}$$

where $k_j = n-j+nN-(N+1)s_j = (N+1)(n-s_j)-j$, $j = 1,2,\dots,n$.

It is known from the definition of $k_j$ that $s_j = 0,1,\dots,(n-1)$ is necessary for $k_j > 0$ if $N > n-1$. The usual method being used [1], (2.16) becomes

$$B_N = \frac{(-1)^{n(n-1)/2}(n!)^n(n-1)!\dots 2!1!}{(N+1)^{(n+1)n/2}(2n-1)!\dots n!} \begin{vmatrix} A_{n-1}^1, & A_{n-1}^2, & \dots, & A_{n-1}^n \\ A_{n-2}^1, & A_{n-2}^2, & \dots, & A_{n-2}^n \\ \cdots & \cdots & \cdots & \cdots \\ A_0^1, & A_0^2, & \dots, & A_0^n \end{vmatrix} + O(1/N) \tag{2.17}$$

where $A_j^k = C_j^{2n} C_n^{n+(N+1)(n-j)-k}$.

A series of skillful calculation related to (2.17) having been made [3], (2.11) is obtained.

Generally, let $u(\theta)$ be an integrable function on $0 \leq \theta \leq 2\pi$ and its Fourier series is $\sum_{p=-\infty}^{\infty} a_p e^{ip\theta}$. Suppose that T is a summation and the kernels of the summation

$$\tau_m = \sum_{p=-\infty}^{\infty} \mu_{mp} a_p e^{ip\theta}$$

are

$$k_m(\theta) = \sum_{p=-\infty}^{\infty} \mu_{mp} e^{ip\theta}. \tag{2.18}$$

Naturally, there is an assumption of the existence of the kernel $k_m(\theta)$, i.e. the convergence of (2.18). If $\tau_m \to s$ for m tending to a limit, then $\sum_{p=-\infty}^{\infty} a_p e^{ip\theta}$ is called T-summable to s.

Let u(U) be an integrable function on $U_n$ and its Fourier series be

$$\sum_f N(f) tr(C_f A_f(U)). \tag{2.19}$$

Again let

$$T_m(V) = B_m^{-1} \det^n \left( \sum_{-\infty}^{\infty} \mu_{mk} V^k \right)$$
$$= B_m^{-1}(k_m(\theta_1)k_m(\theta_2) \ldots k_m(\theta_n))n. \tag{2.20}$$

Then the T-means of (2.19) is

$$\sum_f B_f(m) N(f) tr(C_f A_f(U)), \tag{2.21}$$

where
$$B_f(m) = (cN(f))^{-1} \int_{U_n} \chi_f(\overline{V}) T_m(V) \dot{V}, \tag{2.22}$$

$$B_m = c^{-1} \int_{U_n} \det^n \left( \sum_{-\infty}^{\infty} \mu_{mk} V^k \right) \dot{V}, \tag{2.23}$$

and $e^{i\theta_1}, \ldots, e^{i\theta_n}$ in (2.20) are the characteristic roots of V.

Generally, replacing $k_m(\theta_1) \ldots k_m(\theta_n)$ in (2.20) by kernels $k_m(\theta_1, \theta_2, \ldots, \theta_n)$ of multiple Fourier series, we can give a summation on unitary groups. This is the method from kernel to sum.

**THEOREM 2.4.** [1] Let kernels $k_m(\theta)$ in (2.20) satisfy

1) $k_N(\theta) = O(N^{-\eta})$ for any given $\delta > 0$, where $\delta \leqslant |\theta| \leqslant \pi$,

2) $k_N(\theta) = O(N^{\xi})$ for any $\theta$ where $1 \geqslant \xi > 0$ and $\eta \geqslant \xi$,

3) $\int_{U_m} |T_N(V)| \dot{V} \leqslant H_m$ $(m = 1, 2, \ldots, n)$, where $H_m$ are constants dependent on m only.

Then the T-means (2.21) of Fourier series of u(U) converges to u(U) if u(U) is continuous on $U_n$.

For the summation of Fourier series on unitary groups set up by the method "from sum to kernel", we may begin with giving a corresponding sum of Fourier series on unitary groups by (2.18) [1]

$$\tau_m(U) = \sum_f \mu_{m\ell_1}\mu_{m\ell_2} \cdots \mu_{m\ell_n} N(f)\mathrm{tr}(C_f A_f(U)) \tag{2.24}$$

and call (2.19) T-summable to s of type II if $\tau_m(U) \to s$ when m tends to a limit. The kernels $T^*_m(V)$ of T-summation of type II are

$$T^*_m(V) = \frac{(-i)^{n(n-1)/2}}{(n-1)!\ \dots\ 2!1!} \left(D(e^{i\theta_1},\dots,e^{i\theta_n})\right)^{-1} D\left(\frac{\partial}{\partial\theta_1},\dots,\frac{\partial}{\partial\theta_n}\right)(k_m(\theta_1)\dots k_m(\theta_n)), \tag{2.25}$$

where $e^{i\theta_1}, \dots, e^{i\theta_n}$ are the characteristic roots of V, and

$$D(x_1,x_2,\dots,x_n) = \prod_{1\leq i<j\leq n} (x_i - x_j).$$

Obviously, if we take

$$k_m(\theta_1,\dots,\theta_n) = \sum \mu_{m;\ell_1\dots\ell_n} e^{i(\ell_1\theta_1+\dots+\ell_n\theta_n)}$$

as the summation kernels of multiple Fourier series and rewrite $\mu_{m\ell_1} \cdots \mu_{m\ell_n}$ in (2.24) as $\mu_{m;\ell_1,\dots,\ell_n}$, then we define a summation on unitary groups, the kernels of which can be obtained by changing $k_m(\theta_1)\dots k_m(\theta_n)$ in (2.25) into $k_m(\theta_1,\theta_2,\dots,\theta_n)$.

Take $\mu_{rp} = r^{|p|}$, and then the Abel summation of type II in (2.19)

$$\tau_r(U) = \sum_f r^{|\ell_1|+\dots+|\ell_n|} N(f)\mathrm{tr}(C_f A_f(U)) \tag{2.26}$$

is given (see [2]).

Choose $\mu_{Nk} = A^\alpha_k(N) = A^\alpha_k = \Gamma(\alpha+N-|k|+1)\Gamma(N+1)/(\Gamma(\alpha+N+1)\Gamma(N-|k|+1))$ and then Cesaro $(C,\alpha)$ summation of type II

$$\tau_N(U) = \sum_{N\geq\ell_1\geq\dots\geq\ell_n\geq -N} A^\alpha_{\ell_1} \cdots A^\alpha_{\ell_n} N(f)\mathrm{tr}(C_f A_f(U)) \tag{2.27}$$

is given [see [2]).

The kernel (2.25) corresponding to summations (2.26) and (2.27) takes one-dimensional Poisson kernel and one-dimensional Cesaro kernel respectively as $k_m(\theta)$ respectively.

For Abel and Cesaro summation of type II, the following theorem is valid.

**THEOREM 2.5.** [2] Let u(U) be a function having continuous partial derivatives up to order n(n-1)/2, then the Abel- or Cesaro-summation of type II uniformly converges to u(U).

Wang Shi Kun and Dong Dao Zheng defined Cesaro kernels on rotation groups SO(n)

$$K_N^\alpha(\Gamma) = (B_N^\alpha)^{-1}\det((A_N^\alpha I + \sum_{j=1}^{N} (\Gamma^j+\Gamma'^j) \sum_{r=0}^{N-j} A_r^{\alpha-1})/A_N^\alpha)^{n(n-1)/2}, \tag{2.28}$$

where $\Gamma \in SO(n)$ and $B_N^\alpha$ is a number such that the integral of $K_N^\alpha(\Gamma)$ on SO(n) is equal to 1.

If $u(\Gamma)$ is integrable on SO(n), its Fourier series is

$$u(\Gamma) \sim \sum_m N(m)\mathrm{tr}(C_m A_m(\Gamma)), \tag{2.29}$$

where $A_m(\Gamma)$ are the irreducible representations of SO(n) which take $m = (m_1,\ldots,m_k)$ as its labels, $m_1 \geq \ldots \geq m_k \geq 0$ are integers if $n = 2k+1$ and $m_1 \geq m_2 \geq \ldots \geq m_{k-1} \geq |m_k| \geq 0$ are also integers if $n = 2k$, $N(m) = N(m_1,\ldots, m_k)$ is the order of $A_m(\Gamma)$, and

$$C_m = c^{-1} \int_{SO(n)} u(\Gamma)A_m(\Gamma')\dot{\Gamma}, \tag{2.30}$$

where c is the volume of SO(n) and $\dot{\Gamma}$ is the volume element.

Let $\chi_m(\Gamma) = \mathrm{tr}(A_m(\Gamma))$ and

$$B_m^\alpha = \frac{1}{cN(m)} \int_{SO(n)} \chi_m(\Gamma')K_N^\alpha(\Gamma)\dot{\Gamma}. \tag{2.31}$$

It is easily seen that

$$B^\alpha_{m_1,\ldots,m_k} = B^\alpha_{m_1,\ldots,-m_k},$$

if $n = 2k$, $m_1 \geq \ldots \geq m_k > 0$. Therefore, we only need to calculate the coefficients $B^\alpha_{m_1,\ldots,m_k}$ for $m_1 \geq \ldots \geq m_k \geq 0$.

Cesaro $(C,\alpha)$ means of (2.29) are

$$\Sigma_N^\alpha(\Gamma) = \sum_{(n-1)N\geq m_1} B_m^\alpha\, N(m)\mathrm{tr}(C_m A_m(\Gamma)). \tag{2.32}$$

**THEOREM 2.6.** (Wang Shi Kun and Dong Dao Zheng, see [1]) Let $u(\Gamma)$ be a continuous function on SO(n), $\Gamma \in SO(n)$, then, when $\alpha > (n-2)/(n-1)$, the Fourier series (2.29) of $u(\Gamma)$ is $(C,\alpha)$-summable to itself and, when $u(\Gamma) \in \mathrm{Lip}\, p$,

1) $|\Sigma_N^\alpha(\Gamma) - u(\Gamma)| \leq A_1 N^{-p}$; if $\alpha(n-1)+2-n > p$;

2) $|\Sigma_N^\alpha(\Gamma) - u(\Gamma)| \leq A_2 N^{-p}\log N$, if $\alpha(n-1)+2-n = p$;

3) $|\Sigma_N^\alpha(\Gamma) - u(\Gamma)| \leq A_3 N^{n-2-\alpha(n-1)}$, if $\alpha(n-1)+2-n < p$.

When $\alpha = 1$, the Cesaro summation is just the Fejer summation and its kernels are

$$K_N(\Gamma) = \frac{1}{B_N} \left| \det\left(I + 2 \sum_{j=1}^{N} \Gamma^j \frac{N-j+1}{N+1}\right) \right|^{(n-1)/2} \qquad (2.33)$$

$$B_N = c^{-1} \int_{SO(n)} \left| \det\left(I + 2 \sum_{j=1}^{N} \Gamma^j (N-j+1)(N+1)^{-1}\right) \right|^{(n-1)/2} \dot{\Gamma}.$$

When $n = 2k$, (2.33) becomes

$$\frac{1}{B_N (N+1)^{n(n-1)/2}} \left| \frac{\det(I - \Gamma^{N+1})}{\det(I - \Gamma)} \right|^{(n-1)/2},$$

and

$$B_N = c^{-1}(N+1)^{-n(n-1)/2} \int_{SO(n)} \left| \frac{\det(I - \Gamma^{N+1})}{\det(I - \Gamma)} \right|^{(n-1)/2} \dot{\Gamma}.$$

**THEOREM 2.7.** (Wang Shi Kun and Dong Dao Zheng, see [1]) On SO(2k), the Fejer summation coefficients read

$$B_{m_1 \dots m_k} = \frac{((2k-1)!)^k \prod_{0 \le j < s \le k-1} (s^2 - j^2)}{N(m)(2k-1)! \dots (4k-3)! B_N (N+1)^{k(2k-1)}} \sum_{\substack{s_1=0 \\ e_1 \ge 1-k}}^{4k-2} \dots$$

$$\dots \sum_{\substack{s_k=0 \\ e_k \ge 1-k}}^{4k-2} (-1)^{s_1 + \dots + s_k} C_{s_k}^{4k-2} C_{2k-1}^{3k-2+e_1} \dots C_{s_k}^{4k-2} C_{2k-1}^{3k-2+e_k} \dots$$

$$\dots N((n-1-s_1)(N+1)-m_1, \dots, (n-1-s_k)(N+1)-m_k),$$

where $e_j = (2k-1)N-(N+1)s_j-m_j-k+j$, $j = 1,2,\dots,k$ and $m_1 \ge \dots \ge m_k \ge 0$ and its integral constants are

$$B_N = \frac{((2k-1)!)^k \prod_{0 \le j < s \le k-1} (s^2 - j^2)}{(2k-1)!(2k+1)! \dots (4k-3)!(N+1)^{k(2k-1)}} \sum_{\substack{s_1=0 \\ e_1 \ge 1-k}}^{4k-2} \dots$$

$$\dots \sum_{\substack{s_k=0 \\ e_k \ge 1-k}}^{4k-2} (-1)^{s_1 + \dots + s_k} C_{s_1}^{4k-2} \dots C_{s_k}^{4k-2} C_{2k-1}^{3k-2+e_1} \dots$$

$$\dots C_{2k-1}^{3k-2+e_k} N((n-1-s_1)(N+1), \dots, (n-1-s_k)(N+1)),$$

where $e_j = (2k-1)N-(N+1)s_j-k+j$, $j = 1,2,\dots,k$.

On SO(2k+1) the Fejer summation coefficients are

$$B_m = \frac{((2k)!)^k \frac{1}{2} \ldots (k - \frac{1}{2}) \prod_{0 \leq j < s \leq k-1} ((s + \frac{1}{2})^2 - (j + \frac{1}{2})^2)}{N(m)(2k)!(2k+2)! \ldots (4k-2)! B_N (N+1)^{k(2k-1)}}$$

$$\times \sum_{\substack{s_1=0 \\ e_1 \geq 1-k}}^{4k} \ldots \sum_{\substack{s_k=0 \\ e_k \geq 1-k}}^{4k} (-1)^{s_1+\ldots+s_k} C_{s_1}^{4k} \ldots C_{s_k}^{4k} C_{2k}^{3k-1+e_1} \ldots C_{2k}^{3k-1+e_k}$$

$$\times \prod_{j=1}^{k} ((n-1-s_j)(N+1)-\ell_j - \frac{1}{2})^{-1} N((n-1-s_1)(N+1)-m_1, \ldots, (n-1-s_k)(N+1)-m_k),$$

where $e_j = 2kN-(N+1)s_j-\ell_j$, $\ell_j = m_j+k-j$, $j = 1,2,\ldots,k$ and the integral constants are

$$B_N = \frac{((2k)!)^k \frac{1}{2} \ldots (k - \frac{1}{2}) \prod_{0 \leq j < s \leq k-1} ((s + \frac{1}{2})^2 - (j + \frac{1}{2})^2)}{(2k)!(2k+2)! \ldots (4k-2)!(N+1)^{k(2k-1)}}$$

$$\times \sum_{\substack{s_1=0 \\ e_1 \geq 1-k}}^{4k} \ldots \sum_{\substack{s_k=0 \\ e_k \geq 1-k}}^{4k} (-1)^{s_1+\ldots+s_k} C_{s_1}^{4k} \ldots C_{s_k}^{4k} C_{2k}^{3k-1+e_1} \ldots C_{2k}^{3k-1+e_k}$$

$$\times \prod_{j=1}^{k} ((n-1-s_j)(N+1)-k+j - \frac{1}{2})^{-1} N((n-1-s_1)(N+1), \ldots, (n-1-s_k)(N+1)),$$

where $e_j = 2kN-(N+1)s_j-k+j$, $j = 1,2,\ldots,k$.

The proof of Theorem 2.7 needs the method used in Theorem 2.3 and needs a complicated and skillful calculation. For details, see [1].

He Zhu Qi and Chen Guang Xiao defined Ceasro kernels and Cesaro summation on unitary symplectic groups USP(2n), n = 1,2,... .

Let u(U) be integrable on USP(2n) and its Fourier series is

$$u(U) \sim \sum_f N(f) tr(C_f A_f(U)) \tag{2.34}$$

where $f = (f_1, f_2, \ldots, f_n)$, $f_1 \geq f_2 \geq \ldots \geq f_n \geq 0$ are integers, $A_f(U)$ are the unitary single-valued irreducible representations of USP(2n) which take f as its labels, N(f) are the orders of $A_f(U)$, $\chi_f(U) = tr(A_f(U))$ are the characters of $A_f(U)$, and

$$C_f = c^{-1} \int_{USP(2n)} u(U) A_f(\overline{U}') \dot{U},$$

where c is the volume of USP(2n) and $\dot{U}$ is the volume element.

So, Cesaro means of (2.34) is

$$\tau_N^\alpha(U) = \sum_{(2n+1)N\geq f_1\geq\ldots\geq f_n\geq 0} B_f^\alpha(N)N(f)\mathrm{tr}(C_f A_f(U)) \tag{2.35}$$

where
$$B_f^\alpha(N) = (cN(f))^{-1} \int_{USP(2n)} \chi_f(\overline{V})K_N^\alpha(V)\dot{V} \tag{2.36}$$

and
$$K_N^\alpha(V) = \frac{1}{B_N^\alpha(2A_N^\alpha)^{n(2n+1)}} \left( \frac{\det(\sum_{k=0}^{N} A_{N-k}^{\alpha-1} V^k(I - \overline{V}'^{2k+1}))}{\det(I - \overline{V}')} \right)^{n+1/2} \tag{2.37}$$

is the Ceasro $(C,\alpha)$ kernel. In (2.37), $B_N^\alpha$ are those numbers such that

$$\int_{USP(2n)} K_N^\alpha(V)\dot{V} = 1$$

for $N = 1,2,3,\ldots$

**THEOREM 2.8.** (He Zu Qi and Chen Guang Xiao, see [1]) Let $u(U)$ be continuous on $USP(2n)$, $U \in USP(2n)$, then when $\alpha > (2n-2)/(2n+1)$, the Fourier series (2.34) of $u(U)$ is $(C,\alpha)$-summmable to itself, and when $u(U) \in \mathrm{Lip}\ p$, the following hold

1) $|\tau_N^\alpha(U) - u(U)| \leq A_1 N^{-p}$, if $(2n+1)\alpha-2n+2 > p$;

2) $|\tau_N^\alpha(U) - u(U)| \leq A_2 N^{-p} \log N$, if $(2n+1)\alpha-2n+2 = p$;

3) $|\tau_N^\alpha(U) - u(U)| \leq A_3 N^{2n-2-(2n+1)\alpha}$, if $(2n+1)\alpha-2n+2 < p$.

If $\alpha = 1$, it is just the Fejer summation and its Fejer kernels are

$$\frac{1}{B_N(N+1)^{n(2n+1)}} \left( \frac{\det(I - V^{N+1})}{\det(I - V)} \right)^{2n+1}.$$

Similar to the proof of Theorem 2.3 we can obtain the Fejer summation coefficients and its integral constants.

**THEOREM 2.9.** (He Zu Qi and Chen Guang Xiao, see [1]) The Fejer summation coefficients of Fourier series on the unitary symplectic groups are the following

$$B_f(N) = \frac{(-1)^{n(n+1)/2}((2n+1)!)^n(2n-1)!(2n-3)!\ldots3!}{B_N N(f)(N+1)^{n(2n+1)}(4n)!(4n-2)!\ \ldots\ (2n+2)!}$$

$$\times \sum_{\substack{s_1=0\\k_1\geq 1-n}}^{4n+2} \cdots \sum_{\substack{s_n=0\\k_n\geq 1-n}}^{4n+2} (-1)^{s_1+\ldots+s_n} C_{s_1}^{4n+2}\ldots C_{s_n}^{4n+2}\, C_{k+n-1}^{3n+k_1} \cdots$$

$$\times C_{k_n+n-1}^{3n+k_n} N((n+(2n+1)N-(N+1)s_1-f_1,\ldots,2n-1+(2n+1)N-(N+1)s_n-f_n),$$

where $k_j = (2n+1)N-(N+1)s_j-(f_j+n-j+1)$, $j = 1,2,\ldots,n$, and

$$B_N = \frac{(-1)^{n(n+1)/2}((2n+1)!)^n(2n-1)! \ldots 3!}{(N+1)^{n(2n+1)}(4n)!(4n-2)! \ldots (2n+2)!}$$

$$\times \sum_{\substack{s_1=0 \\ k_1 \geq 1-n}}^{4n+2} \ldots \sum_{\substack{s_n=0 \\ k_n \geq 1-n}}^{4n+2} C_{s_1}^{4n+2} \ldots C_{s_n}^{4n+2} C_{k_1+n-1}^{3n+k_n} \ldots C_{k_n+n-1}^{3n+k_n} (-1)^{s_1+\ldots+s_n}$$

$$\times N(n+(2n+1)N-(N+1)s_1, \ldots, 2n-1+(2n+1)N-(N+1)s_n).$$

Li Shi Xiong and Zheng Xue An defined and discussed Ceasro kernels and Cesaro summation of Fourier series connected with compact Lie groups. To begin with, Cesaro $(C,\alpha)$ kernels are defined on any compact Lie groups whose Lie algebra is one of the compact Lie algebras $(A_n)_u$, $(B_n)_u$, $(C_n)_u$, $(D_n)_u$, $(G_2)_u$, $(F_4)_u$, $(E_6)_u$, $(E_7)_u$, $(E_8)_u$ and $u_n = (A_{n-1})_u \oplus H^1$, $g_2 = (G_2)_u \oplus H^1$, $e_6 = (E_6)_u \oplus H^1$, $\tilde{e}_6 = (E_6)_u \oplus H^2$, $e_7 = (E_7)_u \oplus H^1$ and $H^n$ which is the Lie algebra of torus $T^n$ with dimension n. These Lie algebras are usually called the basic compact Lie algebras. For a general compact Lie group G, the Lie algebra of G can be decomposed either as a direct sum which consists of the basic compact Lie algebras listed above except $(A_n)_u$, $(G_2)_u$, $(E_6)_u$, $e_6$, $(E_7)_u$, or as a direct sum which consists of the basic compact Lie algebras listed above except $H^n$ and at least one of $(A_n)_u$, $(G_2)_u$, $(E_6)_u$, $e_6$, $(E_7)_u$ is included in it. This is called the regular decomposition of a compact Lie algebra. Here the Cesaro kernel of G is just a product or some restriction of the product of Cesaro kernels of several basic compact Lie groups mentioned above, which correspond to the regular decomposition of the Lie algebra of G. We have

**THEOREM 2.10.** (Li Shi Xiong and Zheng Xue An) (1) Let G be a compact Lie group, whose Lie algebra is one of $(A_n)_u$, $(B_n)_u$, $(C_n)_u$, $(D_n)_u$, $(G_2)_u$, $(F_4)_u$, $(E_6)_u$, $(E_7)_u$, $(E_8)_u$, $u_n$, $g_2$, $e_6$, $\tilde{e}_6$ , $e_7$, and again let the critical values $\alpha_0$ corresponding to the above-mentioned basic compact Lie algebras be

$$\alpha_0 = \frac{n}{n+1}, \frac{2n-1}{2n}, \frac{2n-2}{2n+1}, \frac{2n-2}{2n-1}, \frac{2}{3}, \frac{7}{8}, \frac{5}{6}, \frac{7}{8}, \frac{14}{15}, \frac{n-1}{n}, \frac{2}{3}, \frac{5}{6}, \frac{5}{6}, \frac{7}{8}, \tag{2.38}$$

respectively. Then the Cesaro means

$$\tau_N^\alpha(x) = f*K_N^\alpha(x)$$

of Fourier series of any continuous function f(x) on G uniformly converges to f(x) if $\alpha > \alpha_0$, where $x \in G$, $K_n^\alpha(x)$ stand for Cesaro $(C,\alpha)$ kernels on G, * denotes convolution; and if f(x) belongs to Lip p and $\alpha_0 = a/b$, then $\tau_N^\alpha(x)$ satisfy

a) $\left|\tau_N^{\alpha}(x) - f(x)\right| \leqslant A_1N^{-p}$, if $\alpha b-a > p$;

b) $\left|\tau_N^{\alpha}(x) - f(x)\right| \leqslant A_2N^{-p} \log N$, if $\alpha b-a = p$;

c) $\left|\tau_N^{\alpha}(x) - f(x)\right| \leqslant A_3N^{a-\alpha b}$, if $\alpha b-a < p$;

where a and b are given in (2.38).

2) Let G take L as its Lie algebra and the regular decomposition of L be

$$L = L_1 \oplus L_2 \oplus \dots \oplus L_k.$$

By $\alpha_0(L_j)$ we denote the critical values corresponding to $L_j$, $j = 1,2,\dots,k$, and set

$$\alpha_0 = \max_{1 \leqslant j \leqslant k} \{\alpha_0(L_j)\}.$$

Then the Cesaro summation of Fourier series of any continuous function f(x) on G uniformly converges to f(x) if $\alpha > \alpha_0$. Moreover, if $\alpha_0 = a/b$, then a), b), and c) corresponding to 1) are also valid.

The T-summation and T-summation of type II of Fourier series on unitary groups established by the methods "from kernel to sum and from sum to kernel" have similar extensions on compact Lie groups. The related detail is omitted.

## 3. The Cubical Partial Sums of Fourier Series

In this section we consider briefly the definition of the cubical partial sums of Fourier series on unitary groups and its extensions on classical groups and on compact Lie groups.

In the proof of Theorem 3.1, in which the concrete expression for Dirichlet kernels of the cubical partial sums of Fourier series on unitary groups was established, a basic method of studying class functions was set up. The essence of the method is to transform a research problem on class functions to a problem on Fourier series of the functions (such as $g(\theta_1,\dots,\theta_n)$ in (3.6)) on torus which are made by the product of values at the maximum torus of class functions and the Weyl function. This method is also widely applied to research for class functions on classical groups and compact Lie groups. Some researchers abroad such as R. J. Stanton and P. A. Tomas adopted this method in their studies on the almost everywhere convergence of Fourier series of class functions on compact Lie groups.

The cubical partial sum of Fourier series on unitary groups have two forms of extensions on compact Lie groups. One is made by R. J. Stanton and P. A. Tomas They, starting from the convex polyhedron (including the origin as its interior point) on Cartan sub-algebras of Lie algebras of compact Lie groups which is invariant under Weyl groups, defined the polyhedral partial sums, for which one of the fundamental properties for the cubical partial sums defined on unitary groups was used. Another is made by Li Shi Xiong and Zheng Xue An. They, starting from the regular coordinates for the highest weights in a cube or a polyhedron, defined the cubical and polyhedral sums of Fourier series on compact Lie groups, for which another basic property for the cubical partial sums defined on unitary groups was used.

For expressing Dirichlet kernels explicitly, a differential operator was established on unitary groups, by means of which Dirichlet kernels on unitary groups could be simply expressed by Dirichlet kernels of multiple Fourier series. Moreover, when we establish T-summation kernels of type II in section II and when we deduce the integral representations of the spherical means summation, this operator also play an important role. Wang Shi Kun, Dong Dao Zheng, He Zhu Qi, Chen Guang Xiao established corresponding differential operators on rotation groups and unitary symplectic groups respectively. Li Shi Xiong and Zheng Xue An established corresponding differential operators on general compact Lie groups and gave their representations under various systems of coordinates explicitly.

Some researchers abroad such as J. L. Clerc [11] established corresponding differential operators on (semi-simple) compact Lie groups which were expressed as directional derivatives.

When discussing the problem about the central multiplier on compact Lie groups, R. Coifman, G. Weiss [10] and N. J. Weiss [15] established the difference operators similar to the differential operators on unitary groups.

The cubical partial sums of Fourier series (1.9) of an integrable function $u(U)$ on the unitary group $U_n$ are defined by

$$S_N(u,U) = \sum_{N \geq \ell_1 > \dots > \ell_n \geq -N} N(f)\,\mathrm{tr}(C_f A_f(U)) \tag{3.1}$$

where $\ell_1 = f_1+n-1$, $\ell_2 = f_2+n-2$, ..., $\ell_n = f_n$.

Let

$$\mathcal{D}_N(V) = \sum_{N \geq \ell_1 > \dots > \ell_n \geq -N} N(f)\chi_f(V),$$

then

$$S_N(u,U) = u * \mathcal{D}_N(U) = c^{-1}\int_{U_n} u(U\bar{V}')\mathcal{D}_N(V)\dot{V}, \tag{3.2}$$

$\mathcal{D}_N(V)$ is called the Dirichlet kernel.

**THEOREM 3.1.** [2] Let $e^{i\theta_1}, \dots, e^{i\theta_n}$ be the characteristic roots of $V \in U_n$, $d_N(\theta) = \sum_{p=-N}^{N} e^{ip\theta}$, then we have

$$\mathcal{D}_N(V) = \frac{(-i)^{n(n-1)/2}}{(n-1)!\ldots 1!D(e^{i\theta_1},\ldots,e^{i\theta_n})} D\left(\frac{\partial}{\partial\theta_1},\ldots,\frac{\partial}{\partial\theta_n}\right)(d_N(\theta_1)\ldots d_N(\theta_n))$$

$$= \frac{(-i)^{n(n-1)/2}}{(n-1)!\ldots 1!D(e^{i\theta_1},\ldots,e^{i\theta_n})} \det(d_N^{(n-j)}(\theta_k)) \tag{3.3}$$

where $d_N^{(n-j)}(x) = (d/dx)^{n-j} d_N(x)$.

**PROOF.** The function

$$\sum_{N\geq \ell_1\geq\ldots\geq\ell_n\geq -N} \rho_f(r)N(f)\chi_f(V),$$

as Abel-means of $\mathcal{D}_N(V)$, is a class function, hence we only need to consider the following series

$$(D(e^{i\theta_1},\ldots,e^{i\theta_n}))^{-1} \sum_{N\geq \ell_1\geq\ldots\geq\ell_n\geq -N} \rho_f(r)N(f)M_f(e^{i\theta_1},\ldots,e^{i\theta_n}), \tag{3.4}$$

where $M_f(x_1,\ldots,x_n) = \det(x_k^{\ell_j})_{1\leq j,k\leq n}$.

From the definition (1.8) of $\rho_f(r)$, it is easy to see that the series in (3.4)

$$\sum_{N\geq \ell_1\geq\ldots\geq\ell_n\geq -N} \rho_f(r)N(f)M_f(e^{i\theta_1},\ldots,e^{i\theta_n}) \tag{3.5}$$

is the cubical partial sums of the multiple Fourier series of the function

$$g(\theta_1,\ldots,\theta_n) = |(1-re^{i\theta_1})\ldots(1-re^{i\theta_n})|^{-2n}(1-r^2)^{n^2} D(e^{i\theta_1},\ldots,e^{i\theta_n}). \tag{3.6}$$

Thus (3.5) can be expressed as

$$(2\pi)^{-n} \int_0^{2\pi}\ldots\int_0^{2\pi} g(\psi_1,\ldots,\psi_n) \prod_{j=1}^{n} d_N(\psi_j-\theta_j)d\psi_1\ldots d\psi_n. \tag{3.7}$$

In virtue of the skew-symmetry of $g(\psi_1,\ldots,\psi_n)$ under the permutation $(\psi_1,\ldots,\psi_n) \to (\psi_{j_1},\ldots,\psi_{j_n})$, (3.5) can also be expressed as

$$(n!)^{-1}(2\pi)^{-n} \int_0^{2\pi}\ldots\int_0^{2\pi} g(\psi_1,\ldots,\psi_n)P(\psi_1,\ldots,\psi_n;\ \theta_1,\ldots,\theta_n)d\psi_1\ldots d\psi_n$$

$$= \frac{1}{c}\int_{U_n} \frac{(1-r^2)^{n^2}}{|\det(I - rV)|^{2n}} P(V;\theta_1,\ldots,\theta_n)\dot{V}, \tag{3.8}$$

where $P(\psi_1,\ldots,\psi_n;\ \theta_1,\ldots,\theta_n) = \det(d_N(\psi_j-\theta_k))_{1\leq j,k\leq n}$, and when $\theta_1,\ldots,\theta_n$ are given and the characteristic roots of V are $e^{i\psi_1},\ldots,e^{i\psi_n}$, $P(V;\ \theta_1,\ldots,\theta_n)$ is a class function and its value for diagonal matrices is

$$P(\psi_1,\ldots,\psi_n;\ \theta_1,\ldots,\theta_n)\ /\ D(e^{-i\psi_1},\ldots,e^{-i\psi_n}). \tag{3.9}$$

By Theorem 1.1 the value of (3.8) is that of the continuous function of $\psi_1,\ldots,\psi_n$ at point $\psi_1 = \ldots = \psi_n = 0$ when $r \to 1$. By a result in [1], this is

$$(-i)^{n(n-1)/2}((n-1)!\ldots1!)^{-1}\ \det(d_N^{(n-j)}(\theta_k))_{1\leq j,k\leq n}.$$

Thus the conclusion follows from taking limit in (3.4).

For the uniform convergence of the cubical partial sums of Fourier series on unitary groups, the following results are valid.

**THEOREM 3.2.** [4] If $u(U) \in C^{n(n-1)/2+p}(U_n)$ $(0 < p \leq 1)$, then the cubical partial sums $S_N(u,U)$ of its Fourier series converge to $u(U)$ and

$$|S_N(u,U) - u(U)| \leq A\max\{(N^{-1}\log^{n^2}N)^{1/(n+1)},\ N^{-p}\log^{n-1}N\}.$$

Let $u(\Gamma)$ be an integrable function on $SO(n)$, and then the cubical partial sums of its Fourier series (2.29) are defined by

1) $$S_N(u,\Gamma) = \sum_{N\geq \ell_1\geq\ldots\geq|\ell_k|\geq 0} N(m)\mathrm{tr}(C_m A_m(\Gamma)) \tag{3.10}$$

if $n = 2k$ and $\Gamma \in SO(2k+1)$;

2) $$S_N(u,\Gamma) = \sum_{N\geq \ell_1\geq\ldots\geq\ell_k\geq 0} N(m)\mathrm{tr}(C_m A_m(\Gamma)) \tag{3.11}$$

if $n = 2k+1$ and $\Gamma \in SO(2k+1)$, where $m = (m_1,\ldots,m_k)$, $\ell_j = m_j+k-j$, $j = 1,2,\ldots,k$.

The following lemma is needed :

**LEMMA 1.** [8] Let $q_1\ldots,q_k$ be integers such that $q_1 > q_2 > \ldots > q_k \geq 0$, $p_j(q_s)$ be a function dependent only on $q_s$, $j = 1,2,\ldots,k$, and let $N$ be a positive integer, $a$ and $b$ be any real numbers, then

$$\det\Big(\sum_{q=0}^{N} q^{sa-b}\, p_j(q)\Big)_{1\leq s,j\leq k} = \sum_{N\geq q_1>\ldots>q_k\geq 0} \det(q_s^{ja-b})\det(p_j(q_s)).$$

The Dirichlet kernel of the cubical partial sums defined by (3.10) and (3.11) are

$$\mathcal{D}_N(\Gamma) = \sum_{N\geq\ell_1\geq\ldots\geq\ell_n\geq 0} N(m)\sigma_m(\Gamma), \tag{3.12}$$

therefore

$$S_N(u,\Gamma) = u*\mathcal{D}_N(\Gamma)$$

$$= c^{-1}\int_{SO(n)} u(\Gamma W')\mathcal{D}_N(W)\dot{W},$$

where the meaning of $\sigma_m(\Gamma)$ is given in Theorem 1.3. Then by Lemma 1, we have

**THEOREM 3.3.** [8] If $n = 2k$, then

$$\mathcal{D}_N(\Gamma) = \frac{2\det(d_N^{(2s-2)}(\theta_j))_{1\leq s,j\leq k}}{a_{2k}C(k-1,\ldots,0)}; \tag{3.13}$$

and if $n = 2k+1$, then

$$\mathcal{D}_N(\Gamma) = \frac{\det(e_N^{(2s-1)}(\theta_j))_{1\leq s,j\leq k}}{a_{2k+1}S(k-1/2,\ldots,1/2)}, \tag{3.14}$$

where $d_N(\theta) = \sin(N+\frac{1}{2})\theta \ / \ \sin\frac{1}{2}\theta$, $e_N(\theta) = \sin(N+1)\theta \ / \ \sin\frac{1}{2}\theta$, $a_{2k} = 2^{1-k}(2k-2)!\ldots4!2!$, $a_{2k+1} = 2^{-k}i^k(2k-1)!\ldots3!1!$, $C(q_1,\ldots,q_k) = \det(C_{q_s}(\theta_j))$, $C_q(\theta) = 2\cos q\theta$, $S(q_1,\ldots,q_k) = \det(S_{q_s}(\theta_j))$, $S_q(\theta) = 2i\sin q\theta$, $1 \leq j,s \leq k$, and $e^{\pm i\theta_1},\ldots,e^{\pm i\theta_k}$ are the characteristic roots of $\Gamma$.

For the convergence of Fourier series on $SO(n)$, the following theorem is valid.

**THEOREM 3.4.** (Wang Shi Kun and Dong Dao Zheng, see [1]) Let $u(\Gamma)$ be a function defined on $SO(n)$ and $\Gamma \in SO(n)$, moreover, let $u(\Gamma) \in C^{k^2+p}$ if $n = 2k+1$ and $u(\Gamma) \in C^{k(k-1)+p}$ if $n = 2k$, where $0 < p \leq 1$, then the partial sums $S_N(u,\Gamma)$ of Fourier series (2.29) of $u(\Gamma)$ converge to $u(\Gamma)$ and

$$|S_N(u,\Gamma) - u(\Gamma)| \leq A\max\{(N^{-1/(k+1)}(\log N)^{k^2/(k+1)},\ N^{-p}(\log N)^{k-1})\}.$$

The cubical partial sums of Fourier series (2.34) of integrable function $u(U)$ on $USP(2n)$ defined by He Zu Qi and Chen Guang Xiao are

$$S_N(u,U) = \sum_{N\geq \ell_1>\ldots>\ell_n\geq 0} N(f)\mathrm{tr}(C_f A_f(U)) \tag{3.15}$$

where $\ell_k = f_k+n-k+1$.

In the same way, we can obtain

$$S_N(u,U) = c^{-1}\int_{USP(2n)} u(\bar{V}'U)\mathcal{D}_N(V)\dot{V}, \tag{3.16}$$

where

$$\mathcal{D}_N(V) = \sum_{N\geq \ell_1>\ldots>\ell_n\geq 0} N(f)\chi_f(V) \tag{3.17}$$

are Dirichlet kernels.

**THEOREM 3.5.** (He Zu Qi and Chen Guang Xiao, see [1]) Let $u(U)$ be an integrable function on $USP(2n)$, $U \in USP(2n)$, then the partial sums (3.15) of its Fourier series (2.34) can be expressed as (3.16) and

$$\mathcal{D}_N(V) = \frac{(-1)^{n^2}\det(d_N^{(2k-1)}(\theta_j))_{1\leqslant k,j\leqslant n}}{\prod_{k=1}^{n}(2n-2k+1)!\det(\sin(n-p+1)\theta_j)_{1\leqslant p,j\leqslant n}}, \tag{3.18}$$

where $e^{\pm i\theta_1},\ldots,e^{\pm i\theta_n}$ are the characteristic roots of V.

It can be obtained that

**THEOREM 3.6.** (He Zu Qi and Chen Guang Xiao, see [1]) Let u(U) be an integrable function on USP(2n) and $u(U) \in C^{n^2+p}$, $0 < p \leqslant 1$, then $S_N(u,U)$ converges to u(U) uniformly and

$$|S_N(u,U) - u(U)| \leqslant A \max\{(N^{-1}\log^{n^2} N)^{1/(n+1)},\ N^{-p}(\log N)^{n-1}\}.$$

Li Shi Xiong and Zheng Xue An studied the cubical partial sums and polyhedral partial sums of Fourier series on compact Lie groups.

The coordinate representations of the Cartan subalgebra H' of any simple compact Lie algebra are known. As to $(A_{n-1})_u$, we have $H' = \{h_{\lambda_1\ldots\lambda_n} | \sum_1^n \lambda_j = 0\}$, thus the coordinate representation for Cartan subalgebra of $u_n$ are $H = \{h_{\lambda_1\ldots\lambda_n}\}$. The coordinate representation for Cartan subalgebras of other basic compact Lie algebras mentioned in section 2 can also be decided similarly. If L is a compact Lie algebra, then we take the direct sums of the above-mentioned coordinate representations for Cartan subalgebras of those basic compact Lie algebras which are included in the regular decomposition of L (see section 2) as the coordinate representations of the Cartan subalgebra of L. These representations are called the standard coordinate representations for Cartan subalgebra of compact Lie algebras.

Let H be a Cartan subalgebra of a compact Lie algebra L, and the Standard coordinate representation for the points in H be $h_{\lambda_1\ldots\lambda_p}$. Generally speaking, $\lambda_1,\ldots,\lambda_p$ are not necessarily independent of each other. Without loss of generality, assume that $\lambda_1,\ldots,\lambda_q$ is the maximal linearly independent system and that the system of affine coordinates composed by the maximal system is called the regular system on H. Thus any weight $\lambda$ on H can be uniquely expressed as

$$\lambda = f_1\lambda_1 + f_2\lambda_2 + \ldots + f_q\lambda_q,$$

where $f = (f_1,\ldots,f_q)$ are called the regular coordinates for $\lambda$.

Let L be the direct sum of a semi-simple compact Lie algebra L' and the centre $H^k$. Naturally, as a system of vectors on the Cartan sub-algebra H' of L', the system of roots of L' is just the system of vectors on the corresponding Cartan subalgebra $H = H' \oplus H^k$ of L which is called the system of roots for L and the group generated by the reflections with regard to the roots in H is called the Weyl group for L. Besides, $\beta$ denotes the half of the sum of all positive roots. Thus we verified that the equivalent class of all single-valued irreducible representations for a compact Lie group is uniquely determined by the highest weights. Moreover, we explicitly establish the method of calculating the regular coordinates for the highest weights.

Let G be a connected compact Lie group and $\hat{G}$ be the set of the highest weights of all nonequivalent single-valued irreducible representations for G. Let $A_\lambda(g)$, $g \in G$, be the single-valued irreducible unitary representation for G which takes $\lambda$ as its highest weight, and $d_\lambda$ be the order of $A_\lambda(g)$. Let $f(g) \in L(G)$. The Fourier series of $f(g)$ is usually expressed as

$$f(g) \sim \sum_{\lambda \in \hat{G}} d_\lambda \operatorname{tr}(C_\lambda A_\lambda(g)), \tag{3.19}$$

or

$$f(g) \sim \sum_{\lambda \in \hat{G}} d_\lambda f^* \chi_\lambda(g), \tag{3.20}$$

where $C_\lambda = \int_G f(g) A_\lambda(g^{-1}) dg$ and $\int_G dg = 1$, and $\chi_\lambda(g)$ is the character of $A_\lambda(g)$.

If $\lambda \in \hat{G}$ and the regular coordinate for $\lambda + \beta$ is $(\ell_1, \ldots, \ell_q)$ then the cubical partial sums of (3.19) or (3.20) are

$$\begin{aligned} S_N(f;g) &= \sum_{N \geqslant \ell_1, \ldots, \ell_q \geqslant -N} d_\lambda \operatorname{tr}(C_\lambda A_\lambda(g)) \\ &= \sum_{N \geqslant \ell_1, \ldots, \ell_q \geqslant -N} d_\lambda f^* \chi_\lambda(g) \\ &= f^* \mathcal{D}_N(g), \end{aligned}$$

where $\mathcal{D}_N(g)$ is Dirichlet kernels and

$$\mathcal{D}_N(g) - \sum_{N \geqslant \ell_1, \ldots, \ell_q \geqslant -N} d_\lambda \chi_\lambda(g).$$

Assume that $D_1(q)$ is a polyhedron in Euclidean space of dimension q which takes the origin as its interior point and the coefficients of the equations of all faces being integers. Set $D_N(q) = \{x \in E^q,\ x = ty,\ y \in D_1(q),\ 0 \leqslant t \leqslant N\}$. $\lambda + \beta \in D_N(q)$ means that the Regular coordinates for $\lambda + \beta$ belong to $D_N(q)$. Then the polyhedral partial sums of Fourier series of $f(g)$ are

$$S_N(f;g) = \sum_{\lambda+\beta \in D_N(q), \lambda \in \hat{G}} d_\lambda f^* \mathcal{D}_N^*(g),$$

and Dirichlet kernels are

$$\mathcal{D}_N^*(g) = \sum_{\lambda+\beta \in D_N(q), \lambda \in \hat{G}} d_\lambda \chi_\lambda(g).$$

In this connexion, we have the following results.

**THEOREM 3.7.** (Li Shi Xiong and Zheng Xue An) Let G be a compact Lie group, T be a maximal torus of G, dim G = n, dim T = q, m = (n-q)/2 and Lebesgue constants for its Dirichlet kernels be

$$\rho_N(G) = \int_G |\mathcal{D}_N(g)|dg$$

then 1) $\rho_N(G) \doteq A_G N^{[n/3]}(\log N)^s$, where $n \equiv s \bmod 3$, s = 0,1,2, [x] denotes the greatest integer of all integers that are not greater than x, if G takes one of $u_k$, $(B_k)_u$, $(C_k)_u$, $(D_k)_u$ as its Lie algebra;

2) $\rho_N(G) \doteq A_G N^{[(n+1)/3]}(\log N)^s$, where $n \neq 3$, $n+1 \equiv s \bmod 3$, s = 0,1,2; if G takes $(A_k)_u$ as its Lie algebra.

3) $\rho_N(G) \doteq A_G N^5 \log N$, if G takes $(G_2)_u$ as its Lie algebra;

4) $\rho_N(G) \doteq A_G N$, if G takes one of $(A_1)_u$, $(B_1)_u$, $(C_1)_u$ as its Lie algebra;

5) $\rho_N(G) \leqslant A_G N^m(\log N)^s$, where s = 2 for $\tilde{e}_6$ and s = 1 for the others, if G takes one of $g_2$, $(F_4)_u$, $(E_6)_u$, $(E_7)_u$, $(E_8)_u$, $e_6$, $\tilde{e}_6$, $e_7$ as its Lie algebra;

6) $\rho_N(G) = A_G \rho_N(G_1)\rho_N(G_2) \dots \rho_N(G_p)$, if $L = L_1 \oplus L_2 \oplus \dots \oplus L_p$ is the regular decomposition for Lie algebra L of G and $G_k$ is the basic compact Lie group of which Lie algebra is $L_k$, k = 1,2,...,p;

7) Lebesgue constant of the kernel $\mathcal{D}_N^*(g)$ of the polyhedral partial sums for Fourier series on G satisfies

$$\int_G |\mathcal{D}_N^*(g)|dg \leqslant A_G N^m(\log N)^s,$$

where $s \leqslant q$ and s = 1 for those Lie algebras in 1) - 5) and s = 2 only for $\tilde{e}_6$. Moreover, the conclusion in 6) is also valid for general compact Lie groups.

8) The condition for uniform convergence of the cubical and polyhedral partial sums of Fourier series and the estimation for the approximation of the partial sums to the functions can be deduced from Jackson theorem (see Theorem 4.14, 3)).

In Theorem 3.7, $\doteq$ denotes the principal part of $\rho_N(G)$. For Theorem 3.7, 1) - 4), the exact values for constants $A_G$ are already obtained by us.

Usually, the absolute convergence of Fourier series on compact Lie groups is expressed by

$$\sum_{\lambda\in\hat{G}} d_{\lambda_1} \sum_{i,j=1}^{d_\lambda} |c_{ij}^\lambda \ a_{ij}^\lambda(g)| < +\infty \tag{3.21}$$

where $C_\lambda = (c_{ij})_{1\leqslant i,j\leqslant d_\lambda}$, $A_\lambda(g) = (a_{ij}^\lambda(g))_{1\leqslant i,j\leqslant d_\lambda}$.

**THEOREM 3.8.** (Li Shi Xiong and Zheng Xue An) Let G, T, n, q, m be defined as in Theorem 3.7,

1) If $f(g) \in L_2^{k,p}(G)$, and in particular if $f(g) \in C^{k,p}(G)$, where $k = [n/2]$ and $p > n/2 - [n/2]$, $0 < p \leqslant 1$, then the Fourier series for $f(g)$ converges absolutely and uniformly, according to the definition of (3.21).

2) If $f(g) \in L_p^{k,s}(G)$, where k is a non-negative integer, $0 < r < p/(p-1)$, $1 < p \leqslant 2$, $0 < s \leqslant 1$, then Fourier series of $f(g)$ satisfies

$$\sum_{\lambda \in \hat{G}} (d_\lambda \sum_{i,j=1}^{d_\lambda} |c_{ij}^\lambda a_{ij}^\lambda(g)|)^r < +\infty$$

provided $k+s > ((3/p)-1/2)m+q(r^{-1}+p^{-1}-1)$.

3) If $f(g) \in L_p^{k,s}(G)$, where k is a non-negative integer, $0 < r < 2$, $1 < p \leqslant 2$,, $0 < s \leqslant 1$, then Fourier series of $f(g)$ satisfies

$$\sum_{\lambda \in \hat{G}} (d_\lambda \sum_{i,j=1}^{d_\lambda} |c_{ij}|^r)^{1/r} < +\infty$$

provided $k+s > ((3/r)+(3/p)-3)m+q/p$.

4) If $f(g) \in L_p^{k,s}(G)$, $1 < p \leqslant 2$, $0 < r < p/(p-1)$, $0 < s \leqslant 1$, k is non-negative integer, then

$$\sum_{\lambda \in \hat{G}} (d_\lambda \operatorname{tr}(C_\lambda \overline{C_\lambda'}))^r < +\infty$$

provided $k+s > (3/p-3/2)m+q(1/(2r)+1/p-1)$.

The principal results abroad parallel to those on the cubical partial sums of Fourier series in this section and to those on the summations of Fourier series in section 2 are as follows.

In [10], R. Coifman and G. Weiss studied the relation between the central multipliers for Fourier series on compact Lie groups and the multipliers for multiple Fourier series. They proved the following result.

Let H be the Cartan sub-algebra of the Lie algebra of a compact Lie group G, exp be the exponential mapping, and $\varepsilon$ be the unit element of G. If

$$\sum_{\lambda \in \hat{G}} \mathcal{D}(d_\lambda m_\lambda) C_\lambda(\tau) \tag{3.22}$$

defines the bounded multiplier on $L_p(H/\exp^{-1}\varepsilon)$, then

$$\sum_{\lambda \in \hat{G}} m_\lambda d_\lambda \chi_\lambda(g) \tag{3.23}$$

defines the bounded central multiplier on $L^p(G)$, where $p > 1$. In addition, the preceding conditions are also necessary for $p = 1$.

In (3.22), $\tau \in H$,

$$C_\lambda(\tau) = \sum_{\sigma \in W} e^{iB(\lambda, \sigma\tau)}, \tag{3.24}$$

where W denotes the Weyl group, and B( , ) represents the invariant inner product on the Lie algebra for G.

The difference operator $\mathcal{D}$ in (3.22) brings

$$\mathcal{D}(m_\lambda) = (\prod_{j=1}^{m} D_{\alpha_j})m_\lambda, \quad D_{\alpha_j} m_\lambda = m_{\lambda-\alpha_j} - m_\lambda,$$

where $\alpha_1, \alpha_2, \ldots, \alpha_m$ are all positive roots.

R. J. Stanton and P. A. Tomas (see [12] and [13]) discussed the polyhedral partial sums of Fourier series on compact Lie groups defined as follows.

Suppose that R is a closed connex polyhedron which takes the origin as its interior point and is invariant under the transformation of the Weyl group in the Cartan subalgebra H of the Lie algebra of a compact Lie group G, and let $R_t = \{tx | x \in R\}$. They defined the polyhedral partial sums of $f \in L^p(G)$, $p \geq 1$, by

$$S_N f(x) = \sum_{\lambda \in R_N} d_\lambda f * \chi_\lambda(x), \quad x \in G,$$

and proved the following:

1) Let G be a simply connected simple compact Lie group, T be a maximal torus of G, dim G = n, dim T = q and $L_I^p(G)$ be all class functions in $L_p(G)$. If $p > 2n/(n+q)$ and $f \in L_I^p(G)$, then $S_N f(x)$ almost everywhere converges to f.

2) When G, T, n, q are the same as In 1), and $p < 2n/(n+q)$ or $p > 2n/(n-q)$, there exists $f \in L_I^p(G)$ such that $S_N f(x)$ does not converge in the sense of $L^p$ norm.

3) When G, T, n, q are also the same as 1), there exists a number p(R), $2n/(n+q) \leq p(R) \leq (2n-2q+2)/(n-q+2)$ such that $S_N f(x)$ converges in the sense of $L^p$ norm, to f(x) for $p(R) < p < p(R)'$ and $f \in L_I^p(G)$.

4) When G is a simple connected semi-simple compact Lie group, then the results corresponding to 1) - 3) can be composed by combining the results of its simple subgroups.

5) When $p \neq 2$, there exists $f \in L^p(G)$ such that $S_N f(x)$ does not converge in the sense of $L^p$ norm.

6) When $p < 2$, there exists $f \in L_p(G)$ such that $S_N f(x)$ almost everywhere does not converge to f(x).

R. A. Mayer (see [18]) discussed Fourier series for G = SU(2). He proved the following.

1) Let $f \in C^1(G)$, then the Fourier series for f converges uniformly, and there exists $g \in C^1(G)$ such that its Fourier series does not converge absolutely.

2) Let $f \in L^2(G)$ and f belong to class $C^1$ almost everywhere, then the Fourier series of f almost everywhere converges to f. Here f that belongs to

class $C^1$ at one point means that f in a neighborhood of the point is equal to a function in $C^1(G)$.

3) Let $f \in L^1(G)$ and f be equal to zero in a neighborhood of a point $b \in G$. Moreover, Fourier series $\sum_{n=1}^{\infty} P_n f(x)$ for f satisfies $P_n f(b) \to 0$, when $n \to \infty$. Then $\sum_{n=1}^{\infty} P_n f(b)$ converges to zero.

Later, Mayar studied various problems about Fourier series on SU(2) systematically.

In [19], M. E. Taylor discussed the absolute convergence (in the sense of (3.21)) of Fourier series on compact Lie groups and proved : let G be a compact Lie group, dim G = n, and let $s > n/4$ be an integer. If $f \in H_{2s}$ and in particular if $f \in C^{2s}(G)$, then the Fourier series of f converges absolutely and uniformly.

D. L. Ragozin (see 3) of [20]) discussed the problem of the absolute convergence of Fourier series on compact Lie groups in the following sense and the problem of the relation between the convergence and the differentiability of f:

1) $$\sum_{\lambda} d_\lambda \| f * \chi_\lambda(g) \|_{L^p} < +\infty,$$

2) $$\sum_{\lambda} \{d_\lambda \operatorname{tr}(|C_\lambda|^p)\}^{1/p} < +\infty,$$

where the meaning of the related notations is the same as in (3.19) and (3.20), and $\operatorname{tr}(|C_\lambda|^p)$ is defined as follows:

Let $x_1, x_2, \ldots, x_{d_\lambda}$ be the characteristic roots of $C_\lambda \overline{C}_\lambda'$. Then they are non-negative and

$$\operatorname{tr}(|C_\lambda|^p) = \sum_{j=1}^{d_\lambda} x_j^{p/2}.$$

B. Dreseler (see [16] and [17]) studied Lebesgue constants for spherical partial sums of Fourier series on compact Lie groups and proved that the Lebesgue constants are $O(N^{(n-1)/2})$. Moreover, he gave the estimates from above and from below, n being the dimension of the group.

## 4. Summation by Spherical Means

The definition of summation by spherical means in harmonic analysis on unitary groups and the related methods (see [6]) are widely used in the research for harmonic analysis on classical groups and on compact Lie groups.

The spherical means summation of Fourier series on unitary groups, essentially, is such a summation that those terms of Fourier series corresponding to those functions having the same characteristic values of Laplace operator in the representative ring of a unitary group are multiplied by the same coefficient. This can easily be done by taking $\ell_k = f_k + (n-2k+1)/2$ in (4.1) and adding a factor function $\exp(-i(n-1)(\theta_1+\ldots+\theta_n)/2)$ to the integral expression (4.4).

In the research work on the spherical means summation in unitary groups, a method based on the Fourier transformation on Cartan subalgebras was established. As a Cartan sub-algebra, under the invariant inner product, constitutes an Euclidean space, a variety of tools of the Fourier transformation in the Euclidean space can be applied. Some researchers abroad such as R. S. Strichartz (see [14]) adopted a research method, whose basis is the Fourier transformation on the Lie algebra. Comparatively, the former not only can give an explicit expression and rather accurate results but also can give more results to a lot of problems. For example, a wide class of bounded operators on L(G) in Theorem 4.12 (1) which is established by the methods on unitary groups cannot be obtained by the methods on Lie algebras in [14]. But the latter certainly has some advantages over the former in some respects.

Let $u(U)$ be integrable on $U_n$. For Fourier series (1.9), we consider the following sum

$$\sum_m \sum_{\substack{f_1 \geqslant \dots \geqslant f_n \\ \ell_1^2+\dots+\ell_n^2=m,}} N(f)\operatorname{tr}(C_f A_f(U)) \tag{4.1}$$

where $\ell_k = f_k+n-k$, $k = 1,2,\dots,n$.

Let $\phi(t)$ be a function on $0 \leqslant t < \infty$, continuous at $t = 0$ and $\phi(0) = 1$. $\phi(t)$ gives us the means of (4.1) defined as follows:

$$\phi(\sqrt{b}/R)^{-1} \sum_m \phi(\sqrt{m}/R) \sum_{\substack{f_1 \geqslant \dots \geqslant f_n \\ \ell_1^2+\dots+\ell_n^2=m}} N(f)\operatorname{tr}(C_f A_f(U)). \tag{4.2}$$

Obviously, when $u(U)$ is integrable and $R$ is a constant, (4.2) is uniformly convergent for almost all $U \in U_n$, provided

$$\sum_f |\phi(\sqrt{m}/R)N(f)| < +\infty. \tag{4.3}$$

If the limit of (4.2) exists for $R \to \infty$ then Fourier series for $u(U)$ is called $\phi$-summable to a limit. In (4.2), $b = n(n-1)(2n-1)/6$.

Taking

$$H_s^\phi(c) = \frac{1}{c}\int_0^\infty \phi(u/c)u^{s+1}J_s(u)\,du,$$

$$W_s^\phi(c) = H_s^\phi(c)/c^{2s+1},$$

where $J_s(u)$ is the Bessel function of order s of the first kind.

**THEOREM 4.1.** (see [6]). If $u(U)$ is integrable on $U_n$, then (4.2) can be expressed as

$$S_R^\phi(u;U) = \frac{i^{(n-1)/2}R(2\pi)^{-n/2}}{n!(n-1)!\dots1!\phi(\sqrt{b}/R)} \int_{-\infty}^{\infty}\dots\int_{-\infty}^{\infty} \psi_U(e^{i\xi_1},\dots,e^{i\xi_n})D(e^{i\xi_1},\dots,e^{i\xi_n}) \times$$

$$\times D\left(\frac{\partial}{\partial\xi_1},\ldots,\frac{\partial}{\partial\xi_n}\right)H^{\phi}_{(n-2)/2}(|\xi|R)|\xi|^{1-n}d\xi_1 d\xi_2\ldots d\xi_n. \tag{4.4}$$

Here $\phi(t)$ must satisfy the following conditions:

1) $\phi(t)$ is absolutely continuous on any definite interval,

$$2)\quad \int_0^{\infty}|\phi(t)|t^{(n-1)/2}dt < \infty, \tag{4.5}$$

$$3)\quad \frac{\partial^{j_1}}{\partial\xi_1^{j_1}}\cdots\frac{\partial^{j_n}}{\partial\xi_n^{j_n}}\left.\frac{H^{\phi}_{(n-2)/2}(|\xi|R)}{|\xi|^{n-1}}\right|_{|\xi|=\infty} = 0, \tag{4.6}$$

where $0 \leqslant j_1,\ldots,j_n \leqslant n-1$, $j_1 + \ldots + j_n = n(n-1)/2$.

(4.4) can be rewritten as

$$\frac{(-i)^{(n-1)n/2}R(2\pi)^{-n/2}}{n!(n-1)!\ldots 1!\phi(\sqrt{b}/R)}\int_{-\infty}^{\infty}\ldots\int_{-\infty}^{\infty}\psi_U(r^{i\xi_1},\ldots,e^{i\xi_n})D(e^{i\xi_1},\ldots,e^{i\xi_n})$$
$$\times D(\xi_1,\ldots,\xi_n)H^{\phi}_{(n^2-2)/2}(|\xi|R)|\xi|^{1-n^2}d\xi_1 d\xi_2\ldots d\xi_n. \tag{4.7}$$

**PROOF.** We have

$$\frac{1}{c}\int_{U_n}\psi_U(e^{i\xi_1},\ldots,e^{i\xi_n})A_f(\overline{U}')\dot{U}$$
$$= \frac{1}{c}\frac{1}{c}\int_{U_n}\int_{U_n}u(UV\wedge\overline{V}')A_f(\overline{U}')\dot{V}\dot{U}$$
$$= \frac{1}{c}\frac{1}{c}\int_{U_n}\int_{U_n}u(UV\wedge\overline{V}')A_f(V\wedge\overline{V}')A(UV\wedge\overline{V}')'\dot{U}\dot{V}$$
$$= \chi_f(\Lambda)C_f N(f)^{-1},$$

where $\Lambda$ is a diagonal matrix and its diagonal elements are $e^{i\xi_1},\ldots,e^{i\xi_n}$.

From this, we obtain

$$\frac{1}{c}\int_{U_n}S_R^{\phi}(u;U)A_f(\overline{U}')\dot{U} = C_f\phi(\sqrt{m}/R)\phi(\sqrt{b}/R)^{-1}.$$

Thus the Fourier series of (4.4) which are integrable functions on $U_n$ for all $R > 0$ is

$$S_R^{\phi}(u;U) \sim \phi(\sqrt{b}/R)^{-1}\sum_{1}\phi(\sqrt{m}/R)N(f)\mathrm{tr}(C_f A_f(U)). \tag{4.8}$$

By (4.3), the series on the right side of (4.8) is absolutely convergent for almost every $U \in U_n$, thus (4.2) and (4.4) are equal for almost every $u \in U_n$.

As $D\left(\frac{\partial}{\partial\xi_1},\ldots,\frac{\partial}{\partial\xi_n}\right)(H^{\phi}_{(n-2)/2}(|\xi|)|\xi|^{1-n})$ can be calculated by recurrence formula it takes its original sign or the opposite sign under the permutation $(\xi_1,\ldots,\xi_n) \to (\xi_{j_1},\ldots,\xi_{j_n})$ according to the permutation being even or odd.

Thus it is equal to $(-1)^{n(n-1)/2}D(\xi_1,\ldots,\xi_n)H^{\phi}_{(n^2-2)/2}(|\xi|)|\xi|^{1-n^2}$, i.e. (4.4) and (4.7) are equal.

For $\phi(t)$ in the spherical means (4.2), the most interesting examples are the following:

1) $\phi(t) = e^{-t}$, the Poisson-Abel summation,

2) $\phi(t) = e^{-t^2}$, the Gauss-Sommerfeld summation,

3) $\phi(t) = \begin{cases} (1-t^2)^{\delta} & \text{for } 0 \leq t < 1, \\ 0 & \text{for } 1 \leq t, \end{cases}$ the Riesz summation of order $\delta$.

Then, in the Abel-, the Gauss-, and the Riesz-summation of order $\delta$ of Fourier series for $u(U)$ we consider

$$S_R^A(u,U) = \sum_m e^{-\sqrt{m}/R+\sqrt{b}/R} \sum_{\substack{f_1 \geq \ldots \geq f_n \\ \ell_1^2+\ldots+\ell_n^2=m}} N(f)\,\mathrm{tr}(C_f A_f(U)), \tag{4.9}$$

$$S_R^G(u;U) = \sum_m e^{-m/R^2+b/R^2} \sum_{\substack{f_1 \geq \ldots \geq f_n \\ \ell_1^2+\ldots+\ell_n^2=m}} N(f)\,\mathrm{tr}(C_f A_f(U)), \tag{4.10}$$

and

$$S_R^{\delta}(u;U) = \sum_{\substack{f_1 \geq \ldots \geq f_n \\ m = \ell_1^2+\ldots+\ell_n^2 \leq R^2}} (1-b/R^2)^{-\delta}(1-m/R^2)^{\delta}\, N(f)\,\mathrm{tr}(C_f A_f(U)) \tag{4.11}$$

respectively. It is obvious that these three summations satisfy the conditions in Theorem 4.1.

**THEOREM 4.2.** (see [6]). 1) Let $u(U)$ be continuous on $U_n$. Then the Abel means $S_R^A(u,U)$ of the Fourier series for $u(U)$ converges to $u(U)$ uniformly.

2) Let $u(U) \in L^p(U_n)$. Then $S_R^A(u;U) \in L^p(U_n)$ and $S_R^A(u;U)$ converges to $u(U)$ for $R \to \infty$ in the norm of $L^p(U_n)$, where $p \geq 1$; and $\|S_R^A(u;U)\|_p \leq A_0\|u(U)\|_p$.

3) Let $u(U)$ be integrable. Then $S_R^A(u;U)$ converges to $u(U)$ for $R \to \infty$ almost everywhere.

4) Let $u(U) \in \mathrm{Lip}\,\alpha$. Then

$$|S_R^A(u;U) - u(U)| \leq A_1 R^{-\alpha}$$

if $0 < \alpha < 1$, and

$$|S_R^A(u;U) - u(U)| \leq A_2 R^{-1}\log R,$$

if $\alpha = 1$.

**THEOREM 4.3.** (see [6]). 1) If $u(U)$ is continuous on $U_n$, then the Gauss means $S_R^G(u;U)$ of its Fourier series uniformly converges to $u(U)$ for $R \to \infty$, and if the modulus of continuity of $u(U)$ is $\omega(t)$, then

$$|S_R^G(u;U) - u(U)| \leq A_3\omega(R^{-1}),$$

2) If $u(U) \in L^p(U_n)$, then $S_R^G(u;U) \in L^p(U_n)$ and

$$\| S_R^G(u;U) \|_p \leq A_4 \| u(U) \|_p,$$

and $S_R^G(u;U)$ converges to $u(U)$ for $R \to \infty$ in the norm of $L^p(U_n)$, where $p \geq 1$.

3) If $u(U)$ is integrable on $U_n$, then $S_R^G(u;U)$ converges to $u(U)$ for $R \to \infty$ almost everywhere.

**THEOREM 4.4.** (see [6]). If $\delta > (n^2-1)/2$, then

1) $S_R^\delta(u;U)$ converges to $u(U)$ uniformly, for $R \to \infty$ if $u(U)$ is continuous on $U_n$. And if $u(U) \in \text{Lip}\ \alpha$, $0 < \alpha \leq 1$, then

a) $|S_R^\delta(u;U) - u(U)| \leq A_5 R^{-\delta+(n^2-1)/2}$, if $\alpha + (n^2-1)/2 > \delta$;

b) $|S_R^\delta(u;U) - u(U)| \leq A_6 R^{-\alpha}\log R$, if $\alpha + (n^2-1)/2 = \delta$;

c) $|S_R^\delta(u;U) - u(U)| \leq A_7 R^{-\alpha}$, if $\alpha + (n^2-1)/2 < \delta$.

2) If $u(U) \in L^p(U_n)$, $p \geq 1$, then $S_R^\delta(u;U)$ converges to $u(U)$ in the norm of $L^p(U_n)$, for $R \to \infty$, and $\| S_R^\delta(u;U) \|_p \leq A_8 \| u(U) \|_p$.

3) If $u(U)$ is integrable on $U_n$, then $S_R^\delta(u;U)$ converges to $u(U)$ almost everywhere for $R \to \infty$.

In Theorems 4.2, 4.3 and 4.4, the numbers $A_0, A_1, \ldots, A_8$ are independent of $R$.

**THEOREM 4.5.** (see [6]) Let $\phi(t)$ be absolutely continuous on any finite interval, and (4.3), (4.5), and (4.6) be valid for $\phi(t)$. Moreover, we have

$$1) \qquad D\left( \frac{\partial}{\partial \xi_1}, \ldots \frac{\partial}{\partial \xi_n} \right) \frac{H^\phi_{n-1/2}(|\xi|R)}{|\xi|^{n-1}} = O(R^{(n-1)(n+2)/2}),$$

if $0 \leq |\xi| \leq 1/R$, and

$$2) \qquad D\left( \frac{\partial}{\partial \xi_1}, \ldots, \frac{\partial}{\partial \xi_n} \right) \frac{H^\phi_{n/2-1}(|\xi|R)}{|\xi|^{n-1}} = O(R^{-p-1}|\xi|^{-p-n(n+1)/2}),$$

where $p > 0$, if $1/R \leq |\xi| < \infty$.

Then, for $R \to \infty$, $S_R^\phi(u;U)$ uniformly converges to $u(U)$ provided $u(U)$ is continuous.

As in the case of rotation groups, Wang Shikun and Dong Daozheng discussed the summation of Fourier series on rotation groups by spherical means. They proved:

**THEOREM 4.6.** (Wang Shikun and Dong Daozheng see [1]). Let $u(\Gamma)$ be integrable on $SO(n)$ and

$$\sum_m |\phi(\sqrt{b}/R)N(m)| < +\infty, \tag{4.12}$$

where $b = 1^2 + 2^2 + \ldots + (k-1)^2$ if $n = 2k$, and $b = (1/2)^2 + \ldots + (k-1/2)^2$ if $n = 2k+1$, and then the integral representation of the spherical means of Fourier series for $u(\Gamma)$

$$\phi(\sqrt{b}/R)^{-1} \sum_m \phi(\sqrt{q}/R) \sum_{\ell_1^2+\ldots+\ell_k^2=q} N(m)\mathrm{tr}(C_m A_m(\Gamma)) \tag{4.13}$$

is

$$S_R^\phi(u;\Gamma) = \frac{(-i)^{n(n-1)/4-k/2}R(2\pi)^{-k/2}}{\phi(\sqrt{q}/R)a_n k!2^k} \int_{-\infty}^{\infty}\cdots\int_{-\infty}^{\infty} \psi_\Gamma(\xi)G_n(2i\sin\tfrac{1}{2}\xi_1,\ldots,2i\sin\tfrac{1}{2}\xi_k)$$

$$\times G_n\left(\frac{\partial}{\partial\xi_1},\ldots,\frac{\partial}{\partial\xi_k}\right)\left(H^\phi_{k/2-1}(|\xi|R)|\xi|^{1-k}\right)d\xi_1\ldots d\xi_k, \tag{4.14}$$

where $k = [n/2]$, $\ell_j = m_j+k-j+1/2$ if $n = 2k+1$, and $\ell_j = m_j+k-j$ if $n = 2k$, $j = 1,2,\ldots,k$, $G_{2k}(x_1,\ldots,x_k) = D(x_1,\ldots,x_k)$, $G_{2k+1}(x_1,\ldots,x_k) = x_1\ldots x_k D(x_1,\ldots,x_k)$, $a_{2k} = (1/2)D(k-1)^2,\ldots,0^2)$, $a_{2k+1} = (1/2)\ldots(k-1/2)D((k-1)^2,\ldots,0^2)$, and $\phi(t)$ must satisfy the following conditions:

1) $\phi(t)$ is absolutely continuous on any finite interval;

2) $\int_0^\infty |\phi(t)|t^{k/2-1/2}dt < +\infty$;

3) $$\left.\frac{\partial^{j_1}}{\partial\xi_1^{j_1}}\cdots\frac{\partial^{j_k}}{\partial\xi_k^{j_k}}\frac{H^\phi_{k/2-1}(|\xi|R)}{|\xi|^{k-1}}\right|_{|\xi|=\infty} = 0, \text{ where } 0 \leqslant j_1,\ldots,j_k \leqslant n-2.$$

**THEOREM 4.7.** (Wang Shikun and Dong Daozheng, see [1]). By taking the above-mentioned three functions as $\phi(t)$, define the so-called the Abel, the Gauss and the Riesz summations of order $\delta$ of Fourier series on rotation groups respectively. For these three summations the following results are valid where $\delta > n(n-1)/4-1/2$ is needed:

1) These three summations uniformly converge to $u(\Gamma)$ for $R \to \infty$, provided $u(\Gamma)$ is continuous on $SO(n)$.

2) $S_R^A(u;\Gamma)$, $S_R^G(u;\Gamma)$, $S_R^\delta(u;\Gamma)$, in the of $L^p(SO(n))$, $p \geqslant 1$, converge to $u(\Gamma)$, provided $u(\Gamma) \in L^p(SO(n))$.

3) $S_R^A(u;\Gamma)$, $S_R^G(u;\Gamma)$, $S_R^\delta(u;\Gamma)$ almost everywhere converge to $u(\Gamma)$, provided $u(\Gamma)$ is integrable.

**THEOREM 4.8.** (Wang Shikun and Dong Daozheng, see [1]). Let $u(\Gamma)$ be continuous on $SO(n)$, and $\phi(t)$ be absolutely continuous in any finite interval and satisfy

1) $\int_0^{\infty} |\phi(t)| t^{(k-1)/2} dt < +\infty$;

2) $\left. \frac{\partial^{j_1}}{\partial \xi_1^{j_1}}, \ldots, \frac{\partial^{j_k}}{\partial \xi_k^{j_k}} \frac{H^{\phi}_{k/2-1}(|\xi|R)}{|\xi|^{k-1}} \right|_{|\xi|=\infty} = 0$;

3) $G_n\left(\frac{\partial}{\partial \xi_1}, \ldots, \frac{\partial}{\partial \xi_k}\right) \frac{H^{\phi}_{k/2-1}(|\xi|R)}{|\xi|^{k-1}} = O(R^{kn-k^2-1})$, if $0 \leqslant |\xi| \leqslant 1/R$;

4) $G_n\left(\frac{\partial}{\partial \xi_1}, \ldots, \frac{\partial}{\partial \xi_k}\right) \frac{H^{\phi}_{k/2-1}(|\xi|R)}{|\xi|^{k-1}} = O(R^{-p-1}|\xi|^{-p-kn+k^2})$;

if $|\xi| > 1/R$, then $S^{\phi}_R(u;\Gamma)$ uniformly converges to $u(\Gamma)$.

He Zuqi and Chen Guangxiao discussed the summation by spherical means on unitary symplectic groups and obtained the following results.

**THEOREM 4.9.** (He Zuqi and Cheng Guangxiao, see [1]). Let $u(U)$ be integrable on $USP(2n)$ and

1) $\sum_1 |\phi(\sqrt{m}/R) N(f)| < +\infty$,

2) $\phi(t)$ is absolutely continuous in any finite interval,

3) $\int_0^{\infty} |\phi(t)| t^{(n-1)/2} dt < +\infty$,

4) $\left. \left(\frac{\partial}{\partial \xi_1}\right)^{j_1} \cdots \left(\frac{\partial}{\partial \xi_n}\right)^{j_n} \frac{H^{\phi}_{n/2-1}(|\xi|R)}{|\xi|^{n-1}} \right|_{|\xi|=\infty} = 0$, where $1 \leqslant j_1, \ldots, j_n \leqslant 2n-1$.

Then the spherical means of Fourier series for $u(U)$

$$\phi(\sqrt{b}/R)^{-1} \sum_m \phi(\sqrt{m}/R) \sum_{\substack{f_1 \geqslant \ldots \geqslant f_n \geqslant 0 \\ \ell_1^2 + \ldots + \ell_n^2 = m}} N(f) \operatorname{tr}(C_f A_f(U)) \tag{4.15}$$

whose integral representation is

$$S^{\phi}_R(u;U) = \frac{(-i)^{n^2} R(2\pi)^{-n/2}\, 2^{-n}}{n! D(n^2, \ldots, 1^2) n! \phi(\sqrt{b}/R)} \int_{-\infty}^{\infty} \cdots \int_{-\infty}^{\infty} \psi_U(e^{i\xi_1}, \ldots, e^{i\xi_n})$$

$$\times S(n, \ldots, 1) Q\left(\frac{\partial}{\partial \xi_1}, \ldots, \frac{\partial}{\partial \xi_n}\right)(H^{\phi}_{k/2-1}(|\xi|R)|\xi|^{1-n}) d\xi_1 \ldots d\xi_n, \tag{4.16}$$

where $Q(x_1, \ldots, x_n) = x_1 x_2 \ldots x_n D(x_1^2, \ldots, x_n^2)$, $\ell_k = f_k + n + 1 - k$, $k = 1, 2, \ldots, n$, and $b = n(n+1)(2n+1)/6$.

**THEOREM 4.10.** (He Zuqi and Chen Guangxiao, see [1]). By taking the above-mentioned three functions as $\phi(t)$, spherical means $S^A_R(u;U)$, $S^G_R(u;U)$ and $S^{\delta}_R(u;U)$ of Fourier series for $u(U)$ are defined respectively, and the following results are valid (for the Riesz means, the condition $\delta > n^2 + (n-1)/2$ is needed):

1) The three spherical means converge to $u(U)$ for $R \to \infty$ if $u(U)$ is continuous on $USP(2n)$.

2) For $p \geqslant 1$, the three spherical means converge to $u(U)$ in the norm of $L^p(USP(2n))$ if $u(U) \in L^p(USP(2n))$.

3) The three spherical means almost everywhere converge to $u(U)$ for $R \to \infty$ if $u(U)$ is integrable.

**THEOREM 4.11.** (He Zuqi and Chen Guanxiao, see [1]). Let $\phi(t)$ satisfy the conditions in Theorem 4.9. Moreover

$$\left(\frac{d}{udu}\right)^{n^2}\left(\frac{H^{\phi}_{n/2-1}(u)}{u^{n-1}}\right) = O(u^{-(2n^2+n+p)}),$$

where $p > 0$ and $u(U)$ is continuous. Then $S^{\phi}_R(u;U)$ uniformly converges to $u(U)$ for $R \to \infty$.

Li Shixiong and Zheng Xucan discussed the spherical means and the more general means of Fourier series on compact Lie groups.

Let $G$ be a compact Lie group of dimension $n$, $T$ be a maximal torus of dimension $q$ of $G$, $m = (n-q)/2$, $H$ be the Cartan sub-algebra of the Lie algebra of $G$, $B(\ ,\ )$ be the invariant inner product on the Lie algebra of $G$, and $(\ ,\ )^*$ be the special invariant inner product which is called quasi Killing-Cartan form on compact Lie algebras. The related definition can be found in "Fourier analysis on compact Lie groups" to appear in "Advances in Mathematics (in Chinese) [21].

Let $f(g) \in L(G)$. We consider the following means of Fourier series for $f(g)$.

1) Let $\phi(t)$, $H^{\phi}_S(t)$ and $W^{\phi}_S(t)$ define as before and consider the spherical means of Fourier series (3.20) for $f(g)$

$$\sum_{\lambda \in \hat{G}} \phi(|\lambda+\beta|/R)\phi(|\beta|R)^{-1} d_\lambda f * \chi_\lambda(g). \tag{4.17-1}$$

2) Let $\hat{\phi}(h) \in L(H)$, $\hat{\phi}(h)$ be invariant under the transformation of the Weyl group, $h \in H$,

$$\phi(h) = (2\pi)^{-q/2} \int_H \hat{\phi}(y) e^{-iB(h,y)} dy$$

and consider the means

$$\sum_{\lambda \in \hat{G}} \phi\left(\frac{\lambda+\beta}{R}\right)\phi\left(\frac{\beta}{R}\right)^{-1} d_\lambda f^*\chi_\lambda(g), \tag{4.17-2}$$

$$\sum_{\lambda \in \hat{G}} \frac{\sum_\sigma W\phi((\lambda+\beta-\sigma h_o)/R)}{\sum_\sigma W\phi((\beta-\sigma h_o)/R)} d_\lambda f^*\chi_\lambda(g), \tag{4.17-3}$$

$$\sum_{\lambda \in \hat{G}} \frac{\int_G \phi((\lambda+\beta-ad_g\beta)/R)dg}{\int_G \phi((\beta-ad_g\beta)/R)dg} d_\lambda f^*\chi_\lambda(g). \tag{4.17-4}$$

Take $\hat{\phi}(h) = W^{\phi}_{q/2-1}(|h|)$ in (4.17-2). It is obvious that (4.17-2) becomes (4.17-1), where

$$|\lambda+\beta| = |B(\lambda+\beta,\ \lambda+\beta)|^{1/2},$$

and, as a function on H, $\phi(h)$ is invariant under the transformation of the Weyl group. Thus $\phi(h)$ can be uniquely extended as a function on the Lie algebra for G, the values of which are $\phi(ad_g h) = \phi(h)$ for $h \in H$ and $g \in G$.

Let $\alpha_1,\ldots,\alpha_m$ be all positive roots on H; $p(h) = \Pi_{j=1}^{m} B(h,\alpha_j)$; $X_1, X_2,\ldots,X_q$ be the orthonormal basis for $B(\ ,\ )$ in H; $\frac{\partial}{\partial x} = X_1 \frac{\partial}{\partial x_1} + X_2 \frac{\partial}{\partial x_2} + \ldots + X_q \frac{\partial}{\partial x_q}$, and $h = x_1X_1 + x_2X_2 + \ldots + x_qX_q$ be a point in H; again let W denote the Weyl group, $|W|$ be the order of the Weyl group; $\psi_g(h) = \psi_g(\exp h) = \int_G f(gt \exp(h)t^{-1})dt$, $h \in H$, $g,t \in G$, exp be the exponential mapping; $\Delta(h) = \Pi_{j=1}^{m}(2i \sin \frac{1}{2} B(h,\alpha_j))$; $Q(h) = \sum_{\sigma\in W} e^{iB(h_0,\sigma h)}$, $h_0 \in H$. Then the integral expressions for (4.17-1) $\sim$ (4.17-4) are respectively

$$S^{\phi}_{1,R}(f;g) = C_1R^q \int_H \psi_g(-h)\Delta(-h)P(\tfrac{\partial}{\partial x})\{W^{\phi}_{q/2-1}(|h|R)\}dh$$

$$= (-1)^m C_1 R^n \int_H \psi_g(-h)\Delta(-h)P(h)W^{\phi}_{n/2-1}(|h|R)dh;$$

$$S_{2,R}(f;g) = C_2R^q \int_H \psi_g(-h)\Delta(-h)P(\tfrac{\partial}{\partial x})\ \{\hat{\phi}(Rh)\}dh;$$

$$S_{3,R}(f;g) = C_3R^q \int_H \psi_g(-h)\Delta(-h)P(\tfrac{\partial}{\partial x})\ \{\hat{\phi}(Rh)Q(h)\}dh;$$

$$S_{4,R}(f;g) = C_4R^q \int_H \psi_g(-h)|\Delta(-h)|^2P(h)^{-1}P(\tfrac{\partial}{\partial x})\ \{\hat{\phi}(Rh)\}dh,$$

where

$$C_1 = (-i)^m(\phi(|\beta|/R)(2\pi)^{q/2}|W|P(\beta))^{-1},$$

$$C_2 = (-i)^m(\phi(\beta/R)(2\pi)^{q/2}|W|P(\beta))^{-1},$$

$$C_3 = (-i)^m(P(\beta)|W|(2\pi))^{q/2} \textstyle\sum_{\sigma\ W} \phi((\beta-h_0)/R))^{-1},$$

$$C_4 = (-i)^m((2\pi)^{q/2}|W|P(\beta)C_0)^{-1},$$

and $C_0$ depends only on the Lie algebra.

For the means (4.17-1) $\sim$ (4.17-4) and for their corresponding integral expressions $S^{\phi}_{1,R}(f;g) \sim S^{\phi}_{4,R}(f;g)$, the following results are valid.

**THEOREM 4.12.** (Li Shixiong and Zheng Xuean). Let $\hat{\phi}(h)$ have $L^1$ partial derivatives of up to m-times on H. Then the following hold.

1) If $f(g) \in L^p(G)$, $p \geq 1$, for $j = 1,2,3$, then $S^{\phi}_{j,R}(f;g)$ are bounded linear operations on $L^p(G)$ and

$$\|S^{\phi}_{j,R}(f;g)\|_p \leq A(G,\phi,j,R) \|f\|_p,$$

and $S^{\phi}_{j,R}(f;g)$ is regarded as an integrable function for which the Fourier series is just (4.17-j).

2) If $f(g) \in L^p(G)$, $p \geq 1$, and if $P(h)P(\frac{\partial}{\partial x})\{\hat{\phi}(h)\} \in L(H)$, then for $j = 1,2,3,4$, $S^{\phi}_{j,R}(f;g)$ is a bounded linear operator on $L^p(G)$ and

$$\sup_{R>0} \{\|S^{\phi}_{j,R}(f;g)\|_p\} \leq A(G,\phi,j) \|f\|_p,$$

$$\lim_{R\to\infty} \|S^{\phi}_{j,R}(f;g) - f(g)\|_p = 0,$$

and $S^{\phi}_{j,R}(f;g)$ is regarded as an integrable function for which the Fourier series is just (4.17-j).

3) If $|P(h)^{-1}P(\frac{\partial}{\partial x})\{\hat{\phi}(h)\}| \leq A(1 + |h|)^{-n-\varepsilon}$, and $j = 1,2,3,4$, $\varepsilon > 0$, then besides 2) of this theorem, the following results are valid.

a) $S^{\phi}_{j,R}(f;g)$ almost everywhere converges to $f(g) \in L(G)$ for $R \to \infty$.

b) $S^{\phi}_{j,R}(f;g)$ uniformly converges to $f(g)$ for $R \to \infty$ if $f(g)$ is continuous.

c) $\sup_{R>0} |S^{\phi}_{j,R}(f;g)| \leq A(G,\phi,j)(Mf(g) + \int_G |\Delta(t)|^{-1}|f(gt^{-1})|dt)$, where $Mf(g) = \sup_{r>0} \int_{B(g;r)} |f(t)|dt|B(g;r)|^{-1}$, and $B(g;r)$ denotes all $t \in G$ from which the Riemann distance to $g$ is less than $r$;

d) $|\{\sup_R |S^{\phi}_{j,R}(f;g)| > y\}| \leq A(G,\phi,j)y^{-1}\|f\|_{L^1}$.

4) If $\sum_{\lambda\in\hat{G}} |\phi((\lambda+\beta)/R)|d_\lambda < +\infty$, then, in the sense of that

$$\sum_{\lambda\in G} |\phi((\lambda+\beta)/R)d_\lambda f*\chi_\lambda(g)| < +\infty,$$

(4.17-j) absolutely converges for almost every $g \in G$ and $j = 1,2$. And, in the meantime, (4.17-j) is equal to $S^{\phi}_{j,R}(f;g)$ for almost every $g \in G$ and $j = 1,2$, if $\hat{\phi}(h)$ satisfies the first condition, where $f(g) \in L(G)$ besides the above mentioned conditions.

5) (4.17-j) in the sense of 4) absolutely converges for all $R$ and $j = 1,2,3,4$, if $|\phi(h)| \leq C(1 + |h|)^{-n/2-q/2-\varepsilon}$, $\varepsilon > 0$. And, in the meantime, (4.17-j) is equal to $S^{\phi}_{j,R}(f;g)$ for almost every $g \in G$ and $j = 1,2,3$ or $j = 4$, if $\hat{\phi}(h)$ satisfies the first condition or the condition in 3) respectively, besides the above mentioned conditions.

6) From the Poisson summation formula the summation kernels $K^{\phi}_{j,R}(g)$ can be deduced which satisfies

$$S^{\phi}_{j,R}(f;g) = \int_G f(gt^{-1})\, K^{\phi}_{j,R}(t)dt. \tag{4.18}$$

**THEOREM 4.13.** (Li Shixiong and Zheng Xuean). Take $\phi(t) = (1-t^{2k})^{\delta}$, for $0 \leqslant t < 1$ and 0 for $t \geqslant 1$, k being a positive integer. Thus (4.17-1) defines the Riesz summation of order $\delta$ and degree 2k of Fourier series on compact Lie groups. $S^{2k,\delta}_R(f;g)$ denotes $S^{2k,\delta}_{1,R}(f;g)$. When k = 1, it is the usuall Riesz summation denoted by $S^{\delta}_R(f;g)$. Then $S^{2k,\delta}_R(f;g)$ satisfies the following:

1) The conclusion of Theorem 4.12 is valid for $S^{2k,\delta}_R(f;g)$ if $\delta > (n-1)/2$.

2) If f(g) is continuous on G and $\delta > (n-1)/2$ then

$$|S^{2k,\delta}_R(f;g) - f(g)| \leqslant A(G,k,\delta)\omega(f;1/R).$$

3) The saturation order of $S^{2k,\delta}_R$ is $R^{-2k}$.

**THEOREM 4.14.** (Li Shixiong and Zheng Xuean). If $\delta > (n-1)/2$, then

1) $|S^{2k,\delta}_R(f;g) - f(g)| \leqslant A(G,k,\delta,\omega)\, \|f\|_{s,\omega} R^{-s}\omega(1/R)$ if $f(g) \in C^{s,\omega}(G)$, $s \leqslant 2k-1$.

2) $|S^{2k,\delta}_R(f;g) - f(g)| \leqslant A(G,k,\delta)\, \|f\|_{2k} R^{-2k}$ if $f(g) \in C^{2k}(G)$.

3) Let $E_R(f) = \sup_{r \leqslant R} \{\| \sum_{|\lambda+\beta| \leqslant r} d_\lambda tr(C_\lambda A_\lambda(g)) - f(g)\|_\infty\}$, $C_\lambda$ an arbitrary

complex matrix of order $d_\lambda$ and independent of f(g). From 1) and 2) of this theorem, the Jackson Theorem follows:

$$E_R(f) \leqslant A(G,k,\beta)\, \|f\|_{k,\omega} R^{-k}\omega(1/R),$$

if $f(g) \in C^{k,\omega}(G)$. Besides, the Bernstein Theorem can be directly deduced by the unitary representations for compact Lie groups.

4) Let $M > (n-1)/2$, M be an integer, and

$$V^M_R(f;g) = \sum_{k=0}^{M} (-1)^k\, T_{M-k}(4) S^M_{2^{m-k}R}(f;g) \Big/ \prod_{j=0}^{m-1} (4^{M-j}-1),$$

where $T_k(y) = \sum_{j=1}^{C^k_M} y^{x_j}$ and any of $x_1, x_2, \ldots, x_{C^k_M}$ is a sum of k numbers chosen from 1,2,...,M. Then

$$\|V^M_R(f;g) - f(g)| \leqslant (\sup_{R>0} \|V^M_R\| + 1)E_R(f),$$

where $\sup_{R>0} \|V^M_R\| < +\infty$.

**THEOREM 4.15.** (Li Shixiong, Fan Dashan and Zheng Xuean). Take $\delta_0 = (n-1)/2$. Then the kernels of Riesz means of (2k, $\delta$) satisfy

1) $\|K^{2k,\delta_0}_R(g)\|_{L^1} \doteq A(G,k) \log R$, ((see (4.18)).

2) $\|K^{2k,\delta}_R(g)\|_{L^1} \doteq A(G,k,\delta)R^{(n-1)/2-\delta}$, for $0 \leqslant \delta < (n-1)/2$.

3) Let $1 < p \leq 2$, $\delta > (n-1)|1/2-1/p|$, $\dim H > 1$, $f(g) \in L^p(G)$, and then

a) $\| \sup_{R>0} |S_R^{2k,\delta}(f,g)| \|_p \leq A(G,P,k) \|f\|_p$;

b) $S_R^{2k,\delta}(f;g)$ converges to $f(g)$ almost everywhere;

c) $\lim_{R\to\infty} \| S_R^{2k,\delta}(f;g) - f(g) \|_p = 0$.

4) Let $f^*(g;r) = r^{1-n} \int_{B(g;r)} f(y)dy$, where $g,y \in G$. If for almost every $g \in G$ there exists $r_0 > 0$ such that

$$f^*(g;r+2s) + f^*(g;r) - 2f^*(g;r+s) = o(s/\log s)$$

is valid uniformly for $s \leq r \leq r_0$, then the following result of the Salem type is valid : $S_R^{2k,\delta_0}(f;g)$ converges to $f(g)$ (for $R \to \infty$) almost everywhere if $|f|\log^+|f|$ is integral for G being a torus of dimension $n \geq 2$ or if f is integral for G being other compact Lie group. Similarly, we can give the Dini-, the Jordan- and the Lebesgue-test for $S_R^{2k,\delta_0}(f;g)$ on compact Lie groups by use of the function $f^*(g;r)$.

E. M. Stein (see [9]) discussed the following spherical means of Fourier series on compact Lie groups

$$P^t f = \sum_{\lambda \in G} e^{-\mu_\lambda^{1/2} t} d_\lambda f * \chi_\lambda(x)$$

where $x \in G$, $f \in L(G)$, and he proved

1) $\|P^t f\|_p \leq \|f\|_p$, where $t > 0$, $f(x) \in L^p(G)$, $p \geq 1$.

2) $P^t$ is a self-conjugate operator on $L^2(G)$.

3) $f \geq 0$ implies that $P^t f \geq 0$.

4) $\lim_{t\to 0} \frac{P^t f - f}{t} = -(-\Delta)^{1/2} f$,

where

$$(-\Delta)^{1/2} f(x) = \sum_{\lambda \in G} \mu_\lambda^{1/2} d_\lambda f * \chi_\lambda(x), \quad f \in C^\infty(G).$$

5) $u(t;x) \equiv P^t f(x) \in C^\infty(G\times(0,\infty))$, and also $u(t;x)$ satisfies the Laplace equation

$$\left(\frac{\partial^2}{\partial t^2} + \Delta\right)u \equiv 0.$$

6) $u(t;x)$ converges to $f(x)$ for $t \to 0$ in the norm of $L(G)$, where $X_1, X_2, \ldots, X_n$ is a basis of the Lie algebra of G, $(a_{ij}) = (-B(X_i, X_j))^{-1}$.

$$\Delta = \sum_{i,j=1}^{n} a_{ij} X_i X_j, \quad \Delta A_\lambda(x) = -\mu_\lambda A_\lambda(x), \quad \Delta f(x) = \sum_{i,j=1}^{n} a_{ij} X_i X_j f.$$

Let f be a real valued function which belongs to $C^\infty(G)$ and define

$$|\nabla f|^2(x) = \sum_{i,j=1}^{n} a_{ij}(X_i f)(X_j f).$$

If $f \in C^\infty(G\times(0,\infty))$, then

$$|\nabla f|^2(x) = \left( \frac{\partial}{\partial t} f \right)^2 + |\nabla_x f|^2.$$

Then E. M. Stein defined the Littlewood-Paley function of $f \in L^p(G)$ as

$$g(f)(x) = \left( \int_0^\infty t|\nabla u(t;x)|^2 dt\right)^{1/2},$$

and proved the following:

7) Let $f \in L_p(G)$, $1 < P < \infty$. Then $g(f) \in L^p(G)$ and there exists a constant $A_p$ such that

$$\|g(f)\|_p \leqslant A_p\|f\|_p .$$

Conversely, if $\int_G f(x)dx = 0$, then there exists a constant $B_p$ such that

$$\|f\|_p \leqslant B_p\|g(f)\|_p.$$

8) Let the Riesz transformation on G be

$$R_j f = X_j(-\Delta)^{-1/2} f, \; j = 1,2,\ldots,n,$$

where $f \in C^\infty(G)$. Then $R_j$, $j = 1,2,\ldots,n$, are bounded operators on $L^p(G)$ for $1 < p < \infty$, from which follows

$$\|X_i X_j f\|_p \leqslant A_p\|\Delta f\|_p \quad (1 < p < \infty).$$

J. L. Clerc (see [11]) discussed the summation of Fourier series on compact Lie groups by Riesz means of order $\delta$. His main results are as follows:

Let G be a compact Lie group of dimension n and rank q, D(exp h) be Weyl's function of G and

$$S_R^\delta f = \sum_{|\lambda+\beta|<R} (1 - |\lambda+\beta|^2/R^2)^\delta d_\lambda f*\chi_\lambda ;$$

then,

1) $S_R^\delta f \to f$ for $\delta > (n-1)/2$ in the norm of $L^p(G)$, $p \geqslant 1$.
2) $\sup_{R>0} |S_R^\delta f(x)| \leqslant C(Mf(x) + K*|f|(x))$, $\delta > (n-1)/2$.
3) If $\delta > (n-1)/2$, $f \in L(G)$ and m is the Haar measure, then
$m\{\sup_R |S_R^\delta f| > \alpha\} \leqslant \frac{A}{\alpha} \|f\|_1$ ,

and, from this, $S_R^\delta f$ converges to f almost everywhere;

4) If $1 < p \leqslant 2$, $\delta > (n-1)(1/p-1/2)$, then there exists a constant $A_p$ such that

$$\|\sup_R |S_R^\delta f| \|_p \leqslant A_p\|f\|_p .$$

R. S. Strichartz (see [14]) discussed the multiplier transformation on compact Lie algebras and groups.

Let G be a compact Lie group and $\mathfrak{g}$ be its Lie algebra, H be a Cartan subalgebra of $\mathfrak{g}$, $d\mu$ be ad-invariant finite measure on $\mathfrak{g}$. Especially, when $d\mu$ is absolutely continuous, there exists a function $F(x)$, $x \in \mathfrak{g}$, which is integrable and ad-invariant (i.e. $F(h)\,|P(h)|^2$ is integrable on H, $h \in H$ such that $d\mu = F(x)dx$.

R. S. Strichartz proved:

1) If
$$\Phi(x) = \int_{\mathfrak{g}} e^{-iB(x.y)}dy,$$
then
$$\phi(\lambda) = \Phi(\lambda+\beta) \qquad (*)$$
or
$$\phi(\lambda) = \int_G \Phi(\lambda+\beta-ad_y\beta)dy \qquad (**)$$
are bounded operators on L(G).

2) Let $\Phi(x)$ be the same as in 1) and define $\Phi_r(x) = \Phi(x/r)$. Then defines an operator $OP(\Phi)$ on :

$$f \to \int_{\mathfrak{g}} \Phi(y)\hat{f}(x)e^{iB(x.y)}dy, \quad f(y) = (2\pi)^{-n}\int_{\mathfrak{g}} \hat{f}(x)e^{-iB(x.y)}dx,$$

and (*) or (**) defines an operator $op(\phi)$ on G. Then the necessary condition that $OP(\Phi)$ is bounded on $L(\mathfrak{g})$ is that $op(\phi_r)$ is uniformly bounded when $r \to \infty$.

D. L. Ragozin (see [20]), using imbedding method into the Euclidean space, proved the Jackson Theorem, the Bernstein Theorem and other results on compact Lie groups and on compact homogeneous spaces.

As to the harmonic analysis on unitary groups and its extension on classical groups and compact Lie groups, there are many results such as : a variety of theorems of Tauber type, a variety of problems on how to study the harmonic analysis on classical domains through the harmonic analysis on classical groups, and many results on the approximation theory. All these results are omitted, for which the readers are referred to [1] - [6] and other articles.

**REFERENCES**

[1] Gong Sheng (Kung Sun), Harmonic Analysis on Classical Groups (in Chinese), Science Press, Beijing China, 1983.
[2] __________, Acta. Math. Sinica, 10(1960), 239-261 (in Chinese).
[3] __________, ibid 12(1962), 17-31 (in Chinese).
[4] __________, ibid 13(1963), 152-161 (in Chinese).
[5] __________, ibid 13(1963), 323-331 (in Chinese).
[6] __________, ibid 15(1965), 305-325 (in Chinese).
[7] Zhong Jiaqing, Journal of Chinese University of Science and Technology, 9(1979), 31-43.
[8] Gong Sheng, Journal of Chinese University of Science and Technology, 9(1979), 25-30.
[9] Stein, E. M., Annals in Math. Study, Princeton, 1970, No. 63.
[10] Coifman, R. & Weiss. G., Bull. Amer. Math. Soc. 80(1974), 124-126.
[11] Clerc, J. L., Ann. Inst. Fourier, Grenoble, 24(1974), 1:149-172.
[12] Stanton, R. J., Trans. Amer. Math. Soc. 218(1976), 61-81.

[13] Stanton, R. J. & Tomas, P. A., Amer. J. Math. 100(1978), 477-493.
[14] Strichartz, R. S., Trans. Amer. Math. Soc. 193(1974), 99-110.
[15] Weiss, N. J., Amer. J. Math. 94(1972), 103-118.
[16] Dreseler, B., Manuscripta Math. 31(1980), 17-23.
[17] ________, Fourier Analysis and Approximation Theory, Ed. G. Alexits and P. Turan, Vol. I(1976), 327-342.
[18] Mayer, R. A., Duke Math. J. 34(1967), 549-554.
[19] Talor, M. E., Amer. Math. Soc. 19(1968), 1103-1105.
[20] Rayozin, D. L., Trans. Amer. Math. Soc. 150(1970), 41-53.
[21] ________, Math. Ann. 195(1972), 87-94.
[22] ________, ibid, 219(1976), 1-11.
[23] Zheng Xue An, Advances in Math., Vol. 13, 2(1984), 103-118.

Proceedings of the Analysis Conference, Singapore 1986
S.T.L. Choy, J.P. Jesudason, P.Y. Lee (Editors)



# INTERPOLATION OF OPERATORS IN LEBESGUE SPACES WITH MIXED NORM AND ITS APPLICATIONS TO FOURIER ANALYSIS

Satoru IGARI

Mathematical Institute Tôhoku University Sendai,980
Japan

## INTRODUCTION

This note was prepared for the lectures of a workshop and a conference at the National University of Singapore.

Our objective is to discuss an interpolation method of linear operators on functions in a product measure space and apply it to some problems arising in Fourier analysis on Euclidean space.

For a function $f$ on the d-dimensional Euclidean space $R^d$ and $\varepsilon \geq 0$ let $s^\varepsilon(f)$ be the Riesz-Bochner mean of order $\varepsilon$, which is defined by the Fourier transform

$$s^\varepsilon(f)^\wedge(\xi) = (1-|\xi|^2)^\varepsilon \hat{f}(\xi)$$

for $|\xi| < 1$ and $= 0$ otherwise, where

$$\hat{f}(\xi) = \frac{1}{\sqrt{2\pi}^{\,d}} \int_{R^d} f(x)\, e^{-i\xi x} dx.$$

If $d = 1$, $s^\varepsilon$ is a bounded operator of $L^p(R)$ to $L^p(R)$ for $\varepsilon \geq 0$ and $p > 1$ by M.Riesz's theorem. Let $d \geq 2$. Then $s^0$ is bounded on $L^p((R^d)$ if and only if $p = 2$ ( Fefferman[8]) and $s^\varepsilon$ ( $\varepsilon > 0$ ) is unbounded on $L^p(R^d)$ unless $2d/(d+1+2\varepsilon) < p < 2d/(2d-1-2\varepsilon)$ ( Herz[10]). On the otherhand if $d = 2$ and $\varepsilon > 0$, then $s^\varepsilon$ is bounded on $L^4(R^2)$ ( Carleson and Sjölin[3]).

In Chapter III we shall give the following estimate of $s^\varepsilon(f)$, $\varepsilon > 0$, applying Parseval relation and an $L^4(R^2)$-argument due to Carleson-Sjölin[3] and Cordoba[6];

THEOREM. If $d > 2$ and $\varepsilon > 0$, then

$$\int_{R^2} \left(\int_{R^{d-2}} |s^\varepsilon(f)|^2 d\bar{x}\right)^2 d\bar{\bar{x}} \leq C \int_{R^2} \left(\int_{R^{d-2}} |f|^2 d\bar{x}\right)^2 d\bar{\bar{x}}, \qquad (1)$$

for all $f$ in $C_0(R^d)$, where $\bar{x} = (x_0,\ldots,x_{d-2})$ and $\bar{\bar{x}} = (x_{d-2},x_{d-1})$.

Since the operator $s^\varepsilon$ is rotation invariant, we can choose any pair $(x_i,x_j)$, $0 \leq i,j < d$, as a coodinate system of $R^d$ in the inequality (1).

Now we observe that there are $d-2$ variables in the inner integrals in (1) and 2 variables in the outer integrals. Thus it may be speculated that the average $u$ of the exponents of $f$ and $s^{\varepsilon}$ would satisfy

$$1/u = [(d-2)/2 + 2/4]/d = (d-1)/2d,$$

that is, $u = 2d/(d-1)$, which is just the Herz-Pollard bound.

This is motive of our interpolation method. For $f$ in the product measure space $M^{m+n} = M\times M\times\ldots\times M$ and $s, t > 0$ define the mixed norm $\|f\|_{(t,s:p)} = [\int_{M^n}(\int_{M^m}|f|^s dx_0\ldots dx_{m-1})^{t/s}dx_m\ldots dx_{m+n-1}]^{1/t}$, where $p$ denotes the coordinate system of $M^m$. In Chapter II we shall consider the linear operator $T$ which is bounded on the spaces with mixed norm. Under this condition we get some information on the boundedness of $T$ in $L^u(M^{m+n})$, where $1/u = (m/s + n/t)/(m+n)$. Applying our interpolation theorem we get

$$\|s^{\varepsilon}(f)\|_{2d/(d+1)} \leq C\ \|f\|_{2d/(d+1)}$$

for $f$ in $L^{2d/(d+1)}(R^d)$ of product form $f(x) = f_0(x_0)\ldots f_{d-1}(x_{d-1})$, where $\varepsilon > 0$.

In §§ 2 and 3 in Chapter III we show a restriction theorem of Fourier transform and an estimate of Kakeya,s maximal operator for functions of product form applying our interpolation theorem.



## CHAPTER I PRELIMINARIES

### 1.1. Poisson integral and the space $H^2$

For $|z| < 1$ the Poisson kernel $P(z,\theta)$, $-\pi \leq \theta < \pi$, is defined by

$$P(z,\theta) = \Re e\,\frac{e^{i\theta}+z}{e^{i\theta}-z} = \frac{1-r^2}{1-2r\cos(t-\theta)+r^2},\quad z = re^{it}$$

and the conjugate Poisson kernel $Q(z,\theta)$ by

$$Q(z,\theta) = \Im m\, \frac{e^{i\theta}+z}{e^{i\theta}-z} = \frac{2r\sin(t-\theta)}{1-2r\cos(t-\theta)+r^2}.$$

For an integrable function $u$ the Poisson integral $u(z)$ and the conjugate Poisson integral $\tilde{u}$ are defined by

$$u(z) = \int_{-\pi}^{\pi} P(z,\theta)u(e^{i\theta})\frac{d\theta}{2\pi}$$

and

$$\tilde{u}(z) = \int_{-\pi}^{\pi} Q(z,\theta)u(e^{i\theta})\frac{d\theta}{2\pi}, \quad |z| < 1$$

respectively. Suppose that $u$ is real valued. If

$$f(z) = u(z) + i\tilde{u}(z), \tag{1.1}$$

then $f(z)$ is holomorphic in $|z| < 1$ and $\Re e\, f(z) = u(z)$ and $\Im m\, f(0) = 0$.

DEFINITION. Let $p > 0$. A function $f(z)$ holomorphic in $|z| < 1$ is said to belong to the space $H^p$ if

$$\|f\|_{H^p} = \sup_{0<r<1} \left( \int_{-\pi}^{\pi} |f(re^{i\theta})|^p \frac{d\theta}{2\pi} \right)^{1/p} < \infty.$$

A function $f$ in $H^1$ is said to belong to $H^\infty$ if $\|f\|_{H^\infty} = \sup_{|z|<1} |f(z)| < \infty$.

Let $f \in H^2$. Then we have the following properties:

(I) $\lim_{r\to 1} f(re^{i\theta}) = f(e^{i\theta})$ exists a.e.

(II) $\|f(re^{i\cdot}) - f(e^{i\cdot})\|_{L^2} \to 0$ as $r \to 1$.

2.2 The space $N^+$

We have the following ( see e.g.[7])

THEOREM. If $H^2$, then

$$\log|f(z)| \le \int_{-\pi}^{\pi} P(z,\theta) \log|f(e^{i\theta})| \frac{d\theta}{2\pi}, \quad |z| < 1. \tag{1.2}$$

A holomorphic function $f(z)$ in $|z| < 1$ is said to belong to $N$ if

$$\sup_{0<r<1} \int_{-\pi}^{\pi} \log^+|f(re^{i\theta})| \frac{d\theta}{2\pi} < \infty.$$

If $f \in N$, then $f(e^{i\theta}) = \lim f(re^{i\theta})$ exists a.e.

A function $f$ in $N$ is said to belong to $N^+$ if the inequality (1.2) holds for all $|z| < 1$. Thus

$$N \subset N^+ \subset H^2.$$

For details we refer the Duren's book [7].

1.3. Riesz-Thorin interpolatiom theorem

In the following we assume that $(M,\mathfrak{M},\mu)$ and $(N,\mathfrak{N},\nu)$ are

$\sigma$-finite measure spaces. Let $T$ be a linear operator from step functions in $(M,\mathfrak{M},\mu)$ to measurable functions in $(N,\mathfrak{N},\nu)$.

THEOREM(Riesz-Thorin). Let $1 \le p_0,p_1,q_0,q_1 \le \infty$. Suppose that $T$ is a bounded mapping of $L^{p_0}(M)$ to $L^{q_0}(N)$ and of $L^{p_1}(M)$ to $L^{q_1}(N)$ simultaneously and with norm $C_0$ and $C_1$ respectively.

If $0 < \theta < 1$ and

$$\frac{1}{p_\theta} = \frac{1-\theta}{p_0} + \frac{\theta}{p_1} \quad \text{and} \quad \frac{1}{q_\theta} = \frac{1-\theta}{q_0} + \frac{\theta}{q_1},$$

then $T$ is a bounded mapping of $L^{p_\theta}(M)$ to $L^{q_\theta}(N)$ with norm

$$C_\theta \le C_0^{1-\theta} C_1^\theta .$$

Several proofs of Riesz-Thorin theorem are known. Our proof is a slight variant of that of Calderón and Zygmund [2]. It is essentialy same to that of Rochberg and Weiss [13] and it will suggest the method used in Chapter II.

We divide the unit circle $\partial D = [-\pi,\pi)$ into two intervals $I_0, I_1$ of length $2\pi(1-\theta)$ and $2\pi\theta$ respectively. Let $\alpha(z)$ be a function in $H^2$ such that $\Re e\, \alpha(e^{i\theta}) = 1/p_0$ in $\mathrm{int}(I_0)$ and $= p_1$ in $\mathrm{int}(I_1)$, and $\Im m\, \alpha(0) = 0$. For non-zero simple function $w$ in $(M,\mathfrak{M},\mu)$ and complex number $|z| < 1$ define

$$W^z(x) = \|w\|_{p_\theta}^{p_0\gamma(z)} e^{i \arg w(x)} |w(x)|^{p_\theta \alpha(\theta)}$$

if $w(x) \neq 0$ and $= 0$ otherwise, where $\gamma(z) = 1/p_\theta - \alpha(z)$.

For $1 \le p \le \infty$, $p'$ denotes the conjugate index of $p$, that is $1 = 1/p + 1/p'$. Let $\beta(z)$ be a function in $H^2$ which is defined as $\alpha(z)$ for ${}_1q_0'$ and $q_1'$. For given non-zero simple function $f$ in $(N,\mathfrak{N},\nu)$ let

$$F^z(x) = \|f\|_{q_\theta'}^{q_\theta'\delta(z)} e^{i \arg f(x)} |f(x)|^{q_\theta'\beta(z)}$$

if $f(x) \neq 0$ and $= 0$ otherwise, where $\delta(z) = 1/q_\theta' - \beta(z)$.

LEMMA. With the above notations we have

(i)
$$\|w^{e^{i\theta}}\|_{p_j} = \|w\|_{p_\theta}$$
and
$$\|F^{e^{i\theta}}\|_{q_j'} = \|f\|_{q_\theta'} \quad \text{for } e^{i\theta} \in \mathrm{int}(I_j)$$
$j = 0,1$.

(ii)
$$W^0(x) = w(x) \quad \text{and} \quad F^0(x) = f(x).$$

PROOF. For (ii) remark that $\alpha(0) = 1/p_\theta$ and $\beta(0) = 1/q_\theta'$. (i) follows from the definitions of $\alpha$ and $\beta$.

PROOF OF THEOREM. Let $w$ and $f$ be non zero simple functions in $(M,\mathfrak{M},\mu)$ and $(N,\mathfrak{N},\nu)$ respectively and define $W^z$ and $F^z$ as above. Put $\Phi(z) = \int_N TW^z F^z\, d\nu$. Then $\Phi(z) \in H^\infty$, since $\Re e\ \alpha$ is bounded. Thus

$$\log |\Phi(0)| \le \int_{-\pi}^{\pi} \log|\Phi(e^{i\theta})| \frac{d\theta}{2\pi} .$$

By Hölder's inequality and by our assumption we have

$$|\Phi(z)\| \le \|TW^z\|_{q_j} \|F^z\|_{q_j'} \le C_j\|W^z\|_{p_j} \|F^z\|_{q_j'} = C_j\|w\|_{p_\theta}\|f\|_{q_\theta'}$$

for $z = e^{it} \in \text{int}(I_j)$. On the other hand $\Phi(0) = \int Tw\ f\ d\nu$. Thus

$$\log |\int Tw\ f\ d\nu| \le \log(\|w\|_{p_\theta} \|f\|_{q_\theta'}) + (1-\theta)\log C_0 + \theta\log C_1 .$$

Thus

$$|\int Tw\ f\ d\nu| \le C_0^{1-\theta} C_1^{\theta} \|f\|_{q'} \|w\|_{p_\theta} .$$

Taking supremum over $f$ such that $\|f\|_{q'} = 1$, we get the theorem.

## CHAPTER II INTERPOLATION OF OPERATORS IN LEBESGUE SPACES WITH MIXED NORM

### 2.1. Lebesgue spaces with mixed norm

Let $d$ be a positive integer and $(M_j,\mathfrak{M}_j,\mu_j)(j = 0,1,...,d-1)$ be $\sigma$-finite measure spaces. Let $(M^d,\mathfrak{M}^d,\mu^d)$ be the product measure space $\prod_{j=0}^{d-1}(M_j,\mathfrak{M}_j,\mu_j)$. For a subset $p = \{p_0,p_1,...,p_{m-1}\}$ of $\{0,1,...,d-1\}$ let $(M(p),\mathfrak{M}(p),\mu(p)) = \prod_{j\in p}(M_j,\mathfrak{M}_j,\mu_j)$. Thus $d\mu(p)(x_{p_0},...,x_{p_{m-1}}) = d\mu_{p_0}(x_{p_0})...d\mu_{p_{m-1}}(x_{p_{m-1}})$. If $p^c = \{0,1,...,d-1\} - p$, $d\mu^d = d\mu(p)\times d\mu(p^c)$.

Let $1 \le s \le \infty$. $L^s(M^d)$ denotes the space of measurable functions $f$ in $M^d$ such that $\|f\|_s = ( \int_{M^d}|f|^s\, d\mu^d )^{1/s} < \infty$ if $s < \infty$ and $\|f\|_\infty = \text{ess sup}\ |f| < \infty$ if $s = \infty$.

For $1 \le s,t < \infty$ and $p = \{ p_0,...,p_{m-1}\}$, $L^t(M(p^c);L^s(M(p)) = L^t(L^s)$ denotes the space of measurable functions $f$ in $M^d$ such that

$$\|f\|_{(t,s:p)} = \left(\int_{M(p^c)} \left(\int_{M(p)} |f|^s d\mu(p)\right)^{t/s} d\mu(p^c)\right)^{1/t} < \infty.$$

The definition for the cases $s = \infty$ or/and $t = \infty$ will be obvious. In the following we assume that

$$d = m + n,$$

where $m$ and $n$ are positive integer, $1 \le s,t \le \infty$ and $u$ is defined by

$$1/u = [\ m/s + n/t\ ]/d.$$

For $1 \le s \le \infty$, $s'$ denotes the conjugate exponent defined by $1 =$

$1/s + 1/s'$.

2.2 Auxiliary functions ; the case $m \geq n$

Let $P$ be the family of index sets $p$ of $\{0,\ldots,d-1\}$ such that $\mathrm{card}(p) = m$. Let $Q = \{ p \in Q ; 0 \in p \}$ and $R = P - Q$. For $q \in Q$ put $R^q = \{ r \in R ; \mathrm{card}(r \cap q) = m-n \}$.

Divide the unit circle $\partial D$ into $\mathrm{card}(P)$ congruent arcs $I_p$, $p \in P$. Assume $1 \leq s \leq t \leq \infty$. Let $\alpha_o(z)$ and $\alpha_q(z)$ be functions in the Hardy space $H^2$ whose real parts are defined by Table 1 and which satisfy $\Im m\, \alpha_o(0) = \Im m\, \alpha_q(0) = 0$.

Definition of $\alpha_o(z)$ and $\alpha_q(z)$

The case $m \geq n$ Table 1

| | | | $\mathrm{Re}\, \alpha_o(e^{i\theta})$ | $\mathrm{Re}\, \alpha_q(e^{i\theta})$ |
|---|---|---|---|---|
| Q | | $I_{q'}$ | $1/s$ | $0$ |
| | | | ⋮ | ⋮ |
| | | | | $0$ |
| | | $I_q$ | $1/s$ | $1/t - 1/s$ |
| | | | ⋮ | $0$ |
| | | | | ⋮ |
| | | $I_{q''}$ | $1/s$ | $0$ |
| R | $R_q$ | $I_r$ | $1/t$ | $(1/s-1/t)/\mathrm{card}(R^q)$ |
| | | | ⋮ | ⋮ |
| | | | | $(1/s-1/t)/\mathrm{card}(R^q)$ |
| | | | | $0$ |
| | $R - R_q$ | $I_{r'}$ | ⋮ | ⋮ |
| | | | $1/t$ | $0$ |

By the mean value theorem we have

$$\alpha_o(0) = \frac{1}{\mathrm{card}(P)}\left( \sum_{q\in Q} \frac{1}{s} + \sum_{r\in R} \frac{1}{t} \right) = \frac{1}{d}\left( \frac{m}{s} + \frac{n}{t} \right) = \frac{1}{u}, \text{ say.}$$

For simple function $w$ and $f$ on $M^d$ and $|z| < 1$ define $W^z$ and $F^z$ as follows.

$$W^z(x) = A_w(z)e^{i\arg w(x)}|w(x)|^{\alpha_0(z)} \prod_{q\in Q}\left(\int_{M(q)} |w(x)|^u d\mu(q)\right)^{\alpha_q(z)}$$

and

$$F^z(x) = B_f(z)e^{i\arg f(x)}|f(x)|^{u'(1-\alpha_0(z))} \prod_{q\in Q}\left(\int_{M(q)} |f(x)|^{u'} d\mu(q)\right)^{\alpha_q(z)},$$

where

$$A_z(z) = \|w\|_u^{u\gamma(z)}, \quad B_f(z) = \|f\|_{u'}^{-u'\gamma(z)}$$

and

$$\gamma(z) = \frac{m}{n}(\alpha_0(z) - \frac{1}{s}) + (\frac{1}{u} + \frac{1}{t}).$$

LEMMA 1. Let $d = m + n$, $m \geq n \geq 1$ and $1 \leq s \leq t \leq \infty$. Let $w$ and $f$ be non-zero simple functions in $M^d$. Then we have the followings.

(i) $W^0(x) = w(x)$ and $F^0(x) = f(x)$.

(ii) For $p \in P$

$$\|W^z\|_{(t,s:p)} \leq \|w\|_u, \quad z = e^{i\theta} \in \operatorname{int}(I_p).$$

(iii) If $f$ is of the form $f_0(x_0)f_1(x_1)\ldots f_{d-1}(x_{d-1})$, then

$$\|F^z\|_{(t',s':p)} = \|f\|_{u'}, \quad z = e^{i\theta} \in \operatorname{int}(I_p).$$

For a proof see [11].

2.3 Auxiliary functions ; the case $m < n$

Let $d = mk + r$, where $k \geq 2$ and $m \geq r > 0$, so that $n = m(k - 1) + r$. We define a family $P$ of $m$ integers $p^a = \{p_1^a, p_2^a, \ldots, p_m^a\}$, $a = 1,2,\ldots,d$, by four cases : Let $a = mj + b$.

Case A : $0 \leq j < k$ and $0 \leq b < m-r$.

Case B : $0 \leq j < k-1$ and $m-r \leq b < m$.

Case C : $j = k$ and $m-r \leq b < m$.

Case D : $mk \leq a < b$.

Let $a = mj + b$ correspond to the point $(mj+b, b)$ in Fig.2. Then the set $p^a$ is defined as a successive sequence indicated in Fig.2.

Let $\alpha_0(z)$, $\alpha_a(z)$ ( $m \leq a < m(k+1)$) and $\gamma(z)$ be functions in $H^2$ whose real parts are defined by Table 3 and which satisfy

$$\Im m\, \alpha_0(0) = \Im m\, \alpha_a(0) = \Re e\, \gamma(0) = 0.$$

For non-zero simple functions $w$ and $f$ in $M^d$ and $|z| < 1$ define $W^z$ and $F^z$ as follows.

Expression of $p^{mj+b}$

The case $m < n$

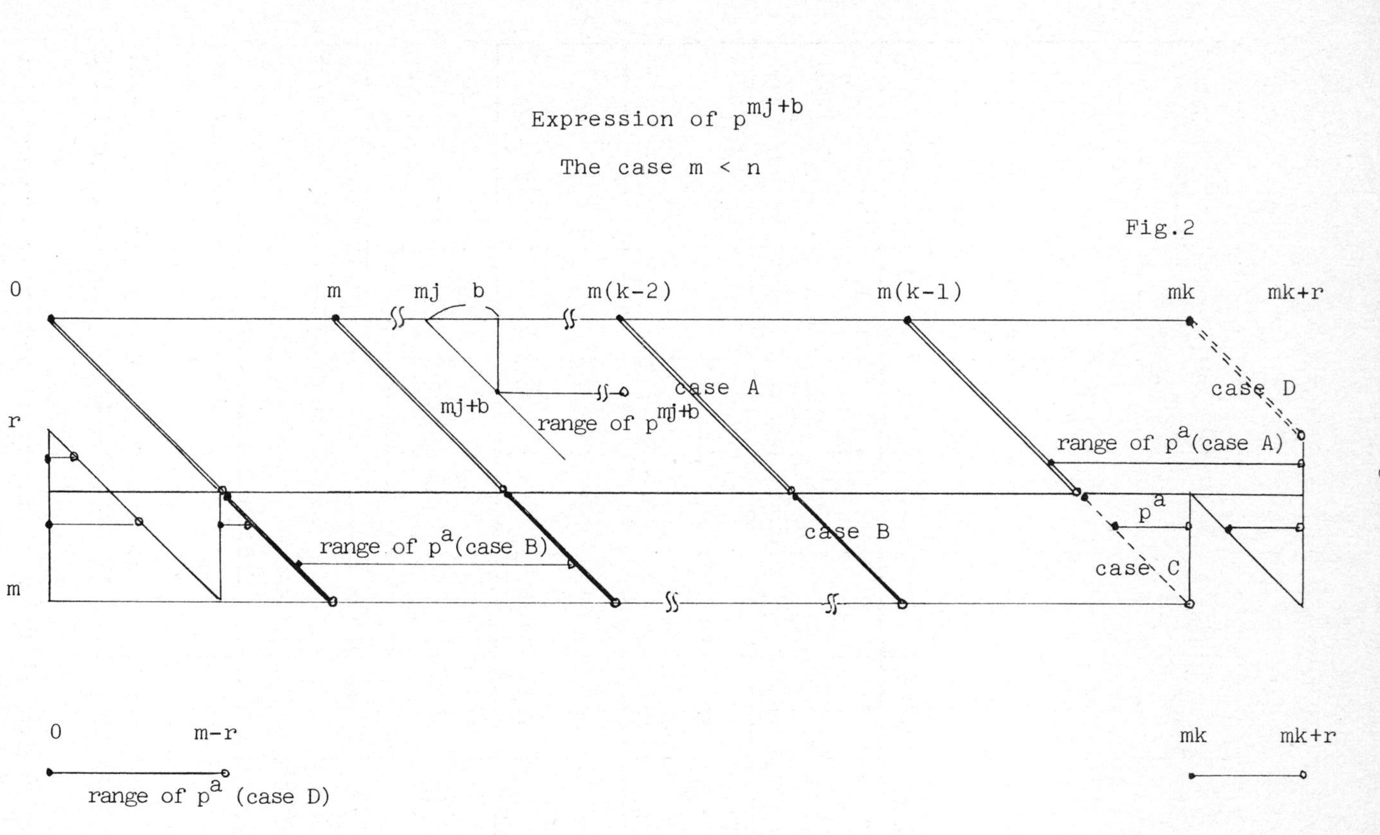


Fig.2

Definition of $\alpha_o(z)$ and $\alpha_q(z)$

The case $m < n$ Table 3

| $I_a$ | $\operatorname{Re}\alpha_o(e^{i\theta})$ | $\operatorname{Re}\ \alpha_{mj+\ell}(e^{i\theta})$ $1\le j<k\ :\ 0\le \ell<m$ | $\operatorname{Re}\ \alpha_{mk+\ell}(e^{i\theta})$ $0\le\ell<m$ | $\operatorname{Re}\ \gamma(e^{i\theta})$ |
|---|---|---|---|---|
| $I_o$ | 1/s | 0 | 0 | 1/u-1/t |
| ⋮ | ⋮ | ⋮ | ⋮ | ⋮ |
| $I_{m-1}$ | 1/s | 0 | 0 | 1/u-1/t |
| $I_m$ | 1/t | 0 | 0 | 1/u-1/t |
| ⋮ | ⋮ | ⋮ | ⋮ | ⋮ |
| $I_{2m-1}$ | 1/t | 0 | 0 | 1/u-1/t |
| ⋮ | ⋮ | ⋮ | ⋮ | ⋮ |
| $I_{m(j-1)}$ | 1/t | 0 | 0 | 1/u-1/t |
| $I_{m(j-1)+\ell}$ | ⋮ | 1/t-1/s | ⋮ | ⋮ |
| $I_{mj-1}$ | 1/t | 0 | 0 | 1/u-1/t |
| $I_{mj}$ | 1/t | 0 | 0 | 1/u-1/t |
| $I_{mj+\ell}$ | ⋮ | 1/s-1/t | ⋮ | ⋮ |
| $I_{m(j+1)-1}$ | 1/t | 0 | 0 | 1/u-1/t |
| ⋮ | ⋮ | ⋮ | ⋮ | ⋮ |
| $I_{m(k-1)}$ | 1/t | 0 | 0 | 1/u-1/t |
| $I_{m(k-1)+\ell}$ | ⋮ | ⋮ | 1/t-1/s | ⋮ |
| $I_{mk-1}$ | 1/t | 0 | 0 | 1/u-1/t |
| $I_{mk}$ | 1/t | 0 | (1/s-1/t)/r | (1/u-1/t)-(1/s-1/t)m/r |
| ⋮ | ⋮ | ⋮ | ⋮ | ⋮ |
| $I_{mk+r-1}$ | 1/t | 0 | (1/s-1/t)/r | (1/u-1/t)-(1/s-1/t)m/r |

$$W^z(x) = A_w(z)|w(x)|^{u\alpha_o(z)} e^{i\ \arg w(x)}$$
$$\times \prod_{\ell=o}^{m-1} \prod_{j=1}^{k} \left(\int |w(x)|^u d\mu (P^{\ell} \cup P^{m+\ell} \cup \ldots \cup P^{m(j-1)+\ell})\right)^{\alpha_{mj+\ell}(z)}$$

if $w(x) \neq 0$ and $= 0$ otherwise, where $A_w(z) = \|w\|_u^{u\gamma(z)}$, and

$$F^z(x) = B_f(z)\ |f(x)|^{u'(1-\alpha_o(z))} e^{i \arg f(x)}$$
$$\times \prod_{\ell=o}^{m-1} \prod_{j=1}^{k} \left(\int |f(x)|^{u'} d\mu(p^{\ell}\cup p^{m+\ell}\cup \ldots \cup p^{m(j-1)+\ell})\right)^{\alpha_{mj+\ell}(z)},$$

if $f(x) \neq 0$ and $= 0$ otherwise, where $B_f(z) = \|f\|_{u'}^{-u'\gamma(z)}$.

Let $P = \{ p^a \}$ and divide the unit circle $\partial D$ into card(P) congruent arcs $I_{p_a}$. Then we get a similar lemma as in §2.2. 2.4.

2.4. Interpolation theorems

Let $d = m + n$. Let $P$ be the family of $m$ integers defined in §2.2 or §2.3 according to $m \geq n$ or $m < n$ and $I_p$, $p \in P$, be arcs of the preceeding sections. Let $( M,\mathfrak{M},\mu )$ and $( N,\mathfrak{N},\nu )$ be $\sigma$-finite measure spaces. $( M(p),\mathfrak{M}(p),\mu(p))$, $( N^d,\mathfrak{N}^d,\nu^d)$, etc. will denote the spaces defined in §2.2.

THEOREM 1. Let $T$ be a linear operator of simple functions in $M^d$ to measurable functions in $N$. Let $v(e^{i\theta})$ be a measurable function in $\partial D$ such that $1 \leq v(e^{i\theta}) \leq \infty$. Define $v$ by

$$\frac{1}{v} = \int_{\partial D} \frac{1}{v(e^{i\theta})} \frac{d\theta}{2\pi}.$$

(i) Let $1 \leq u_o \leq u_1 \leq \infty$ and

$$1/u = ( m/u_o + n/u_1)/d.$$

Suppse that

$$\|Tw\|_{v(e^{i\theta})} \leq C(e^{i\theta})\|w\|_{(u_1,u_o:p)} \tag{2.1}$$

for simple functions $w$ and $e^{i\theta} \in \mathrm{int}(I_p)$, $p \in P$, where $C(e^{i\theta})$ is a measurable function in $\partial D$. Then

$$\|Tw\|_u \leq C\ \|w\|_u, \tag{2.2}$$

where

$$C = \exp \int_{\partial D} \log C(e^{i\theta})\ \frac{d\theta}{2\pi}.$$

(ii) Let $1 \leq u_1 \leq u_o \leq \infty$. Then under the assumption (2.1) we have (2.2) for all functions $w$ of the product form $w_o(x_o)w_1(x_1) \ldots w_{d-1}(x_{d-1})$.

THEOREM 2. Let $T$ be a linear operator of simple functions in $M^d$ to measurable functions in $N^d$.

(i) Let $1 \leq u_o \leq u_1 \leq \infty$ and $1 \leq v_1 \leq v_o \leq \infty$. Suppose that

$$\|Tw\|_{(v_1,v_o:p)} \leq C_p\|w\|_{(u_1,u_o:p)} \tag{2.3}$$

for all $w$ and $p \in P$.

If

$$1/u = ( m/u_o + n/u_1)/d \quad \text{and} \quad 1/v = ( m/v_o + n/v_1)/d,$$

then

$$\|Tw\|_v \le C \|w\|_u, \tag{2.4}$$

where

$$C = (\prod_{p\in P} C_p)^{1/\mathrm{card}(P)}.$$

(ii) Let $1 \le u_1 \le u_0 \le \infty$ and $1 \le v_1 \le v_0 \le \infty$. Then under the assumption of (2.3) we have (2.4) for all functions $w$ of product form.

For a proof we refer to Igari [11].

## CHAPTER III APPLICATIONS OF INTERPOLATION THEOREMS TO SOME PROBLEMS ARISING IN FOURIER ANALYSIS

In this chapter we shall show three examples of applications of the interpolation theorems given in Chapter II. They are closely related to the spherical summation problem of multiple Fourier transform. We omit the proofs since the proofs of Theorems 3 and 6 in §§3.1 and 3.2 are given in [11], and that of Theorem 8 will appear in Igari [12].

3.1. Estimates of Riesz-Bochner means

Let $P$ be the family of all subsets $p \subset \{0,1,\dots,d-1\}$ such that $\mathrm{card}(p) = d-2$. Using the notations in Chapter II we have

THEOREM 3. If $\varepsilon > 0$, then

$$\int_{R^2}\Big(\int_{R^{d-2}}|s^\varepsilon(f)|^2 dx(p)\Big)^2 dx(p^c) \le C \int_{R^2}\Big(\int_{R^{d-2}}|f|^2 dx(p)\Big) dx(p^c) \tag{3.1}$$

for all $f \in C_c^\infty(R^d)$ and $p \in P$.

Theorem 3 implies that

$$\|s^\varepsilon(f)\|_{(4,2:p)} \le C \|f\|_{(4,2:p)}.$$

By duality

$$\|S^\varepsilon(f)\|_{(4/3,2:p)} \le C \|f\|_{(4/3,2:p)}.$$

Applying Theorem 2(ii) we get

$$\|s^\varepsilon(f)\|_{2d/(d+1)} \le C \|f\|_{2d/(d+1)}$$

for $f$ of the product form $f_0(x_0)f_1(x_1)\dots f_{d-1}(x_{d-1})$.

By an interpolation theorem for multilinear operators ( see [1] )

THEOREM 4. Let $\varepsilon > 0$ and $2d/(d+1)) \le u \le 2$. Then

$$\|s^\varepsilon(f)\|_u \le C \|f\|_u$$

for $f$ of product form.

3.2. Restriction problem of Fourier transform to the unit sphere

Suppose that

$$\left(\int_{S^{d-1}} |\hat{f}(\xi)|^2 d\sigma(\xi)\right)^{1/2} \le C \left(\int_{R^d} |f(x)|^p dx\right)^{1/p}, \quad (3.2)$$

where $d\sigma$ is an area element on $S^{d-1}$ and $f \in C_c^\infty(R^d)$. Put $p_0(\lambda) = 2d/(d+1+2\lambda)$. By Fefferman's remark ( see [9]), if $p_0(\lambda) < p < 2$, then

$$\|s^\lambda(f)\|_p \le C \|f\|_p .$$

(3.2) holds for $1 \le p < 2d/(d+1)$ for radial functions. Tomas [14] pproved that (3.2) is valid for general $f$ if $1 \le p < 2(d+1)/(d+3)$, but it fails for $p > 2d(d+1)/(d+3)$.

In [11] we showed the following

THEOREM 5. If $d \ge 2$, $1 \le u \le 2d/(d+1)$ and $f \in C_c^\infty(R^d)$, then

$$\left(\int_{S^{d-1}} |\hat{f}(\xi)|^2 |\xi_0| d\sigma(\xi)\right)^{1/2} \le \frac{1}{\sqrt{2\pi}} \int_{R^1} \left(\int_{R^{d-1}} |f|^2 dx_1 \ldots dx_{d-1}\right)^{1/2} dx_0 . \quad (3.3)$$

Let $P$ be the family of $d-1$ indices in $\{ 0,1,\ldots,d-1 \}$ and $I_p$, $p \in P$, be disjoint arcs in $\partial D$ of length $2\pi/d$. Let $\delta_p(z)$ be functions in $H^2$ such that $\Re e\ \delta_p(e^{i\theta}) = 1$ a.e. in $I_p$ and $= 0$ otherwise, and $\Im m\ \delta_p(0) = 0$. Then $\delta_p(0) = 1/d$. Define a mapping $T^z$ by

$$T^z f(\xi) = \hat{f}(\xi) \prod_{j=o}^{d-1} |\xi_j|^{\delta_j(z)} .$$

Applying an analytic operator version of Theorem 1(ii) with $M = R$, $N = S^{d-1}$ and $u_0 = 2$, $u_1 = 1$, we get

THEOREM 6. If $d \ge 2$, $1 \le u \le 2d/(d+1)$ and $f$ in $C_c^\infty(R^d)$ is of product form

$$\left(\int_{S^{d-1}} |\hat{f}(\xi)|^2 |\xi_0 \ldots \xi_{d-1}|^{1/d} d\sigma(\xi)\right)^{1/2} \le \frac{1}{\sqrt{2\pi}} \|f\|_u . \quad (3.4)$$

3.3. Kakeya's maximal function

Let $\Re$ be a family of on-empty bounded open sets in $R^d$. For a locally integrable function $f$ on $R^d$ the maximal operator $M_\Re$ related to $\Re$ is defined by

$$M_\Re f(x) = \sup_{x \in R \in \Re} \frac{1}{|R|} \int_R |f| dx.$$

When $\Re$ is the family of all open balls in $R^d$, $M_\Re f$ is Hardy-Littlewood maximal function. For given $N > 2$ and $a > 0$ let $\Re$

be the family of all rectangles in $R^d$ with size $a\times\ldots\times a\times aN$, but with arbitrary direction.

When $d = 2$, the operator $M_{\mathfrak{R}}$ has arisen in the work of Fefferman [9] and Cordoba [5] to estimate Riesz-Bochner operator.

In fact Cordoba [6] proved that when $d = 2$

$$\|M_{\mathfrak{R}}f\|_2 \le C(\log N)^{1/2} \|f\|_p.$$

Recently, Christ, Duoandikoexea and Rubio de Francia [4] showed that if $d \ge 3$ and $1 < p \le (d+1)/2$,

$$\|M_{\mathfrak{R}}f\|_p \le C(\log N)^{\beta}N^{d/p-1}\|f\|_p.$$

for some constant $\beta > 0$.

In Igari [12] we have shown the following. For $x = (x_0,x_1,\ldots,x_{d-1})$ denote $\bar{x} = (x_0,\ldots,x_{d-3})$ and $\bar{\bar{x}} = (x_{d-2},x_{d-1})$.

THEOREM 7. There exists a constant $C$ such that

$$\int_{R^2}(\sup_{\bar{x}}|M_{\mathfrak{R}}f(\bar{x},\bar{\bar{x}})|)^2 d\bar{\bar{x}} \le C \log^3 N \int_{R^2}(\sup_{\bar{\bar{x}}}|f(\bar{x},\bar{\bar{x}})|)^2 d\bar{\bar{x}}. \quad (3.5)$$

(3.5) implies that

$$\|M_{\mathfrak{R}}f\|_{(2,\infty:p)} \le C \log^{3/2}N \|f\|_{(2,\infty:p)},$$

where $p = (d-2,d-1)$.

Thus by Theorem 2(ii) we get

THEOREM 8. Let $d \ge 3$. Then there exists a constant $C$ such that

$$\|M_{\mathfrak{R}}f\|_d \le C \log^{3/2}N \|f\|_d$$

for $f$ in $L^d(R^d)$ of product form.

REFERENCES

[1] J.Berg and J.Löfström, Interpolation Spaces, An Introduction, Springer-Verlag, Berlin,Heidelberg/New York, 1976.

[2] A.P.Calderon and A.Zygmund, On the theorem of Hausdorff-Young and its extensions, Ann.Math.Studies,25(1950),166-188.

[3] L.Carleson and P.Sjölin, Oscillatory integrals and multiplier Problem for the disk, Studia Math.,44(1972),287-299.

[4] M.Christ,J.Duoandikoetxea and J.L.Rubio de Francia, Maximal operators related to the Radon transform and the Calderon-Zygmund method of rotations, Duke Math.J.,53(1986),189-209.

[5] A.Cordoba, The Kakeya maximal function and the spherical summation multipliers, Amer.J.Math.,99(1977),1-22.

[6] A.Cordoba, The multiplier problem for the polygon, Annales

of Math.,lo5(1977),581-588.

[7] P.Duren, Theory of $H^p$ Spaces, Acad.Press,New York/London, 1970.

[8] C.Fefferman, The multiplier problem for the ball, Annales of Math.,94(1974),330-336.

[9] C.Fefferman, A note on spherical summation multipliers, Israel J.Math.,15(1973),44-52.

[10] C.Herz, On the mean inversion of Fourier and Hankel transforms, Proc.Nat.Acad.Sci.USA,40(1954),996-9.

[11] S.Igari, Interpolation of operators in Lebesgue spaces with mixed norm and its applications to Fourier analysis,Tohoku Math.J.,38(1986),469-490.

[12] S.Igari, On Kakeya,s maximal function, Proc.Japan Acad.

[13] R.Rochberg and G.Weiss, Analytic families of Banach spaces and some of their uses, Recent Progress in Fourier Analysis, ed.by I.Peral and J.L.Rubio de Francia, North-Holland 1985, 173-201.

[14] P.A.Tomas, A restriction theorem for the Fourier transform, Bull.Amer.Math.Soc.,81(1975),477-478.

Proceedings of the Analysis Conference, Singapore 1986
S.T.L. Choy, J.P. Jesudason, P.Y. Lee (Editors)



# SHIFT INVARIANT MARKOV MEASURES AND THE ENTROPY MAP OF THE SHIFT*

CHOO-WHAN KIM

Department of Mathematics and Statistics
Simon Fraser University
Burnaby, B.C.
CANADA V5A 1S6

Let $\Omega = \times_0^\infty\{0,1,\ldots,s-1\}$, $s \geq 2$, and let $T$ be the shift on $\Omega$. Let $P(\Omega,T)$ be the compact convex set of all shift invariant normalized Borel measures on $\Omega$, and let $M(\Omega,T)$ be the set of all Markov measures in $P(\Omega,T)$. Let $M_1(\Omega,T)$ be the set of all Markov measures in $M(\Omega,T)$ that are induced by irreducible stochastic matrices, and let $M_2(\Omega,T)$ be the set of all Markov measures in $M(\Omega,T)$ that are induced by reducible recurrent stochastic matrices. We show that both $M(\Omega,T)$ and $M_2(\Omega,T)$ are compact, connected nonconvex subsets of $P(\Omega,T)$, and $M_1(\Omega,T)$ is an open, connected, strongly nonconvex dense subset of $M(\Omega,T)$ such that $M(\Omega,T) = M_1(\Omega,T) \cup M_2(\Omega,T)$. We also show that the entropy map of the shift is an affine upper semicontinuous function from $P(\Omega,T)$ onto $[0,\log s]$ and is a continuous function on $M(\Omega,T)$ which maps $M_1(\Omega,T)$ onto $[0, \log s]$.

## 1. INTRODUCTION

Throughout this paper, $s$ will denote a fixed but arbitrary integer such that $s \geq 2$. Let $S = \{0,1,\ldots,s-1\}$ be endowed with the discrete topology, let $\Omega = \times_0^\infty S$ be endowed with the product topology, and let $\mathcal{B}$ be the $\sigma$-algebra of Borel sets in $\Omega$. Note that $\Omega$ is a compact metrizable space. Each element $\omega \in \Omega$ is a sequence $(\omega_n)_{n\geq 0}$ where $\omega_n \in S$. For each $n \geq 0$, let $x_n: \Omega \to S$ be the continuous surjection defined by $x_n(\omega) = \omega_n$. The transformation $T: \Omega \to \Omega$ defined by $(T\omega)_n = \omega_{n+1}$ for each $n \geq 0$ is a continuous surjection and is called the shift on $\Omega$.

Let $P(\Omega)$ denote the set of all probability measures on $(\Omega,\mathcal{B})$ endowed with the weak* topology. Then $P(\Omega)$ is a compact convex metrizable space. For each $\mu \in P(\Omega)$, let $T\mu \in P(\Omega)$ be such that $(T\mu)(E) = \mu(T^{-1}E)$ for each $E \in \mathcal{B}$. Define $P(\Omega,T) = \{\mu \in P(\Omega): T\mu = \mu\}$. Note that $P(\Omega,T)$ is a compact convex subset of $P(\Omega)$.

Let $M(s \times s)$ be the set of all $s \times s$ stochastic matrices $P = (p_{ij})_{ij \in S}$ and let $\Pi$ be the set of all probability vectors $p = (p_i)_{i \in S}$. A vector $p \in \Pi$ is called positive if $p_i > 0$ for all $i \in S$, denoted by $p > 0$. Let $\Pi^+ = \{p \in \Pi: p > 0\}$. A vector $p \in \Pi$ is called a stationary distribution of a matrix $P \in M(s \times s)$ if $pP = p$, i.e., $\sum_{i=0}^{s-1} p_i p_{ij} = p_j$ for each $j \in S$. For

*Research supported by NSERC Canada.

each $P \in M(s \times s)$, let $\Pi(P)$ denote the set of all stationary distributions of $P$, i.e., $\Pi(P) = \{p \in \Pi\colon pP = p\}$. Note that, for each $P \in M(s \times s)$, $\Pi(P) \neq \phi$ and $P$ is irreducible iff $\Pi(P) = \{p\}$ for some $p \in \Pi^+$. For any $P \in M(s \times s)$ and any $p \in \Pi$, there exists, by the Kolmogorov existence theorem, a unique measure in $P(\Omega)$, denoted by $\mu_{pP}$, such that

$$\mu_{pP}(x_k = i_k,\ 0 \le k \le n) = p_{i_0} p_{i_0 i_1}, \ldots, p_{i_{n-1} i_n}$$

for each $n \ge 0$ and each sequence $i_0, \ldots, i_n$ in $S$. The measure $\mu_{pP}$ is called the Markov measure induced by $P$ and $p$ or the $(p,P)$ Markov measure. Let $M(\Omega)$ denote the set of all Markov measures, i.e., $M(\Omega) = \{\mu_{pP}\colon p \in \Pi,\ P \in M(s \times s)\}$, and let $M(\Omega,T) = \{\mu_{pP} \in M(\Omega)\colon pP = p\}$. Then we have $M(\Omega,T) = M(\Omega) \cap P(\Omega,T)$. Elements of $M(\Omega,T)$ are called the shift invariant Markov measures. Let $M_1(\Omega,T)$ be the set of all Markov measures in $M(\Omega,T)$ that are induced by irreducible stochastic matrices, and let $M_2(\Omega,T)$ be the set of all Markov measures in $M(\Omega,T)$ that are induced by reducible recurrent stochastic matrices.

For each $p = (p_i) \in \Pi$, we define the stochastic matrix $P = (p_{ij})_{i,j \in S}$ by $p_{ij} = p_j$ for all $i,j \in S$, then $pP = p$. In this case the $(p,P)$ Markov measure is called the $p$ Bernoulli measure, denoted by $\mu_p$. For each Bernoulli measure $\mu_p$, we have $\mu_p(x_k = i_k,\ 0 \le k \le n) = p_{i_0}, \ldots, p_{i_n}$ for each $n \ge 0$ and each sequence $i_0, \ldots, i_n$ in $S$. Let $B(\Omega,T)$ denote the set of all Bernoulli measures, and let $B(\Omega,T)^+ = \{\mu \in B(\Omega,T)\colon p > 0\}$. Note that $B(\Omega,T) \subset M(\Omega,T)$.

In Section 2, we show that $M(s \times s)$ is a compact convex set and the set of all irreducible matrices in $M(s \times s)$ is an open, convex dense subset of $M(s \times s)$. On the other hand, the set of all reducible matrices in $M(s \times s)$ is a compact connected subset of $M(s \times s)$. The notion of strongly nonconvex set is introduced in Definition 2.15.

In Section 3, we show that $B(\Omega,T)$ is a compact, connected strictly non-convex set, and $B(\Omega,T)^+$ is an open, connected, strongly nonconvex dense subset of $B(\Omega,T)$.

The main results of this paper are stated in Section 3. We show that both $M(\Omega,T)$ and $M_2(\Omega,T)$ are compact, connected nonconvex sets and $M_1(\Omega,T)$ is an open, connected, strongly nonconvex dense subset of $M(\Omega,T)$. We also have $M(\Omega,T) = M_1(\Omega,T) \cup M_2(\Omega,T)$.

In Section 5, the results of Sections 3 and 4 are used to show that the entropy map of the shift is a continuous function on $M(\Omega,T)$ which maps both $M_1(\Omega,T)$ and $B(\Omega,T)$ onto $[0, \log s]$.

For background information on Markov measures and on Markov chains we refer the reader to Billingsley [1], Chung [2], Denker et al [3], Feller [4], and Walters [7].

2. PRELIMINARIES

Given $\mu \in P(\Omega)$, the state space $E$ of the stochastic process $\{x_n\}_{n \geq 0}$ defined on the probability space $(\Omega, \mathcal{B}, \mu)$ is defined by $E = \{i \in S: \mu(x_n = i) > 0$ for some $n \geq 0\}$. Clearly, $E$ is a nonempty subset of $S$.

Proposition 2.1. Let $\mu \in P(\Omega)$. The following assertions are equivalent:

(i) $\mu$ is the Markov measure induced by $P \in M(s \times s)$ and $p \in \Pi$.

(ii) The process $\{x_n\}_{n \geq 0}$ defined on $(\Omega, \mathcal{B}, \mu)$ is a Markov chain with the state space $E$, the stationary transition matrix $(P_{ij})_{i,j \in E}$ and the initial distribution $(p_i)_{i \in E}$.

Proof. The implication (i) $\Rightarrow$ (ii) follows from Theorem 1 of Chung [2, p.7]. Note that $p_i = \mu(x_0 = i)$ for all $i \in E$ and $p_i = 0$ for all $i \in S - E$, so that $\Sigma_{i \in E}\, p_i = 1$. Let $i \in E$. Then there exists an $n \geq 0$ such that $\mu(x_n = i) > 0$ so that $p_{ij} = \mu(x_{n+1} = j | x_n = i)$ for all $j \in S$. It follows that $\Sigma_{j \in E}\, p_{ij} = 1$ for each $i \in E$, so that $(p_{ij})_{i,j \in E}$ is a stochastic matrix.

(ii) $\Rightarrow$ (i): Suppose (ii) holds. Let $F = S - E$. Define the stochastic matrix $P' = (P'_{ij})_{i,j \in S}$ by $p'_{ij} = p_{ij}$ for each $(i,j) \in E \times E$, $p'_{ij} = 0$ for each $(i,j) \in E \times F$, $p'_{ij} = \delta_{ij}$ for each $(i,j) \in F \times S$ where $\delta_{ij}$ denotes Kronecker's delta. Let $p = (p_i)_{i \in S}$ be the probability vector defined by $p_i = \mu(x_0 = i)$ for each $i \in S$. Then $\mu$ is the $(p,P')$ Markov measure.

We state without proofs the following three propositions.

Proposition 2.2. Let $\mu \in P(\Omega)$. Then the following assertions are equivalent:

(i) $\mu \in P(\Omega,T)$.

(ii) The process $\{x_n\}_{n \geq 0}$ defined on $(\Omega, \mathcal{B}, \mu)$ is stationary.

Proposition 2.3. Let $\mu \in P(\Omega,T)$. Then the following assertions are equivalent:

(i) $\mu$ is the $(p,P)$ Markov measure where $pP = p$.

(ii) The process $\{x_n\}_{n \geq 0}$ defined on $(\Omega, \mathcal{B}, \mu)$ is a stationary Markov chain with the state space $E$, the stationary transition matrix $(p_{ij})_{i,j \in E}$ and the stationary initial distribution $(p_i)_{i \in E}$ where $E \subset S$.

Proposition 2.4. Let $\mu \in M(\Omega,T)$. Then the following assertions are equivalent:

(i) $\mu \in B(\Omega,T)$.

(ii) The process $\{x_n\}_{n \geq 0}$ defined on $(\Omega, \mathcal{B}, \mu)$ is a sequence of independent, identically distributed random variables with finite mean.

Let $C(\Omega)$ denote the usual Banach space of all continuous real functions on $\Omega$ and let $1_A$ denote the indicator function of a set $A \subset \Omega$. Let $\mathcal{S}$ be the collection of all thin cylinder sets: $Z(i_0,\ldots,i_n) = (x_k = i_k,\ 0 \leq k \leq n)$ where $n \geq 0$ and $i_0,\ldots,i_n \in S$, together with the empty set $\phi$. It is easily seen that $\mathcal{S}$ is a countable base for the product topology of $\Omega$ and is a semialgebra which generates the $\sigma$-algebra $\mathcal{B}$. Note also that each set in $\mathcal{S}$ is a clopen set, i.e., a set which is both closed and open, so that $1_A \in C(\Omega)$ for all $A \in \mathcal{S}$. Using the Stone-Weierstrass theorem, we show readily that the family of all linear combinations of indicator functions of sets in $\mathcal{S}$ is dense in $C(\Omega)$.

By the preceding remark, we obtain at once the following propositions.

Proposition 2.5. Let $\mu, \mu_n \in P(\Omega)$ where $n = 1,2,\ldots$. Then the following assertions are equivalent:

(i) $\mu_n \to \mu$.

(ii) $\lim_n \mu_n(A) = \mu(A)$ for each $A \in \mathcal{S}$.

Proposition 2.6. Let $\mu_n$ be the $(p_n, P_n)$ Markov measures where $p_n = (p_i(n))_{i \in S}$, $P_n = (p_{ij}(n))_{i,j \in S}$, $n = 1,2,\ldots$. Let $\mu$ be the $(p,P)$ Markov measure where $p = (p_i)_{i \in S}$, $P = (p_{ij})_{i,j \in S}$.

(i) If $\lim_n p_i(n) = p_i$ each $i \in S$, and $\lim_n p_{ij}(n) = p_{ij}$ for each $i.j \in S$, then $\mu_n \to \mu$.

(ii) If $\mu_n \to \mu$, then $\lim_n p_i(n) = p_i$ for each $i \in S$, and $\lim_n p_{ij}(n) = p_{ij}$ for each $i \in S$ with $p_i > 0$ and each $j \in S$.

Prepostion 2.7. Let $\mu$ be the $p$ Bernoulli measure, and let $\mu_n$ be the $p_n$ Bernoulli measures where $n = 1,2,\ldots$. Then $\mu_n \to \mu$ iff $\lim_n p_i(n) = p_i$ for each $i \in S$.

For brevity, we shall denote $(p_{ij})_{i,j \in S}$ by $(p_{ij})$. Let $P = (p_{ij}) \in M(s \times s)$. The matrix $P$ is called positive if $p_{ij} > 0$ for all $i,j$. Define $P^0 = I$, the unit matrix, i.e., $I = (\delta_{ij})$. For each positive integer $n$, the $n$-th power of the matrix $P$ is denoted by $P^n = (p_{ij}^{(n)})$ where $p_{ij}^{(1)} = p_{ij}$. Note that $P^n \in M(s \times s)$ for all $n \geq 0$. An element of $S$ is called a state. A state $i$ is called recurrent if $\sum_{n=1}^{\infty} p_{ii}^{(n)} = \infty$ and transient if $\sum_{n=1}^{\infty} p_{ii}^{(n)} < \infty$. A state $i$ is called absorbing if $p_{ii} = 1$. We say that the matrix $P$ is recurrent if all of its states are recurrent. In general, the matrix $P$ has at least one recurrent state.

Definition 2.8. Define

$M_1(s \times s) = \{P \in M(s \times s):\ P$ is irreducible$\}$,

$M_2(s \times s) = \{P \in M(s \times s):\ P$ is reducible recurrent$\}$,

$M_3(s \times s) = M(s \times s) - M_1(s \times s) - M_2(s \times s)$,

$M_r(s \times s) = M_2(s \times s) \cup M_3(s \times s)$.

Note that $M_3(s \times s)$ denotes the set of all matrices in $M(s \times s)$ having transient states and $M_r(s \times s)$ denotes the set of all reducible matrices in $M(s \times s)$.

We shall always assume that $M(s \times s)$ is endowed with the relative product topology of $[0,1]^n$ with $n = s^2$. Then we obtain the following proposition, the easy proof of which we leave to the reader.

Theorem 2.9. $M(s \times s)$ is a compact convex set.

Theorem 2.10. $M_r(s \times s)$ is a compact connected subset of $M(s \times s)$.

Proof. Suppose $P_n \to P$ where $P_n \in M_r(s \times s)$ and $P \in M(s \times s)$. If $P \in M_1(s \times s)$, then, by Gantmacher [5, p. 51], $(I + P)^{s-1}$ is a positive matrix. Let $A_n = I + P_n$ and $B = I + P$. We see readily that $\{A_n\}_{n \geq 1}$ are also reducible and $A_n^{s-1} \to B^{s-1}$ as $n \to \infty$. Since $B^{s-1}$ is positive, there exists a positive integer $n'$ such that $A_n^{s-1}$, $n \geq n'$, are positive, a contradiction. Therefore, $P \in M_r(s \times s)$ and $M_r(s \times s)$ is a compact subset of $M(s \times s)$.

To prove the connectedness of $M_r(s \times s)$, let $\Lambda$ be the collection of all proper nonempty subsets $E$ of $S$. We may assume that $\Lambda = \{E_n : 1 \leq n \leq 2^s - s\}$. For each $E_n \in \Lambda$, let

$$M_r(s \times s, E_n) = \{(p_{ij}) \in M(s \times s) : p_{ij} = 0 \text{ for each } (i,j) \in E_n \times E_n'\}$$

where $E_n' = S - E_n$. It is plain that each $M_r(s \times s, E_n)$ is convex and contains the unit matrix I. Consequently, $M_r(s \times s) = \bigcup_{n=1}^{2^s - 2} M_r(s \times s, E_n)$ is connected.

Remark 2.11. $M_2(2 \times 2)$ consists of the $2 \times 2$ unit matrix, so that it is compact convex. Assume $s \geq 3$. Then $M_2(s \times s)$ is not convex as the following example shows. Define $P, Q \in M_2(s \times s)$ by

$$p_{ij} = \frac{1}{s-1} \text{ for all } i,j \in \{0,1,\ldots,s-2\},\ p_{s-1,s-1} = 1,$$

$$q_{00} = 1,\ q_{ij} = \frac{1}{s-1} \text{ for all } i,j \in \{1,2,\ldots,s-1\}.$$

We obtain, for each $c \in (0,1)$, $cP + (1-c)Q \in M_1(s \times s)$. On the other hand, we show, by a modification of the proof of Theorem 2.10, $M_2(s \times s)$ is connected. We also note that $M_2(s \times s)$ is not closed. Define $P_n \in M_2(s \times s)$ by

$$p_{ij}(n) = \delta_{ij} \text{ for } 0 \leq i \leq s-3,\ 0 \leq j \leq s-1,\ p_{s-2,s-2}(n) = 1 - \frac{1}{n+1},$$

$$p_{s-2,s-1}(n) = 1/n+1,\ p_{s-1,s-2}(n) = p_{s-1,s-1}(n) = 1/2$$

where $n = 1,2,\ldots$. Then $P_n$ converges to the matrix $P \in M_3(s \times s)$ defined by

$$p_{ij} = \delta_{ij} \text{ for } 0 \leq i \leq s-2,\ 0 \leq j \leq s-1,\ p_{s-1,s-2} = p_{s-1,s-1} = \frac{1}{2}.$$

Similarly, we prove that $M_3(s \times s)$ is connected and is not convex. To prove $M_3(s \times s)$ is not closed, let $P_n \in M_3(s \times s)$ be such that

$$p_{ij}(n) = \delta_{ij} \text{ for } 0 \leq i \leq s-2, 0 \leq j \leq s-1, p_{s-1,s-2}(n) = \frac{1}{n}, p_{s-1,s-2}(n) = 1 - \frac{1}{n}.$$

where $n = 1,2,\ldots$. Then $P_n \to I \in M_2(s \times s)$. Note also that $M_r(s \times s)$ is not convex.

Proposition 2.12. The set of all positive stochastic matrices is dense in $M(s \times s)$. In particular, $M_1(s \times s)$ is a dense subset of $M(s \times s)$.

Proof. (i) Suppose $P = (p_{ij}) \in M_1(s \times s)$. For each $i \in S$, there is $j_i \in S$ such that $p_{ij_i} > 0$. Let $r$ be a positive integer such that $p_{ij_i} > \frac{1}{r}$ for all $i$. Define the matrices $P_n = (p_{ij}(n))_{i,j}$, $n = 1,2,\ldots$, by

$$p_{ij_i}(n) = p_{ij_j} - \frac{1}{r+n},$$

$$p_{ik}(n) = p_{ik} + \frac{1}{(s-1)(r+n)} \quad \text{for all } k \in S - \{j_i\}$$

where $i \in S$. Then $\{P_n\}_{n \geq 1}$ are positive stochastic matrices and converges to $P$.

(ii) Suppose $P = (p_{ij}) \in M_2(s \times s)$. Then there exists a (unique) partition $\{C_k\}_{1 \leq k \leq m}$, $2 \leq m \leq s$, of $S$ such that each matrix $P_k = (p_{ij})_{i,j \in C_k}$ is irreducible stochastic. Choose $i_k, j_k \in C_k$ such that $p_{i_k j_k} > 0$, $1 \leq k \leq m$. Let $r$ be a positive integer such that $p_{i_k j_k} > \frac{1}{r}$ for all $k$. Define the matrices $Q_n = (q_{ij}(n))_{i,j}$ by

$$q_{i_k j_k}(n) = p_{i_k j_k} - \frac{1}{r+n},$$

$$q_{i_1 j_2}(n) = q_{i_2 j_3}(n) = \ldots = q_{i_{m-1} j_m} = q_{i_m j_1} = \frac{1}{r+n},$$

$$q_{ij}(n) = p_{ij} \quad \text{elsewhere}$$

where $n = 1,2,\ldots$. Then $Q_n \in M_1(s \times s)$ and $Q_n \to P$. By (i), we obtain a sequence $\{P_n\}_{n \geq 1}$ of positive stochastic matrices which converges to $P$.

(iii) Suppose $P = (p_{ij})_{i,j} \in M_3(s \times s)$. Using the canonical form of the matrix $P$ (see Seneta [6, p. 15]), together with a modification of the argument in (ii), we obtain a sequence $\{Q_n\}_{n \geq 1}$ in $M_1(s \times s)$ which converges to $P$. Using (i) again, we complete the proof.

It is plain $M_1(s \times s)$ is a convex subset of $M(s \times s)$, so that, by theorem 2.10 and Proposition 2.12, we obtain the following proposition.

Theorem 2.13. $M_1(s \times s)$ is an open, convex, dense subset of $M(s \times s)$.

Definition 2.14. For any $i_0, \ldots, i_{n-1} \in S$ where $n \geq 1$, let $[i_0, \ldots, i_{n-1}]$ denote the point $(\omega_m)_{m \geq 0} \in \Omega$ such that $\omega_{kn+m} = i_m$ for all $k \geq 0$ and all $m \in \{0,1,\ldots,n-1\}$. For each point $\omega \in \Omega$, let $\varepsilon_\omega$ denote the unit mass at $\{\omega\}$, i.e., $\varepsilon_\omega(A) = 1_A(\omega)$ for all $A \subset \Omega$.

For convenience, we introduce the following concept.

Definition 2.15. A subset $E \subset P(\Omega)$ is called strongly nonconvex if $c\mu + (1-c)\nu \in P(\Omega) - E$ for all $\mu, \nu \in E$ with $\mu \neq \nu$ and all $c \in (0,1)$.

We shall always assume that every subset of $P(\Omega)$ discussed in the rest of this paper is endowed with the relative weak* topology of $P(\Omega)$. For concepts and notation not explained in this paper, we refer to the standard works: see for example Billingsley [1], Denker et al [3], and Walters [7].

## 3. BERNOULLI MEASURES

We begin by proving the following basic lemmas.

<u>Lemma 3.1</u>. The mapping $p \to \mu_p$ from $\Pi$ onto $B(\Omega,T)$ is a homeomorphism, and $B(\Omega,T)$ is a compact connected set.

<u>Proof</u>. It is plain that $\Pi$ is a compact convex subset of the space $[0,1]^S$, so that it is a compact connected set. On the other hand, we see readily that the mapping $p \to \mu_p$ is a bijection between $\Pi$ and $B(\Omega,T)$ and is continuous by Proposition 2.7. Therefore, the mapping is a homeomorphism, and $B(\Omega,T)$ is a compact connected set.

<u>Lemma 3.2</u>. Let $p = (p_i)_{i \in S}$ and $q = (q_i)_{i \in S}$ be probability vectors such that $p \neq q$, and let $A(p) = \{i \in S: p_i > 0\}$, $A(q) = \{i \in S: q_i > 0\}$. Let $\mu$ and $\nu$ be the p-Bernoulli measure and the q-Bernoulli measure.

(i) If $A(p) \cap A(q) = \phi$, then $c\mu + (1-c)\nu \in M(\Omega,T) - B(\Omega,T)$ for all $c \in (0,1)$.

(ii) If $A(p) \cap A(q) \neq \phi$, then $c\mu + (1-c)\nu \in P(\Omega,T) - M(\Omega,T)$ for all $c \in (0,1)$.

<u>Proof</u>. Let $c \in (0,1)$ be arbitrary and $\rho = c\mu + (1-c)\nu$. Then $\rho \in P(\Omega,T)$, so that, by Proposition 2.2, $\{x_n\}_{n \geq 0}$ is a stationary process on $(\Omega,\mathcal{B},\rho)$. Let $A_1 = A(p)$ and $A_2 = A(q)$. Clearly both $A_1$ and $A_2$ are nonempty sets. Define $E = A_1 \cup A_2$, $B = S - A_1$. For each $i \in S$, define $r_i = \rho(x_0 = i)$, that is, $r_i = cp_i + (1-c)q_i$. We see readily that $r_i > 0$ iff $i \in E$, so that $r = (r_i)_{i \in E}$ is a positive probability vector. Define the stochastic matrix $R = (r_{ij})_{i,j \in E}$ by $r_{ij} = \rho(x_1 = j | x_0 = i)$, that is, $r_{ij} = (cp_ip_j + (1-c)q_iq_j)/r_i$. It is easily seen that $\Sigma_{i \in E} r_i r_{ij} = r_j$ for each $j \in E$, that is, $rR = r$.

(i) Assume $A_1 \cap A_2 = \phi$. Then we obtain easily that

$$r_{ij} = p_j \text{ for each } (i,j) \in A_1 \times E, \quad r_{ij} = q_j \text{ for each } (i,j) \in A_2 \times E.$$

Let $n \geq 0$ and let $i_0,\ldots,i_n \in S$ be such that $\rho(x_k = i_k, 0 \leq k \leq n) > 0$, equivalently, either $\{i_0,\ldots,i_n\} \subset A_1$ or $\{i_0,\ldots,i_n\} \subset A_2$. If $\{i_0,\ldots,i_n\} \subset A_1$, then, for each $j \in E$,

$$\rho(x_{n+1} = j | x_k = i_k, 0 \leq k \leq n) = \rho(x_{n+1} = j | x_n = i_n) = p_j = r_{i_n j}.$$

If $\{i_0,\ldots,i_n\} \subset A_2$, then, for each $j \in E$,

$$\rho(x_{n+1} = j | x_k = i_k, 0 \leq k \leq n) = \rho(x_{n+1} = j | x_n = i_n) = q_j = r_{i_n j}.$$

By Proposition 2.3, we obtain $\rho \in M(\Omega,T)$. Since

$$\rho(x_{n+1} = j | x_k = i_k, 0 \leq k \leq n) = p_j \neq cp_j = \rho(x_{n+1} = j)$$

for any $n \geq 0$ and any $i_0,\ldots,i_n,j \in A_1$, it follows from Proposition 2.4 that $\rho \in M(\Omega,T) - B(\Omega,T)$.

To prove (ii), we shall consider the following two cases.

Case 1. $A_1 \cap A_2 \neq \phi$ and $B \cap A_2 = \phi$, or equivalently, $A_2 \subset A_1$. Since $p \neq q$, there exist $i \in A_2$, $j \in A_1$ such that $i \neq j$, $p_i \neq q_i$, $p_j \neq q_j$.

Suppose $\rho \in M(\Omega,T)$. Then we have $\rho(x_2 = j|x_0 = i, x_1 = i) = r_{ij}$, that is,

$$(cp_1^2/t_i)p_j + ((1-c)q_i^2/t_i)q_j = (cp_i/r_i)p_j + ((1-c)q_i/r_i)q_j$$

where $t_i = cp_1^2 + (1-c)q_i^2$, so that $cp_i^2/t_i = cp_i/r_i$. It follows $p_i = q_i$, a contradiction. Consequently, $\rho \in P(\Omega,T) - M(\Omega,T)$.

Case 2. $A_1 \cap A_2 \neq \phi$ and $B \cap A_2 \neq \phi$. Choose $i \in A_1 \cap A_2$, $j \in B \cap A_2$. Clearly $i \neq j$. If $\rho \in M(\Omega,T)$, then $\rho(x_2 = j|x_0 = j, x_1 = i) = r_{ij}$, or equivalently, $q_j = (1-c)q_iq_j/(cp_i + (1-c)q_i)$ so that $cp_iq_j = 0$, a contradiction. Hence $\rho \in P(\Omega,T) - M(\Omega,T)$. Thus the proof is complete.

From Lemma 3.2 we obtain

Proposition 3.3. Both $B(\Omega,T)$ and $B(\Omega,T)^+$ are strictly nonconvex sets.

The next theorem follows immediately from Lemma 3.1 and Proposition 3.3.

Theorem 3.4. $B(\Omega,T)$ is a compact, connected strongly nonconvex set.

We shall now prove the following result.

Theorem 3.5. $B(\Omega,T)^+$ is an open, connected, strictly nonconvex dense subset of $B(\Omega,T)$.

Proof. It is easily seen that $\Pi^+$ is a convex open subset of $\Pi$, so that, by Lemma 3.1, $B(\Omega,T)^+$ is a connected open subset of $B(\Omega,T)$. By Proposition 3.3, it remains to show $B(\Omega,T)^+$ is dense in $B(\Omega,T)$. Let $\mu_p$ be any measure in $B(\Omega,T) - B(\Omega,T)^+$. Then the set $E = \{i: p_i > 0\}$ is a nonempty proper subset of $S$. Let card $E = t$, $r = s - t$ and $i_0 \in E$. There exists a positive integer $n_0$ such that $p_i > 1/n_0$ for all $i \in E$. Define the probability vector $p_n \in \Pi^+$, $n > n_0$, by

$$p_{i_0}(n) = p_{i_0} - \frac{1}{n}, \quad p_i(n) = \frac{1}{rn} \quad \text{for all } i \in S - E,$$

$$p_i(n) = p_i \quad \text{for all } i \in E - \{i_0\}.$$

We have then $p_n \to p$, equivalently $\mu_{p_n} \to \mu_p$.

Remark 3.6. An immediate consequence of Lemmas 3.1 and 3.2 is that the homeomorphism $p \to \mu_p$ from $\Pi$ onto $B(\Omega,T)$ is not affine, i.e.,

$c\mu_p + (1-c)\mu_q \neq \mu_{cp + (10c)q}$ for all $p,q \in \Pi$ with $p \neq q$ and all $c \in (0,1)$.

## 4. SHIFT INVARIANT MARKOV MEASURES

The principal aims of this section is to prove analogues of Theorems 2.9, 2.10 and 2.13 for shift invariant Markov measures. We begin with the following definition.

Definition 4.1. Let $M_1(\Omega,T)$ be the set of all Markov measures in $M(\Omega,T)$ that are induced by irreducible matrices, i.e.,

$$M_1(\Omega,T) = \{\mu_{pP} \in M(\Omega,T): P \in M_1(s \times s), pP = p\}.$$

Similarly we define

$M_n(\Omega,T) = \{\mu_{pP} \in M(\Omega,T): P \in M_n(s \times s),\ p \in \Pi(P)\}$ where $n = 2,3$,

$M_r(\Omega,T) = M_2(\Omega,T) \cup M_3(\Omega,T)$.

Note that $M(\Omega,T) = M_1(\Omega,T) \cup M_r(\Omega,T)$.

Theorem 4.2. $M(\Omega)$ is a compact, connected nonconvex subset of $P(\Omega)$.

Proof. It is clear that $\Pi \times M(s \times s)$ is a compact convex subset of $[0,1]^n$ where $n = s + s^2$. By part (i) of Proposition 2.6, the mapping $(p,P) \to \mu_{pP}$ from $\Pi \times M(s \times s)$ onto $M(\Omega)$ is a continuous surjection, so that $M(\Omega)$ is a compact connected subset of $P(\Omega)$. By part (ii) of Lemma 3.2, $M(\Omega)$ is not convex.

Lemma 4.3. $M(\Omega,T)$ is a compact nonconvex subset of $M(\Omega)$.

Proof. It follows from Theorem 4.2, together with $M(\Omega,T) = M(\Omega) \cap P(\Omega,T)$, that $M(\Omega,T)$ is compact. By part (ii) of Lemma 3.2, $M(\Omega,T)$ is not convex.

Lemma 4.4. $M_1(\Omega,T)$ is homeomorphic with $M_1(s \times s)$ and is connected.

Proof. For each $P \in M_1(s \times s)$, let $p = f(P)$ be the unique stationary distribution of $P$. By an elementary argument, we show that the mapping $f: M_1(s \times s) \to \Pi$ is continuous, so that, by part (i) of Proposition 2.6, the mapping $F: M_1(s \times s) \to M(\Omega,T)$ defined by $F(P) = \mu_{pP}$ where $p = f(P)$ is also continuous. To prove $F$ is an injection, suppose $F(P) = F(Q)$ for some $P,Q \in M_1(s \times s)$, that is, $\mu_{pP} = \mu_{qQ}$, $p = f(P)$, $q = f(Q)$ for some $P,Q \in M_1(s \times s)$. Then we obtain at once $p = q$. Since both $p$ and $q$ are positive, we also get $P = Q$. Note that $M_1(\Omega,T) = \{F(P): P \in M_1(s \times s)\}$. It follows from part (ii) of Proposition 2.6 that the mapping $F$ is a homeomorphism between $M_1(s \times s)$ and $M_1(\Omega,T)$. Since $M_1(s \times s)$ is a convex set, $M_1(\Omega,T) = F(M_1(s \times s))$ is connected. This completes the proof.

By a partition of $S$, we shall always mean a finite collection $\{C_k\}_{k=1}^n$ of pairwise disjoint nonempty subsets $C_k$ of $S$ such that $S = \bigcup_{k=1}^n C_k$ where $2 \le n \le s$.

Lemma 4.5. $M_2(\Omega,T)$ is connected.

Proof. Let $\Lambda_n$ be the finite collection of all n-set partition $\alpha = \{C_k\}_{k=1}^n$ of $S$ where $2 \le n \le s$.

Assume $2 \le n \le s-1$. Let $\alpha = \{C_k\}_{k=1}^n$ be any element of $\Lambda_n$. Let $M_1(C_k)$ be the set of all irreducible stochastic matrices $(p_{ij})_{ij \in C_k}$. Let $\Omega_k = \times_{i=0}^{\infty} S_i$ where $S_i = C_k$ for all $i$, let $T_k$ be the shift on $\Omega_k$, and let $M_1(\Omega_k,T_k)$ be the set of all shift invariant Markov measures defined on $\Omega_k$ that are induced by matrices in $M_1(C_k)$. Each $\mu_k$ of $M_1(\mu_k,T_k)$ is of the form $\mu_{p_k P_k}$ where $P_k \in M_1(C_k)$ and $p_k$ is the unique stationary distribution of $P_k$. We shall always extend each measure $\mu_k$ to a unique measure in $P(\Omega)$ by $\mu_k(\Gamma) = \mu_k(\Gamma \cap \Omega_k)$ for each $\Gamma \subseteq \Omega$. By Lemma 4.4, $M_1(\Omega_k,T_k)$ is connected. Let $\Pi_n$ be the convex set of all n-dimensional probability vectors $c = (c_k)_{k=1}^n$. Define the connected set $X_{n,\alpha}$ by

$$X_{n,\alpha} = \Pi_n \times M_1(\Omega_1, T_1) \times \ldots \times M_1(\Omega_n, T_n) .$$

Define the mapping $F_n : X_{n,\alpha} \to M_2(\Omega, T)$ by

$$F_n(c, \mu_1, \ldots, \mu_n) = \sum_{k=1}^{n} c_k \mu_k .$$

It is straightforward to show that the mapping $F_n$ is continuous, so that $F_n(X_{n,\alpha})$ is connected.

It is evident that $\Lambda_s$ denotes the partition of S into one-point sets. Let $M_2(\Omega,T: \Lambda_s)$ denote the set of all Markov measures induced by the unit matrix. Then $M_2(\Omega,T:\Lambda_s) = \{ \sum_{k=0}^{s-1} c_k \varepsilon_{[k]} : 0 \le c_k \le 1, \sum_{k=0}^{s-1} c_k = 1\}$. Clearly $M_2(\Omega,T: \Lambda_s)$ is convex.

We now have

$$M_2(\Omega,T) = \sum_{n=1}^{s-1} (\cup_{\alpha \in \Lambda_n} F_n(X_{n,\alpha})) \cup M_2(\Omega,T: \Lambda_s) .$$

To prove the connectedness of $M_2(\Omega,T)$, it is enough to show that $M_2(\Omega,T: \Lambda_s)$ is not separated from any of $F_n(X_{n,\alpha})$ where $2 \le n \le s-1$, $\alpha \in \Lambda_n$. Let $\alpha = \{C_k\}_{k=1}^n \in \Lambda_n$ where $2 \le n \le s-1$. Define the sequence $P_m = (p_{ij}(m)) \in M_2(s \times s)$, $m \ge 1$, as follows:

If $C_k = \{i\}$, then $p_{ii}(m) = 1$ .

If $C_k = \{i_1, \ldots, i_j\}$ where $i_1 < i_2 < \ldots < i_j$ , then

$$p_{ii}(m) = 1 - \frac{1}{m+1} \quad \text{for all } i \in C_k ,$$

$$p_{i_1 i_2}(m) = p_{i_2 i_3}(m) = \ldots = p_{i_{j-1} i_j}(m) = p_{i_j i_1}(m) = \frac{1}{m+1} .$$

Then we have $P_m \to I$ . Let p be the uniform probability vector on S , and let $\mu_m$ be the $(p,P_m)$ Markov measures. We see readily that $\mu_m \in F_n(x_{n,\alpha})$, $m \ge 1$, and $\mu_m \to \frac{1}{s} \sum_{k=0}^{s-1} \varepsilon_{[k]} \in M_2(\Omega,T: \Lambda_s)$.

<u>Lemma 4.6</u>. $M_r(\Omega,T) = M_2(\Omega,T)$.

<u>Proof</u>. Let $P = (p_{ij}) \in M_2(s \times s)$. Then there is a unique partition $\{C_k\}_{k=1}^n$, $2 \le n \le s$, of S such that each $P_k = (p_{ij})_{i,j \in C_k}$ is an irreducible stochastic matrix. Let $p_k = (p_i(k))_{i \in C_k}$ be the unique probability vector such that $p_k P_k = p_k$ and let $\pi_k \in \Pi$ be the extension of $p_k$ . We see easily that $\Pi(P)$ is the convex hull of $\{\pi_k\}_{1 \le k \le n}$, i.e.,

$$\Pi(P) = \{ \sum_{k=1}^{n} c_k \pi_k : 0 \le c_k \le 1, \sum_{k=1}^{n} c_k = 1\} .$$

Let $\mu_k$ be the $(\pi_k, P)$ Markov measure, i.e., $\mu_k = \mu_{\pi_k P}$ . Let $M_2(\Omega,T: P)$ denote the set of all Markov measures in $M(\Omega,T)$ that are generated by the matrix P , i.e., $M_2(\Omega,T: P) = \{\mu_{pP} : P \in \Pi(P)\}$. Then $M_2(\Omega,T: P)$ is the convex hull of $\{\mu_k\}_{1 \le k \le m}$,

$$M_2(\Omega,T\colon P) = \{\sum_{k=1}^{m} c_k\mu_k : 0 \le c_k \le 1,\ \sum_{k=1}^{m} c_k = 1\}\ .$$

Therefore, we obtain

$$M_2(\Omega,T) = \cup\{M_2(\Omega,T\colon P) : P \in M_2(s \times s)\}.$$

Suppose $P = (p_{ij}) \in M_3(s \times s)$. Then $S$ is uniquely partitioned into $\{C_1,\ldots,C_{n-1},D\}$, $2 \le n \le s$, such that $D$ is the nonempty set of transient states and each $P_k = (p_{ij})_{i,j \in C_k}$ is an irreducible stochastic matrix. Let $p_k$ be the stationary distribution of the matrix $P_k$ and let $\pi_k \in \Pi$ be the extension of $p_k$. If $M_3(\Omega,T\colon P)$ denotes the set of all Markov measures in $M(\Omega,T)$ that are induced by the matrix $P$, then

$$M_3(\Omega,T\colon P) = \{\sum_{k=1}^{n-1} c_k\mu_k : 0 \le c_k \le 1,\ \sum_{k=1}^{n-1} c_k = 1\}$$

where $\mu_k$ denotes the $(\pi_k,P)$ Markov measure. Define the matrix $P' = (p'_{ij})$ by $p'_{ij} = p_{ij}$ for each $(i,j) \in (S-D) \times S$ and $p'_{ij} = \delta_{ij}$ for each $(i,j) \in D \times S$. It follows that $P' \in M_2(s \times s)$ and

$$M_3(\Omega,T\colon P) \subset M_2(\Omega,T\colon P') \subset M_2(\Omega,T).$$

Consequently, $M_3(\Omega,T) \subset M_2(\Omega,T)$ so that $M_r(\Omega,T) = M_2(\Omega,T)$.

We now prove the following analogue of Theorem 2.10. See also Remark 2.11.

Theorem 4.7. $M_2(\Omega,T)$ is a compact, connected nonconvex set.

Proof. It follows from Lemma 4.5, together with Lemma 3.2, that $M_2(\Omega,T)$ is a connected nonconvex set. To prove the compactness of $M_2(\Omega,T)$, suppose $\mu_n \to \mu$ where $\mu_n \in M_2(\Omega,T) = M_r(\Omega,T)$ and $\mu \in M(\Omega,T)$. Let $\mu_n$ be the $(p_n,P_n)$ Markov measure with $P_n \in M_r(s \times s)$, $p_n \in \Pi(P_n)$, and let $\mu$ be the $(p,P)$ Markov measure with $P \in M(s \times s)$, $p \in \Pi(P)$. Suppose $\mu \in M_1(\Omega,T)$. By Lemma 4.4, there is a unique pair $(p,P) \in \Pi \times M_1(s \times s)$ such that $pP = p$ and $\mu = \mu_{pP}$. It follows from part (ii) of Lemma 2.6, together with Theorem 2.10, that $P_n \to P \in M_r(s \times s)$, a contradiction. Therefore we must have $\mu \in M_2(\Omega,T)$.

The next result is an analogue of Proposition 2.12.

Proposition 4.8. $M_1(\Omega,T)$ is dense in $M(\Omega,T)$.

Proof. It is enough to show that, for each $\mu \in M_2(\Omega,T)$, there is a sequence $\{\mu_n\}$ in $M_1(\Omega,T)$ such that $\mu_n \to \mu$. Let $\mu$ be the $(\pi,P)$ Markov measure in $M_2(\Omega,T)$. Suppose that $\{C_k\}_{k=1}^m$ is a partition of $S$ such that each $P_k = (p_{ij})_{i,j \in C_k}$ is an irreducible stochastic matrix. Let $p_k$ be the unique stationary distribution of the matrix $P_k$, and let $\pi_k \in \Pi$ be the extension of $p_k$. Then $\mu = \sum_{k=1}^{m} c_k\mu_{\pi_k P}$ where $0 \le c_k$, $\sum_{k=1}^{m} c_k = 1$.

(i) Assume that $0 < c_k < 1$ for all $k \in \{1,2,\ldots,m\}$. For each $k$, choose $i_k, j_k \in C_k$ such that $p_{i_k j_k} > 0$. Let $n_0$ be a positive integer such that $p_{i_k j_k} > 1/n_0$ for all $k$. Let $t_1 = 1$ and let $t_k \in (0,1]$,

$2 \le k \le m$, that will be determined later. For each $n > n_0$, define the irreducible stochastic matrix $P_n = (p_{ij}(n))_{i,j \in S}$ by

$$p_{i_k j_k}(n) = p_{i_k j_k} - \frac{t_k}{n}, \quad p_{i_k j_{k+1}}(n) = \frac{t_k}{n} \quad \text{for } 1 \le k \le m-1,$$

$$p_{i_m j_1}(n) = \frac{t_m}{n}, \quad p_{i_m j_m}(n) = p_{i_m j_m} - \frac{t_m}{n}$$

$$p_{ij}(n) = p_{ij} \quad \text{elsewhere.}$$

Let $p_n = (p_i(n))_i$ be the unique stationary distribution of the matrix $P_n$. We shall show that the distributions $p_n$ are independent of $n$. That is, there is a probability vector $p = (p_i)_i$ such that $p = p_n$ for all $n > n_0$. Since $p_n P_n = p_n$, we obtain

$$p_{j_1}(n) = \Sigma_{i \in C_1} p_i(n) p_{ij_1} + (p_{i_m}(n) t_m - p_{i_1}(n)) \frac{1}{n},$$

$$p_j(n) = \Sigma_{i \in C_1} p_i(n) p_{ij} \quad \text{for } j \in C_1 - \{j_1\},$$

so that

$$\Sigma_{j \in C_1} p_j(n) = \Sigma_{j \in C_1} \Sigma_{i \in C_1} p_i(n) p_{ij} + (p_{i_m}(n) t_m - p_{i_1}(n)) \frac{1}{n}$$

$$= \Sigma_{i \in C_1} p_i(n) + (p_{i_m}(n) t_m - p_{i_1}(n)) \frac{1}{n}.$$

Therefore, we get $p_{i_1}(n) = p_{i_m}(n) t_m$. Similarly, we also obtain $p_{i_k}(n) t_k = p_{i_{k+1}}(n) t_{k+1}$ for $1 \le k \le m-1$, so that

$$p_{i_1}(n) = p_{i_k}(n) t_k \quad \text{for } 2 \le k \le m.$$

From the above equations, together with the equations

$$p_j(n) = \Sigma_{i \in C_k} p_i(n) p_{ij} \quad \text{for all } j \in C_k - \{j_k\},\ k = 1,2,\ldots,m$$

$$\sum_{j=0}^{s-1} p_j(n) = 1,$$

there exist $a_{ik} > 0$ such that

$$p_i(n) = a_{ik} p_{i_k}(n) = a_{ik} p_{i_1}(n)/t_k$$

for all $i \in C_k - \{i_k\}$, $k = 1,\ldots,m$. Note that each $a_{ik}$ is independent of $n$. Then we obtain

$$1 = \sum_{k=1}^{m} \Sigma_{i \in C_k} p_i(n) = \sum_{k=1}^{m} ((A_k + 1)/t_k) p_{i_1}(n)$$

where $A_k = \Sigma_{i \in C_k - \{i_k\}} a_{ik}$. Define the probability vector $p = (p_i)_{i \in S}$ by

$$p_{i_1} = \left(\sum_{k=1}^{m} ((A_k + 1)/t_k)\right)^{-1}, \quad p_{i_k} = p_{i_1}/t_k \quad \text{for } 2 \le k \le m,$$

$$p_i = (a_{ik}/t_k) p_{i_1} \quad \text{for } i \in C_k - \{i_k\},\ 1 \le k \le m.$$

Then we obtain $p = p_n$ for all $n > n_0$.

Since $pP_n = p$ and $P_n \to P$ as $n \to \infty$, we obtain $pP = P$, $\mu_{pP_n} \in M_1(\Omega,T)$ and $\mu_{pP_n} \to \mu_{pP} \in M_2(\Omega,T)$ as $n \to \infty$. It remains to choose $\{t_k\}_{2 \le k \le m}$ such that $p = \sum_{k=1}^{m} c_k\pi_k$. Using $c_k = \Sigma_{i \in C_k} p_i$, $1 \le k \le m$, we obtain $p_{i_1} = c_1/(A_1+1)$ and

$$t_k = \frac{c_1(A_k+1)}{c_k(A_1+1)}, \quad 1 \le k \le m .$$

(ii) We may assume without loss of generality that $0 < c_k$ for $1 \le k \le q < m$ and $c_k = 0$ for $q+1 \le k \le m$. Put $u = m-q$. Define

$$\mu_r = \sum_{k=1}^{q-1} c_k\mu_{\pi_k P} + c_q(1 - \frac{1}{r})\mu_{\pi_q P} + \frac{c_q}{ur}\left(\sum_{k=q+1}^{m} \mu_{\pi_k P}\right)$$

where $r = 2,3,\ldots$. Then $\mu_r \in M_2(\Omega,T)$ and $\mu_r \to \mu$ as $r \to \infty$. By (i), there is a sequence in $M_1(\Omega,T)$ which converges to $\mu_r$. This completes the proof.

We obtain at once from Lemmas 4.3 and 4.4, together with Proposition 4.8, the following analogue of Theorem 2.9.

Theorem 4.9. $M(\Omega,T)$ is a compact, connected nonconvex set.

We are now in a position to prove the following analogue of Theorem 2.13.

Theorem 4.10. $M_1(\Omega,T)$ is an open, connected, strongly nonconvex dense subset of $M(\Omega,T)$.

By Lemma 4.4, Theorem 4.7 and Proposition 4.8, it remains to show the following proposition.

Proposition 4.11. Let $P$ and $Q$ be any matrices in $M_1(s \times s)$ with stationary distributions $p$ and $q$, respectively. Let $\mu$ be the $(p,P)$ Markov measures and let $\nu$ be the $(q,Q)$ Markov measure. Then the following assertions are equivalent:

(i) $P = Q$.

(ii) $\mu = \nu$.

(iii) $c\mu + (1-c)\nu \in M(\Omega,T)$ for all $c \in (0,1)$.

(iv) $c\mu + (1-c)\nu \in M(\Omega,T)$ for some $c \in (0,1)$.

The proof of this proposition follows from the following two lemmas.

Lemma 4.12. Let $P = (p_{ij})$ and $Q = (q_{ij})$ be any matrices in $M_1(s \times s)$ with stationary distributions $p = (p_i)$ and $q = (q_i)$, respectively. Let $\mu$ be the $(p,P)$ Markov measure and $\nu$ the $(q,Q)$ Markov measure. If $c\mu + (1-c)\nu \in M(\Omega,T)$ for some $c \in (0,1)$, and $p_{i'j'} \ne q_{i'j'}$ for some $i',j' \in S$, then

(i) $p_{i_0}p_{i_0i_1}\cdots p_{i_{n-1}i'}q_{i'j'} = q_{i_0}q_{i_0i_1}\cdots q_{i_{n-1}i'}p_{i'j'}$ for all $n \ge 1$ and all $i_0,\ldots,i_{n-1} \in S$,

(ii) $p_i p_{ii'} q_{i'} = q_i q_{ii'} p_{i'}$ for all $i \in S$ ,

(iii) $p_i p_{ij} p_{ji'} q_{i'} = q_i q_{ij} q_{ji'} p_{i'}$ for all $i,j \in S$ ,

(iv) $i' \neq j'$ .

Proof. Let $\rho = c\mu + (1-c)\nu$ where $0 < c < 1$ . Clearly $\rho \in P(\Omega,T)$. Note that both $p$ and $q$ are positive probability vectors. Define the positive probability vector $r = (r_i)_{i \in S}$ by $r_i = \rho(x_0 = i) = cp_i + (1-c)q_i$, and define the stochastic matrix $R = (r_{ij})_{i,j \in S}$ by $r_{ij} = (cp_i p_{ij} + (1-c)q_i q_{ij})/r_i$ . It is easily seen that $R \in M_1(s \times s)$ and $rR = r$ .

Suppose that $\rho \in M(\Omega,T)$, equivalently, $\rho \in M_1(\Omega,T)$, and $p_{i'j'} \neq q_{i'j'}$ for some $i',j' \in S$. By Proposition 2.3, we obtain $\rho(x_{n+1} = j' | x_0 = i_0,\dots,x_{n-1} = i_{n-1}, x_n = i') = r_{i'j'}$, for any $n \geq 1$ and any $i_0,\dots,i_{n-1} \in S$ such that $\rho(x_0 = i_0,\dots,x_{n-1} = i_{n-1}, x_n = i') > 0$ , or equivalently,

$$\frac{cp_{i_0} p_{i_0 i_1} \cdots p_{i_{n-1} i'} p_{i'j'} + (1-c) q_{i_0} q_{i_0 i_1} \cdots q_{i_{n-1} i'} q_{i'j'}}{cp_{i_0} p_{i_0 i_1} \cdots p_{i_{n-1} i'} + (1-c) q_{i_0} q_{i_0 i_1} \cdots q_{i_{n-1} i'}} = \frac{cp_{i'} p_{i'j'} + (1-c) q_{i'} q_{i'j'}}{cp_{i'} + (1-c) q_{i'}}$$

for any $n \geq 1$ and any $i_0,\dots,i_{n-1} \in S$ such that $p_{i_0 i_1} \cdots p_{i_{n-1} i'} + q_{i_0 i_1} \cdots q_{i_{n-1} i'} > 0$ . Since $p_{i'j'} \neq q_{i'j'}$ , we obtain

$$\frac{cp_{i_0} p_{i_0 i_1} \cdots p_{i_{n-1} i'}}{cp_{i_0} p_{i_0 i_1} \cdots p_{i_{n-1} i'} + (1-c) q_{i_0} q_{i_0 i_1} \cdots q_{i_{n-1} i'}} = \frac{cp_{i}}{cp_{i'} + (1-c) q_{i'}}$$

for any $n \geq 1$ and any $i_0,\dots,i_{n-1} \in S$ such that $p_{i_0 i_1} \cdots p_{i_{n-1} i'} + q_{i_0 i_1} \cdots q_{i_0 i_1} \cdots q_{i_{n-1} i'} > 0$ , or equivalently,

$$p_{i_0} p_{i_0 i_1} \cdots p_{i_{n-1} i'} q_{i'} = q_{i_0} q_{i_0 i_1} \cdots q_{i_{n-1} i'} p_{i'}$$

for any $n \geq 1$ and any $i_0,\dots,i_{n-1} \in S$ such that $p_{i_0 i_1} \cdots p_{i_{n-1} i'} + q_{i_0 i_1} \cdots q_{i_{n-1} i'} > 0$ . Thus (i) holds.

Both (ii) and (iii) follow from (i). On the other hand, we get from (ii) that $p_{i'i'} p_{i'} q_{i'} = q_{i'i'} p_{i'} q_{i'}$ so that, since $p_{i'} > 0$ and $q_{i'} > 0$ , $p_{i'i'} = q_{i'i'}$ . Thus, (iv) holds.

Lemma 4.13. Let $P = (p_{ij})$ and $Q = (q_{ij})$ be any two distinct matrices in $M_1(s \times s)$ with stationary distributions $p = (p_i)$ and $q = (q_i)$, respectively. Let $\mu$ be the $(p,P)$ Markov measure and $\nu$ the $(q,Q)$ Markov measure. If $c\mu + (1-c)\nu \in M(\Omega,T)$ for some $c \in (0,1)$, then

$$p_{ij} = \frac{p_j q_i}{p_i q_j} q_{ij} \quad \text{for all } i,j \in S .$$

<u>Proof</u>. Suppose $\rho = c\mu + (1-c)\nu$, $0 < c < 1$, and $\rho \in M(\Omega,T)$, equivalently, $\rho \in M_1(\Omega,T)$. Define $A_1 = \{j \in S: p_{jk} \neq q_{jk}$ for some $k \in S\}$ and $B = S - A_1$. Since $P \neq Q$, we get $A_1 \neq \phi$. It follows from part (ii) of Lemma 4.12 that

$$p_{ij} = \frac{p_j q_i}{p_i q_j} q_{ij} \quad \text{for all } (i,j) \in S \times A_1 .$$

If $A_1 = S$, then we are done.

Suppose $A_1 \neq S$, equivalently, $B \neq \phi$. Since $P$ is irreducible, the matrix $(p_{ij})_{i \in B,\, j \in A_1}$ is a nonzero matrix. Define

$A_2 = \{j \in B: p_{jk} > 0$ for some $k \in A_1\}$. Clearly, $A_2 \neq \phi$ and $A_1 \cap A_2 = \phi$. Let $j$ be any state in $A_2$ and let $k \in A_1$ be such that $p_{jk} > 0$. Again using part (ii) of Lemma 4.12, we get

$$0 < q_{jk} = p_{jk} = \frac{p_k q_j}{p_j q_k} q_{jk} ,$$

so that $p_j q_k = p_k q_j$. By part (iii) of Lemma 4.12, we get

$$p_i p_{ij} p_{jk} q_k = q_i q_{ij} q_{jk} p_k \quad \text{for all } i \in S ,$$

so that

$$p_{ij} = \frac{p_k q_i}{p_i q_k} q_{ij} = \frac{p_j q_i}{p_i q_j} q_{ij} \quad \text{for all } i \in S .$$

Therefore, we get

$$p_{ij} = \frac{p_j q_i}{p_i q_j} q_{ij} \quad \text{for all } (i,j) \in S \times A_2 .$$

If $S = A_1 \cup A_2$, then we are done. Otherwise, by repeating the above procedure finitely many times, we obtain a partition $\{A_k\}_{1 \le k \le n}$ of $S$ such that the equations hold on each of the sets $S \times A_k$, $1 \le k \le n$. This completes the proof.

<u>Proof of Proposition 4.11</u>. Since the implications (i) ⇒ (ii) ⇒ (iii) ⇒ (iv) hold trivially, it remains to show (iv) ⇒ (i). To this end, assume (iv) holds. Suppose $P \neq Q$. Then by Lemma 4.13, we obtain $p_{ij} = p_j q_i q_{ij} / p_i q_j$ for all $i,j \in S$. Since $\sum_{j=0}^{s-1} p_{ij} = 1$ for each $i \in S$, we obtain

$$\sum_{j=1}^{s-1} p_j \left(\frac{q_i}{q_j} q_{ij}\right) = p_i \quad \text{for all } i \in S .$$

If we define the matrix $Q' = (q'_{ij})_{i,j \in S}$ by

$$q'_{ij} = \frac{q_j}{q_i} q_{ji} \quad \text{for all } i,j \in S ,$$

then $Q'$ is an irreducible stochastic matrix. We also have

$$\sum_{j=0}^{s-1} p_j q'_{ji} = p_i , \quad \sum_{j=0}^{s-1} q_j q'_{ji} = q_i$$

for all $i \in S$. By the uniqueness of stationary distributions of $Q'$, we have $p = q$ so that $p_{ij} = \frac{p_j q_i}{p_i q_j} q_{ij} = q_{ij}$ for all $i,j \in S$, a contradiction.

Remark 4.14. We give an application of Proposition 4.11. A sequence $\{p_n\}_{n \geq 0}$ of functions $p_n: \times_0^n S \to [0,1]$ is said to satisfy the consistency conditions if

(i) $\sum_{i=0}^{s-1} p_0(i) = 1$,

(ii) $\sum_{i=0}^{s-1} p_{n+1}(i_0,\ldots,i_n,i) = p_n(i_0,\ldots,i_n)$ for all $n \geq 0$ and all $i_0,\ldots,i_n \in S$.

Any pair $(p,P) \in \Pi \times M(s \times s)$ gives rise to such a sequence $\{p_n\}_{n \geq 0}$ by the formula $p_n(i_0,\ldots,i_n) = p_{i_0} p_{i_0 i_1} \cdots p_{i_{n-1} i_n}$. There are sequences $\{p_n\}_{n \geq 0}$ satisfying the consistency conditions that do not arise from any pair $(p,P) \in \Pi \times M(s \times s)$.

Let $(p_{ij})_{i,j \in S}$ and $(q_{ij})_{i,j \in S}$ be any distinct irreducible stochastic matrices with stationary distributions $(p_i)_{i \in S}$ and $(q_i)_{i \in S}$, respectively. Let $c \in (0,1)$ be arbitrary. Define

$$p_n(i_0,\ldots,i_n) = cp_{i_0} p_{i_0 i_1} \cdots p_{i_{n-1} i_n} + (1-c) q_{i_0} q_{i_0 i_1} \cdots q_{i_{n-1} i_n}$$

for all $n \geq 0$ and all $i_0,\ldots,i_n \in S$. Clearly the sequence $\{p_n\}_{n \geq 0}$ satisfy the consistency conditions. By Proposition 4.11, the sequence $\{p_n\}_{n \geq 0}$ is not generated by any pair $(p,P) \in \Pi \times M(s \times s)$. By the Kolmogorov existence theorem, the sequence $\{p_n\}_{n \geq 0}$ defines a unique measure $\tau \in P(\Omega,T) - M(\Omega,T)$ such that $\tau(x_0 = i_0,\ldots,x_n = i_n) = p_n(i_0,\ldots,i_n)$ for all $n \geq 0$ and all $i_0,\ldots,i_n \in S$.

## 5. THE ENTOPY MAP OF THE SHIFT

The object of this section is to show the entropy map of the shift is a surjection from $B(\Omega,T)$ onto $[0, \log s]$. We start with basic properties of the entropy map.

In the sequel, $\xi$ denotes the partition of $\Omega$ defined by $\xi = \{(x_0 = i): i \in S\}$. By the Kolmogorov-Sinai theorem, we get $h_\mu(T) = h_\mu(T,\xi)$ for all $\mu \in P(\Omega,T)$. The mapping $\mu \to h_\mu(T)$ defined on $P(\Omega,T)$ is called the entropy map of the shift $T$, or simply, the entropy map. It is well-known (see Billingsley [1], Denker et al [3], and Walters [7]) that

$$h_\mu(T) = - \Sigma_i p_i \log p_i \quad \text{or} \quad \Sigma_{i,j} p_i p_{ij} \log p_{ij}$$

according as $\mu$ is the $p$ Bernoulli measure or the $(p,P)$ Markov measure. All logarithms are natural ones. We state without proof the following lemma (see Denker et al [3, Proposition 10.13] or Walters [7, Theorem 8.1]).

<u>Lemma 5.1</u>. The entropy map $h_\mu(T)$ is an affine function from $P(\Omega,T)$ into $[0,\infty)$.

<u>Lemma 5.2</u>. The entropy map $h_\mu(T)$ is an upper semicontinuous function on $P(\Omega,T)$ and $0 \le h_\mu(T) \le \log s$ for all $\mu \in P(\Omega,T)$.

<u>Proof</u>. Let $n$ be any positive integer. Then we have, for each $\mu \in P(\Omega,T)$,

$$0 \le H_\mu(\bigvee_0^{n-1} T^{-j}\xi) = - \sum_{i_0} \cdots \sum_{i_{n-1}} \mu(x_j = i_j,\ 0 \le j \le n-1) \log (x_j = i_j,\ 0 \le j \le n-1) \le n \log s .$$

Suppose $\mu_k \to \mu$ in $P(\Omega,T)$. By Proposition 2.5, we obtain $H_{\mu_k}(\bigvee_0^{n-1} T^{-j}\xi) \to H_\mu(\bigvee_0^{n-1} T^{-j}\xi)$ as $k \to \infty$. Therefore, for each fixed $n \ge 1$, the mapping $\mu \to H_\mu(\bigvee_0^{n-1} T^{-j}\xi)$ is a continuous function of $\mu$.

Since $\{\frac{1}{n}H_\mu(\bigvee_0^{n-1} T^{-j}\xi)\}_{n \ge 1}$ is a nonincreasing sequence of continuous functions of $\mu$, $h_\mu(T) = \lim_n \frac{1}{n}H_\mu(\bigvee_0^{n-1} T^{-j}\xi)$ is an upper semicontinuous function of $\mu$. Note that $0 \le h_\mu(T) \le \log s$ for all $\mu$ in $P(\Omega,T)$.

We denote by $\lambda$ the Bernoulli measure induced by the uniform probability vector on $S$.

<u>Lemma 5.3</u>. Let $\mu \in P(\Omega,T)$. Then the following assertions are equivalent:

(i) $h_\mu(T) = \log s$.

(ii) $\mu = \lambda$.

<u>Proof</u>. (i) $\Rightarrow$ (ii): Suppose (i) holds, equivalently, $h_\mu(T,\xi) = \log s$. From the proof of Lemma 5.2, we get $H_\mu(\bigvee_0^{n-1} T^{-j}\xi) = \log s^n$ for all $n \ge 1$, so that, by Corollary 4.2.1 of Walters [7].

$$\mu(x_j = i_j,\ 0 \le j \le n-1) = s^n = \lambda(x_j = i_j,\ 0 \le j \le n-1)$$

for all $n \ge 1$. By the uniqueness theorem of finite measures, we obtain $\mu = \lambda$.

The implication (ii) $\Rightarrow$ (i) is evident, and the proof is complete.

<u>Lemma 5.4</u>. Let $\mu \in B(\Omega,T)$. Then the following assertions are equivalent:

(i) $h_\mu(T) = 0$.

(ii) $\mu$ is the $(\delta_{ij})_i$ Bernoulli measure for some $j \in S$.

<u>Proof</u>. Clearly (ii) $\Rightarrow$ (i). To prove (i) $\Rightarrow$ (ii), let $\mu$ be the $p$ Bernoulli measure. Suppose $h_\mu(T) = 0$, i.e., $- \Sigma_i p_i \log p_i = 0$. Then we have $p_i \log p_i = 0$ for all $i$, so that $p_i = 0$ or $1$ for all $i$. Since $\Sigma_i p_i = 1$, there is a unique $j \in S$ such that $p_i = \delta_{ij}$ for all $i$. Therefore $\mu$ is the $(\delta_{ij})_i$ Bernoulli measure.

The next lemma is a generalization of Lemma 5.4.

<u>Lemma 5.5</u>. Let $\mu \in M(\Omega,T)$. Then the following assertions are equivalent:

(i) $h_\mu(T) = 0$

(ii) There exist a permutation matrix $P$ and a stationary distribution $p$ of $P$ which induce the measure $\mu$.

<u>Proof</u>. (i) $\Rightarrow$ (ii): Suppose (i) holds. Let $\mu$ be the $(p,P)$ Markov measure where $P = (p_{ij}) \in M(s \times s)$, $p = (p_i) \in \Pi(P)$. Let $E = \{i \in S: p_i > 0\}$ and $F = S - E$. Note that $P' = (p_{ij})_{i,j \in E}$ is stochastic, and $p' = (p_i)_{i \in E}$ is a positive probability vector with $p'P' = p'$. An elementary calculation yields

$$h_\mu(T) = \Sigma_{i \in E}(- \Sigma_{j \in E} p_{ij} \log p_{ij})p_i = 0 ,$$

so that

$$- \Sigma_{j \in E} p_{ij} \log p_{ij} = 0$$

for each $i \in E$, equivalently,

$$p_{ij} = 0 \text{ or } 1 \text{ for all } i,j \in E .$$

Since $\Sigma_{j \in E} p_{ij} = 1$ for each $i \in E$, there is, for each $i \in E$, a unique $\varphi(i) \in E$ such that

$$p_{ij} = \delta_{\varphi(i)j} \quad \text{for all } j \in E .$$

Then the mapping $\varphi: E \to E$ must be surjective. If not, there is $j \in E$ such that $j \neq \varphi(i)$ for all $i \in E$, equivalently,

$$p_{ij} = 0 \quad \text{for all } i \in E .$$

We have $0 < p_j = \Sigma_{i \in E} p_i p_{ij} = 0$, a contradiction. Since $E$ is a finite set, the mapping $\varphi$ is bijective, so that $P'$ is a permutation matrix. Note that

$$p_{ij} = \delta_{\varphi(i)j} \quad \text{for each } i \in E \text{ and each } j \in S .$$

If $F = \phi$, then $P = P'$, so that we are done.

Suppose $F \neq \phi$. Define the permutation matrix $Q = (q_{ij})_{i,j \in S}$ by

$$q_{ij} = p_{ij} \quad \text{for each } (i,j) \in E \times S ;$$
$$q_{ij} = \delta_{ij} \quad \text{for each } (i,j) \in F \times S .$$

We obtain easily $pQ = p$. Let $\nu$ denote the $(p,Q)$ Markov measure. It remains to show that, for each $n \geq 0$,

$$\mu(x_0 = i_0,\ldots,x_n = i_n) = \nu(x_0 = i_0,\ldots,x_n = i_n),$$

equivalently,

$$p_{i_0} p_{i_0 i_0} \cdots p_{i_{n-1} i_n} = p_{i_0} q_{i_0 i_1} \cdots q_{i_{n-1} i_n}$$

for all $i_0,\ldots,i_n \in S$. We see at once the equations hold for $n = 0,1$. Suppose the equations hold for some $n \geq 1$. Let $i_0,\ldots,i_{n+1}$ be any states. If $\mu(x_j = i_j,\ 0 \leq j \leq n) = 0$, then $\mu(x_j = i_j,\ 0 \leq j \leq n+1) = 0 = \nu(x_j = i_j,\ 0 \leq j \leq n+1)$. If $\mu(x_j = i_j,\ 0 \leq j \leq n) > 0$, then $i_j \in E$ for $0 \leq j \leq n$, so that $p_{i_n i_{n+1}} = q_{i_n i_{n+1}}$ and $\mu(x_j = i_j,\ 0 \leq j \leq n+1) = \nu(x_j = i_j,\ 0 \leq j \leq n+1)$. By induction, we obtain $\mu = \nu$.

(ii) $\Rightarrow$ (i): Suppose that $\mu$ is the $(p,P)$ Markov measure where $P = (p_{ij})_{i,j \in S}$ is a permutation matrix with a stationary distribution

$p = (p_i)_{i \in S}$. Let $\varphi: S \to S$ be the bijection such that $p_{ij} = \delta_{\varphi(i)j}$ for all $i,j \in S$. Then we have

$$\Sigma_{j \in S}\, p_{ij} \log p_{ij} = 0 \quad \text{for each} \quad i \in S ,$$

so that $h_\mu(T) = \Sigma_{i \in E}(-\Sigma_{j \in E}\, p_{ij} \log p_{ij})p_i = 0$ where $E = \{i \in S: p_i > 0\}$. This completes the proof.

From the proof of Lemma 5.5, the following lemma is obvious.

Lemma 5.6. Let $\mu \in M_1(\Omega,T)$. Then the following assertions are equivalent:

(i) $h_\mu(T) = 0$ .

(ii) $\mu$ is induced by a unique irreducible permutation matrix in $M_1(s \times s)$.

The following example illustrate $h_\mu(T) = 0$ for some $\mu \in P(\Omega,T) - M(\Omega,T)$.

Example 5.7. Assume $s \geq 3$ . Let $P,Q \in M_1(s \times s)$ be distinct irreducible permutation matrices. Let $\mu$ and $\nu$ be the shift invariant Markov measures induced by $P$ and $Q$ , respectively. It follows from Lemmas 5.1 and 5.6, together with Proposition 4.11, that

$c\mu + (1-c)\nu \in P(\Omega,T) - M(\Omega,T)$ and $h_{c\mu + (1-c)\nu}(T) = 0$ for all $c \in (0,1)$.

Example 5.8. Assume $S = \{0,1\}$. Let $\mu$ be the Markov measure induced by the $2 \times 2$ unit matrix, together with the probability vector $(0,1)$, i.e., $\mu = \varepsilon_{[0]}$. Let $\nu$ be the $(q,Q)$-Markov measure where

$$Q = \begin{bmatrix} 0 & 0 \\ 1 & 0 \end{bmatrix} , \quad q = (\tfrac{1}{2}, \tfrac{1}{2}) .$$

It is straightforward to show that $c\mu + (1-c)\nu \in P(\Omega,T) - M(\Omega,T)$ for all $c \in (0,1)$. By Lemma 5.1, we also obtain $h_{c\mu + (1-c)\nu}(T) = 0$ for all $c \in (0,1)$.

We are now in a position to prove the main results of this section.

Theorem 5.9. The entropy map $h_\mu(T): P(\Omega,T) \to [0, \log s]$ is an affine, upper semicontinuous surjection.

Proof. By Lemmas 5.1 and 5.2, it remains to show that the entropy map is surjective. Let $\mu = \varepsilon_{[0]}$. Define the function $f(t)$ by

$$f(t) = h_{t\lambda + (1-t)\mu}(T), \quad 0 \leq t \leq 1 .$$

It follows from Lemmas 5.1, 5.3 and 5.4 that

$$f(t) = t(\log s), \quad 0 \leq t \leq 1 .$$

For any $u \in (0, \log s)$, let $c = u/\log s$ and $\nu = c\lambda + (1-c)\mu$. Using part (ii) of Lemma 3.2, we get $\nu \in P(\Omega,T) - M(\Omega,T)$ and $u = f(c) = h_\nu(T)$. This completes the proof.

We give an application of Theorem 4.9.

Theorem 5.10. The entropy map $h_\mu(T): M(\Omega,T) \to [0, \log s]$ is a continuous surjection.

Proof. Suppose $\mu_n \to \mu$ in $M(\Omega,T)$. Let $\mu$ be the $(p,P)$ Markov measure and let $\mu_n$ be the $(p_n,P_n)$ Markov measure. By Proposition 2.6, we have $\lim_n p_i(n) = p_i$ for each $i \in S$ and $\lim_n p_{ij}(n) = p_{ij}$ for each $i,j \in S$

provided $p_i > 0$ . If $p_i = 0$, then

$$0 \le - p_i(n)p_{ij}(n)\log p_{ij}(n) \le p_i(n)/e$$

for each $j$ , so that $\lim_n p_i(n)p_{ij}(n)\log p_{ij}(n) = 0$ for each $j$ . We have then

$$\lim_n h_{\mu_n}(T) = -\lim_n \sum_{i=0}^{s-1}\sum_{j=0}^{s-1} p_i(n)p_{ij}(n)\log p_{ij}(n) = -\sum_{i=0}^{s-1}\sum_{j=0}^{s-1} p_i p_{ij}\log p_{ij} = h_\mu(T).$$

Therefore, $h_\mu(T)$ is continuous on $M(\Omega,T)$. It follows from Theorem 4.9, together with Lemmas 5.3 and 5.5, that the entropy map is a surjection between $M(\Omega,T)$ and $[0, \log s]$.

We obtain from Theorem 4.10, together with Theorem 5.10 and Lemmas 5.3 and 5.6, the following theorem.

Theorem 5.11. The entropy map $h_\mu(T)\colon M_1(\Omega,T) \to [0, \log s]$ is a continuous surjection.

Using Theorem 3.4, together with Theorem 5.10 and Lemmas 5.3 and 5.4, we obtain the next theorem.

Theorem 5.12. The entropy map $h_\mu(T)\colon B(\Omega,T) \to [0, \log s]$ is a continuous surjection.

Theorem 3.5 has the following application.

Theorem 5.13. The entropy map $h_\mu(T)\colon B(\Omega,T)^+ \to (0, \log s]$ continuous surjection.

Proof. By Theorem 3.5, together with Theorem 5.12 and Lemma 5.3, it is enough to show that there is a sequence $\{\mu_n\}$ in $B(\Omega,T)^+$ such that $h_{\mu_n}(T) \to 0$ as $n \to \infty$ . Define the probability vectors $p_n = (p_i(n))$ in $\Pi$ by

$$p_1(n) = 1 - \frac{1}{n+1}\ , \ p_i(n) = \frac{1}{(s-1)(n+1)} \quad \text{for } i = 1,2,\ldots,s-1$$

where $n = 1,2,\ldots$ . Let $\mu_n$ be the $p_n$ Bernoulli measures. Then we have

$$h_{\mu_n}(T) = -\left(1 - \frac{1}{n+1}\right)\log\left(1 - \frac{1}{n+1}\right) - \frac{1}{n+1}\log\frac{1}{n+1} - \frac{1}{n+1}\log\frac{1}{s-1} \to 0$$

as $n \to \infty$ .

We may use anyone of Theorems 5.10, 5.11 and 5.12 to show that the entropy map $h_\mu(T)\colon P(\Omega,T) \to [0, \log s]$ is surjective. We conclude this section by observing that the entropy map is not continuous on $P(\Omega,T)$. We illustrate this assertion with a modification of an example in Walters [7, p. 184].

Example 5.14. Define $F(T^n) = \{\omega \in \Omega\colon T^n\omega = \omega\}$ and $\mu_n = \frac{1}{s^n}\sum_{\omega \in F(T^n)} \varepsilon_\omega$ where $n = 1,2,\ldots$ . Note that card $F(T^n) = s^n$ and $\mu_n \to \lambda$ as $n \to \infty$ . It is straightforward to show that $\mu_n \in P(\Omega,T) - M(\Omega,T)$ and $h_{\mu_n}(T) = 0$ for all all $n$ . Consequently, we have $h_{\mu_n}(T) \not\to h_\lambda(T) = \log s$ as $n \to \infty$ .

REFERENCES

[1] Billingsley, P., Ergodic Theory and Information (Wiley, 1965).

[2] Chung, K.L., Markov chains with Stationary Transition Probabilities (Springer, 2nd ed., 1967).
[3] Denker, M., Grillenberger, C. and Sigmund, K., Ergodic Theory on Compact Spaces (Springer, Lecture Notes in Math. 527, 1976).
[4] Feller, W., An Introduction to Probability Theory and Its Applications, Vol. 1 (Wiley, 3rd ed. 1968).
[5] Gantmacher, F.R., The Theory of Matrices, Vol. 2 (Chelsea, 1959).
[6] Seneta, E., Non-negative Matrices and Markov Chains (Springer, 2nd ed. 1981).
[7] Walters, P., An Introduction to Ergodic Theory (Springer, 1982).

Proceedings of the Analysis Conference, Singapore 1986
S.T.L. Choy, J.P. Jesudason, P.Y. Lee (Editors)



# TRANSLATION INVARIANT OPERATORS AND MULTIPLIERS OF BANACH-VALUED FUNCTION SPACES

Hang-Chin Lai and Tsu-Kung Chang

Institute of Mathematics, National Tsing Hua University, Hsinchu, Taiwan, Republic of China

Let $G$ be a locally compact abelian group, and $A$ be a commutative Banach algebra and $X$ a Banach $A$-module. In this paper, we investigate the invariant operators from a Banach-valued function space defined on $G$ into another Banach-valued function space, and characterize the space of all invariant operators as the following isometrically isomorphic relations under some appropriate conditions:

(i) $(L^1(G,Y), L^1(G,X)) \cong \Xi(Y, M(G,X))$;

(ii) $(L^1(G,Y), L^p(G,X)) \cong \Xi(Y, L^p(G,X))$, $1 < p < \infty$;

(iii) $\mathrm{Hom}_{L^1(G,A)}(L^1(G,A), L^1(G,X)) \cong \mathrm{Hom}_A(A, M(G,X))$;

(iv) $\mathrm{Hom}_{L^1(G,A)}(L^1(G,A), L^p(G,X)) \cong \mathrm{Hom}_A(A, L^p(G,X))$, $1 < p < \infty$

where $(E(G,Y), G(G,X))$ denotes the space of all invariant operators of $E$ to $F$, $\Xi(Y,Z)$ is the space of bounded linear operators from $Y$ to $Z$, and $\mathrm{Hom}_A$ means the $A$-module homomorphisms. Moreover (i) and (ii) with $Y = A$ coincide with (iii) and (iv) respectively if and only if $A = \mathbb{C}$, the complex field. This means that any invariant operator of a Banach function space is a multiplier if and only if $A \cong \mathbb{C}$

## 1. INTRODUCTION AND PRELIMINARIES

Let $G$ be a locally compact abelian group with Haar measure $dt$, $A$ be a commutative Banach algebra and $X$ a Banach space. Denote by $L^1(G,A)$ the space of all Bochner integrable $A$-valued functions defined on $G$ which is a commutative Banach algebra under convolution, and $L^p(G,X)$ the space of all $X$-valued measurable functions defined on $G$ whose $p$-power of $X$-norm are integrable over $G$ which is a Banach space for each $p$, $1 \leq p < \infty$.

Subject Classification (AMS 1980): 20B05, 43A22, 46G10.
Key Words and Phrases: Banach module, homomorphism, invariant operator, multiplier, Bochner integral, vector measure, Radon Nikodym property.

For two Banach spaces $X$ and $Y$, a bounded linear operator $T$ from a Banach function space $E(G,Y)$ to another Banach function space $F(G,X)$ is said to be an *invariant operator* if $T$ commutes with translation operator $\tau_a (a \in G)$. Through out this paper, the space of all invariant operators from $E(G,Y)$ to $F(G,X)$ is denoted by

$$(E(G,Y), F(G,X)).$$

Our purpose of this paper is to characterize the space of invariant operators under some appropriate conditions.

If $V$ and $W$ are A-module, it is known (see Rieffel [11] cf. also Lai [7], [8]) that

$$(1.1) \quad \mathrm{Hom}_A(V,W^*) \cong (V \otimes_A W)^*$$

in which a linear operator $T \in \mathrm{Hom}_A(A,W^*)$ corresponding to a continuous linear functional $\psi$ on $V \otimes_A W$ is given by

$$(Tv)(w) = \psi(v \otimes w) \quad \text{for} \quad v \in V,\ w \in W.$$

Here $\mathrm{Hom}_A(V,W^*)$ is the space of all A-module homomorphism from $V$ to $W^*$, that is, each $T \in \mathrm{Hom}_A(V,W^*)$ satisfies

$$T(av) = aT(v) \quad \text{for all} \quad a \in A,\ v \in V,$$

where $T$ is a continuous linear operator from $V$ to $W^*$; $V \otimes_A W$ denotes the A-module tensor product space of $V$ and $W$, that is $V \hat{\otimes}_\gamma W/K$, where $K$ is the closed linear subspace of the projective tensor product space $V \hat{\otimes}_\gamma W$ generated by the element of the form:

$$av \otimes w - v \otimes aw \quad \text{for any} \quad a \in A,\ v \in V,\ w \in W.$$

If $X$ is a Banach A-module, then $L^p(G,X)$, $1 \leq p < \infty$ is a Banach $L^1(G,A)$-module. In [7] and [8], Lai characterized various spaces of module homomorphisms as well as Banach-valued function spaces defined on a locally compact abelian group $G$ under certain appropriate conditions. The module homomorphism space is generally called the *multiplier space*. It is well known that, in the scalar-valued function spaces over $G$, a bounded linear operator is a multiplier if and only if it is an invariant operator. For example,

$$(1.2) \quad \mathrm{Hom}_{L^1(G)}(L^1(G), L^1(G)) \cong (L^1(G), L^1(G)) \cong M(G)$$

where $M(G)$ denotes the space of bounded regular measures on $G$, that is, if $T$ is a bounded linear operator on $L^1(G)$ then the following statements are equivalent:

(a) $T(f * g) = Tf * g = f * Tg$ for all $f,g \in L^1(G)$

(b) $\tau_s T = T\tau_s$ for $s \in G$ where $\tau_s f(t) = f(ts^{-1}) = f(t-s)$

(c) there is a unique measure $\mu \in M(G)$ such that

$$Tf = \mu * f \quad \text{for any} \quad f \in L^1(G).$$

Moreover it is also known that

(1.3) $$\mathrm{Hom}_{L^1(G)}(L^1(G), L^p(G)) \cong (L^1(G), L^p(G)) \cong L^p(G),$$

$1 < p < \infty$ and the relationship between both sides of $\cong$ is given by the following equivalent statement: Let $T$ be a bounded linear operator of $L^1(G)$ to $L^p(G)$. Then

$\tau_s T = T\tau_s$ if and only if there exists a function $g \in L^p(G)$ such that

$$Tf = f * g \quad \text{for all} \quad f \in L^1(G).$$

However, in the Banach-valued function spaces, an invariant operator need not be a multiplier. In [12], Tewari, Dutta and Vaidya proved the following theorem.

<u>THEOREM A</u> ([12]: Theorem 3). If the dimension $\dim A > 1$, then there is a bounded linear invariant operator $T$ of $L^1(G,A)$ such that

$$T \notin \mathrm{Hom}_A(L^1(G,A), L^1(G,A)).$$

Using this result, they disprove the Akinyele's results about the equivalence between the multiplier and invariant operator on $L^1(G,A)$. In [12], they proved that

(1.4) $$\mathrm{Hom}_A(L^1(G,A), L^1(G,A)) = M(G,A)$$

provided $A$ has an identity of norm 1. This result is extended by Lai [8] as in the following theorems.

<u>THEOREM B</u> ([8, Theorem 9]). Let $A$ be a commutative Banach algebra with an identity of norm 1 and $X$ a Banach A-module. Then the following statements are equivalent:

(a) $T \in \mathrm{Hom}_{L^1(G,A)}(L^1(G,A), L^1(G,X))$

(b) there exists a unique $\mu \in M(G,X)$ such that

$$Tf = f * \mu \quad \text{for all} \quad f \in L^1(G,A).$$

Moreover

(1.5) $$\mathrm{Hom}_{L^1(G,A)}(L^1(G,A), L^1(G,X)) \cong M(G,X)$$

Evidently if $X = A$, (1.5) is reduced to (1.4).

THEOREM C( 8, Theorem 6 ). Let $A$ be a commutative Banach algebra with identity of norm 1 and $X$ an A-module. If the topological dual and bidual spaces $X^*$, $X^{**}$ of $X$ have the Radon Nikodym property in the wide sense with respect ot $G$, then the following statements are equivalent:

(a) $T \in \mathrm{Hom}_{L^1(G,A)}(L^1(G,A), L^p(G,X))$, $1 < p < \infty$

(b) there exists a unique $g \in L^p(G,X)$ such that

$$Tf = f * g \quad \text{for all} \quad f \in L^1(G,A).$$

Moreover,

$$(1.6) \quad \mathrm{Hom}_{L^1(G,A)}(L^1(G,A), L^p(G,X)) \cong L^p(G,X), \ 1 < p < \infty.$$

Recently, Quek [10, Theorem 9] proved that if $X$ has the wide Radon Nikodym property then the isometric isomorphism of (1.6) in Theorem C holds. Quek's result imporves Theorem C since if $X$ is embedded as a closed subspace of $X^{**}$ in the norm topology, the Theorem 2 of Diestel and Uhl[1; p.81] implies that $X$ has the wide Radon Nikodym property whenever $X^{**}$ has.

As the remark in [12: p.229] indicates, it would be interesting to characterize the set of all bounded linear invariant operators on various Banach-valued function spaces over $G$. In this paper, we shall characterize the spaces $(L^1(G,Y), L^p(G,X))$, for $1 \leq p < \infty$ and establish the relationships in the case $Y = A$ between $\mathrm{Hom}_{L^1(G,A)}(L^1(G,A), L^p(G,X))$ and invariant operators. Moreover, since Theorem A shows that not every invariant operator is a multiplier, the question arises naturally under what conditions every invariant operator in $(L^1(G,A), L^p(G,X))$ is a multiplier. We will prove that a necessary and sufficient condition is that $A = \mathbb{C}$, the complex field.

## 2. INVARIANT OPERATORS

We will prove first that if $A = \mathbb{C}$ in (1.5) and (1.6) then every invariant operator of $L^1(G)$ to $L^p(G,X)$, $1 \leq p < \infty$ will be a multiplier.

THEOREM 1. Let $X$ be a Banach space. A bounded linear operator $T$: $L^1(G) \to L^p(G,X)$, $1 \leq p < \infty$ is an invariant operator if and only if it is a multiplier.

Proof. If $T$ is an invariant operator, then for any $x^* \in X^*$, the topological dual space, define a mapping

$$T_{x^*} : L^1(G) \longrightarrow L^p(G)$$

by

$$T_{x^*}f = x^* \circ Tf \quad \text{for all} \quad f \in L^1(G),$$

it is a bounded linear invariant operator.

Indeed, $T_{x^*}$ is clearly linear whenever $T$ is, and

$$\begin{aligned} \|T_{x^*}f\|_p = \|x^* \circ Tf\|_p &\leq \|x^*\| \, \|Tf\|_{pX} \\ &\leqq \|x^*\| \, \|T\| \, \|f\|_1 \end{aligned}$$

shows the continuity of $T_{x^*}$ where $\| \; \|_{pX}$ is the norm of $L^p(G,X)$ (see Lai [8]). Now let $\tau_s$, $s \in G$ be a translation operator, we then have

$$\begin{aligned} \tau_s(T_{x^*}f)(t) &= T_{x^*}F(ts^{-1}) \\ &= x^* \circ Tf(ts^{-1}) \\ &= x^* \circ \tau_s(Tf)(t) \\ &= x^* \circ T(\tau_s f)(t) \\ &= T_{x^*}(\tau_s f)(t) \qquad \text{for all} \quad f \in L^1(G) \quad \text{and} \quad t \in G, \end{aligned}$$

that is,

$$\tau_s T_{x^*} = T_{x^*}\tau_s.$$

This shows that $T_{x^*}$ is invariant whenever $T$ is.

It is known that (see (1.2) for $p = 1$ and (1.3) for $1 < p < \infty$) the invariant operators and multipliers are equivalent in the case of scalar-valued function spaces. It follows that for any $x^* \in X^*$,

$$\begin{aligned} x^* \circ T(f * g) &= T_{x^*}(f * g) \\ &= f * T_{x^*}g \\ &= f * (x^* \circ Tg) \\ &= x^* \circ (f * Tg) \quad \text{for all} \quad f, g \in L^1(G). \end{aligned}$$

Thus

$$\text{(a)} \quad x^* \circ T(f * g) = x^* \circ (f * Tg) \qquad \text{for all} \quad f, g \in L^1(G).$$

Now if we take $f, g \in C_c(G)$, the continuous functions with compact support, then the support of $f * Tg$ is contained in a compact subset $K = \text{supp } f \subset G$, and so from Corollary 7 of Diestel and Uhl [1, p.48], we see that the identity (a) implies

(b) $T(f * g) = f * Tg$

almost every where as an element in $L^p(G,X)$ for all $f,g \in C_c(G)$. Since $C_c(G)$ is dense in $L^1(G)$, (b) holds in $L^p(G,X)$ for all $f,g \in L^1(G)$. Hence

$$T \in \text{Hom}_{L^1(G)}(L^1(G), L^p(G,X)).$$

The "if part" of theorem is known. Therefore the proof is complete. Q.E.D.

Applying Theorem 1, we can establish the following theorem for invariant operators.

<u>THEOREM 2</u>. Let $X$ and $Y$ be two Banach space. Then the following two statements are equivalent

(i) $T \in (L^1(G,Y), L^1(G,X))$

(ii) there exists a unique $L \in \Xi(Y, M(G,X))$, the bounded linear operator of $Y$ to $M(G,X)$, such that

$$T(f \otimes y) = f * Ly \quad \text{for all} \quad f \in L^1(G),\ y \in Y.$$

Moreover

(2.1) $(L^1(G,Y), L^1(G,X) \cong \Xi(Y, M(G,X))$.

<u>Proof</u>. (i) $\Longrightarrow$ (ii).

Let $T \in (L^1(G,Y), L^1(G,X))$. Since $L^1(G,Y) = L^1(G) \hat{\otimes}_\gamma Y$,

$$\begin{aligned} \|f \otimes y\|_\gamma &= \|f\|_1 \|y\|_Y \\ &= \|fy\|_{1Y} \\ &= \int_G \|F(t)y\|_Y dt \\ &= (\int_G |f(t)|dt) \|y\|_Y, \end{aligned}$$

we can write $T(f \otimes y) = T(fy)$ for all $f \in L^1(G)$, $y \in Y$. For each $y \in Y$, we define $T_y: L^1(G) \to L^1(G,X)$ by

$$T_y f = T(fy) \quad \text{for all} \quad f \in L^1(G).$$

Evidently, $T_y$ is translation invariant whenever $T$ is. So that $T_y \in (L^1(G), L^1(G,X))$. Applying Theorem 1, we see that $T_y$ is a multiplier. That is,

$$T_y \in \text{Hom}_{L^1(G)}(L^1(G), L^1(G,X)).$$

It follows from Theorem B, by taking $A = \mathbb{C}$, that there exists a $\mu_y \in M(G,X)$ such that

$$T_y f = f * \mu_y \quad \text{for all} \quad f \in L^1(G)$$

and $\|T_y\| = \|\mu_y\|$. Note that $\|T_y\| \leq \|y\|_Y \|T\|$. Thus the mapping $Y \to M(G,X)$ defined by

$$L: Y \longrightarrow \mu_y$$

is bounded linear such that

$$T(fy) = f * L(y) \quad \text{with} \quad \|L\| \leq \|T\|.$$

(ii) $\Longrightarrow$ (i).

Conversely if $L \in \Xi(Y, M(G,X))$, we defined a mapping

$$T_L^1: L^1(G) \times Y \to L^1(G,X)$$

by

$$T_L^1(f,y) = f * L(y) \quad \text{for all} \quad f \in L^1(G),\ y \in Y.$$

Then $T_L^1$ is a bilinear continuous operator, and by the universal property of tensor product, there exists a linear map $T_L$,

$$T_L: \quad L^1(G) \hat{\otimes}_\gamma Y = L^1(G,Y) \longrightarrow L^1(G,X)$$

such that

$$T_L(f \otimes y) = f * L(y) \quad \text{for all} \quad f \in L^1(G),\ y \in Y$$

and satisfying

$$\|T_L\| \leqq \|L\|.$$

This $T_L$ is translation invariant since

$$\begin{aligned} \tau_s T_L(f \otimes y) &= \tau_s T(fy) \\ &= \tau_s(f * L(y)) \\ &= \tau_s f * L(y) \\ &= T_L(\tau_s fy) \\ &= T_L \tau_s(fy) \qquad \text{for all} \quad s \in G,\ y \in Y,\ f \in L^1(G). \end{aligned}$$

Hence $T_L \in (L^1(G,Y), L^1(G,X))$. By the first paragraph in the proof, we obtain $\|T_L\| = \|L\|$.

Finally, the one-one correspondence between $(L^1(G,Y), L^1(G,X))$ and $\Xi(Y, M(G,X))$ is obvious. Therefore we obtain

$$(L^1(G,Y), L^1(G,X)) \cong \Xi(Y, M(G,X))$$

and the proof is completed. Q.E.D.

According to Theorem C with $A = \mathbb{C}$ and Theorem 1, the invariant operators of $L^1(X,Y)$ to $L^p(G,X)$ for $1 < p < \infty$ can be characterized as good as the proof of Theorem 1. We state it as in the following theorem.

THEOREM 3. Let X and Y be Banach spaces. If X has the wide Radon Nikodym property with respect to G, then the following two statements are equivalent:

(i) $T \in (L^1(G,Y),\ L^p(G,X))$

(ii) there exists $L \in \Xi(Y,\ L^p(G,X))$, $1 < p < \infty$ such that

$$T(f \otimes y) = T(fy) = f * L(y) \quad \text{for all} \quad f \in L^1(G),\ y \in Y.$$

Moreover

$$(L^1(G,Y),\ L^p(G,X)) \cong \Xi(Y,\ L^p(G,X)).$$

REMARK 1. (i) Note that if $Y = \mathbb{C}$. Then Theroem 2 and 3 reduce to Theorem 1.

(ii) If $Y = \mathbb{C} = X$ then Theorems 2 and 3 are coincide with the usual multipliers, that is,

$$(L^1(G),\ L^1(G)) \cong \mathrm{Hom}_{L^1(G)}(L^1(G),\ L^1(G)) \cong M(G)$$

$$(L^1(G),\ L^p(G)) \cong \mathrm{Hom}_{L^1(G)}(L^1(G),\ L^p(G)) \cong L^p(G),$$

## 3. MULTIPLIERS OF VECTOR-VALUED FUNCTION SPACE

Let $Y = A$, in Theorems 2 and 3, be a commutative Banach algebra. Then we have the following characterizations.

THEOREM 4. Let A be a commutative Banach algebra (need not have identity) and X a Banach A-module. Then

$$(3.1) \quad \mathrm{Hom}_{L^1(G,A)}(L^1(G,A),\ L^1(G,X)) \cong \mathrm{Hom}_A(A,\ M(G,X)).$$

Proof. We have known that a multiplier operator $T \in \mathrm{Hom}_{L^1(G,A)}(L^1(G,A),\ L^1(G,X))$ is an invariant operator, thus $T \in (L^1(G,A),\ L^1(G,X)$. According to Theorem 2 with $Y = A$, there exists a unique $L \in \Xi(A,\ M(G,X))$ such that

(a) $T(fa) = f * L(a)$ for all $f \in L^1(G)$, $a \in A$.

Here $L(a) \in M(G,X)$ and $f * L(a)$ is an X-valued Bochner integrable function over G since $L^1(G)$ acts on $M(G,X)$ under convolution, and $f * L(a)$ vanishers on the singular part of $M(G,X)$. Hence it is an element of $L^1(G,X)$ and the relationship between T and L in (a) is well posed.

Moreover, for $f,\ g \in L^1(G)$, $a,\ b \in A$,

$$T(fa * gb) = T((f * g)ab) = (f * g) * L(ab)$$

and

$$T(fa * gb) = fa * T(gb) = (f * g) * aL(b)$$

we then have

$$L(ab) = aL(b) \quad \text{for all} \quad a, b \in A.$$

This shows that $L$ is an A-module homomorphism, that is

$$L \in \mathrm{Hom}_A(A, M(G,X)).$$

Conversely, for $L \in \mathrm{Hom}_A(A, M(G,X))$, we define

(b) $T_L(fa) = f * L(a)$ for all $f \in L^1(G)$, $a \in A$.

Then $fa \in L^1(G,A)$, $T_L$ is a bounded linear mapping from $L^1(G,A)$ to $L^1(G,X)$ since $L^1(G) * M(G,X)$ is laid in the space of $L^1(G,X)$. We want to show that $T$ is an $L^1(G,A)$-module homomorphism. In fact, for any $f \in L^1(G)$ and $a \in A$, $fa \in L^1(G,A) = L^1(G) \hat{\otimes}_\gamma A$, thus each $h \in L^1(G) \hat{\otimes}_\gamma A$ can be written by $h = gb$ for some $g \in L^1(G)$ and $b \in A$. Then

$$\begin{aligned} T_L(h * fa) &= T_L(gb * fa) \\ &= T_L((f * g)ba) \\ &= g * f * L(ab) \\ &= (g * f) * bL(a) \\ &= (bg) * (f * L(a)) \\ &= h * T_L(fa) \end{aligned}$$

and $T_L$ is an $L^1(G,A)$-module homomorphism.

It is easy to show $||T|| = ||L||$ for $T$ and $L$ in the relations of (a) and (b). Therefore the isometric isomorphism of (3.1) is proved. Q.E.D.

Under the same argument as in Theorem 4, we have the following theorem.

<u>THEOREM 5</u>. Let $A$ be a commutative Banach algebra and $X$ an A-module. Suppose that $X$ has the wide Radon Nikodym property with respect to $G$. Then

$$\text{(3.2)} \quad \mathrm{Hom}_{L^1(G,A)}(L^1(G,A), L^p(G,X)) \cong \mathrm{Hom}_A(A, L^p(G,X))$$

for $1 < p < \infty$.

<u>REMARK 2</u>. If $A$ has an identity of norm 1 in Theorems 4 and 5, then

(3.1) is isometrically isomorphic to $M(G,X)$, and

(3.2) is isometrically isomorphic to $L^p(G,X)$.

## 4. NECESSARY CONDITION FOR AN INVARIANT OPERATOR TO BE A MULTIPLIER

This section gives a main characterization for an invariant operator to be a multiplier in Banach function sapces. Although a multiplier is an invariant operator, the converse is not true, for example one can consult Theorem A. The following theorem gives a slight extension of Theorem A.

THEOREM 6. Let A be a commutative Banach algebra of dimension larger than 1 that has an identity of norm 1, X be an A-module. Then there exists a bounded linear invariant operator T of $L^1(G,A)$ to $L^p(G,X)$ such that

$$T \notin \mathrm{Hom}_{L^1(G,A)}(L^1(G,A),\ L^p(G,X)) \quad \text{for} \quad 1 \leq p < \infty.$$

Proof. We only prove the case when $p = 1$. If $1 < p < \infty$, the proof of theorem is the same, *mutatis mutandis*, as the proof in case $p = 1$.

Let $\chi$ be any nonzero character on A, that is, a multiplicative linear funcitonal on A. Define $\phi$: A $\to$ A by

$$\phi(a) = \chi(a)e \quad \text{for all} \quad a \in A,$$

where e is the unit of A. Then $\phi$ is a bounded linear operator on A. Since $\dim A > 1$, $\{\chi(a)e|a \in A\} \subsetneqq A$. Thus there is an element $b \in A$ such that

$$\phi(b) \neq b = b\phi(e).$$

Let $\{e_\alpha\}$ be an approximate identity of $L^1(G)$ having compact support in G. For each $\alpha$ and $L \in \mathrm{Hom}_A(A,X)$, we define a mapping $\mu_\alpha$: $A \to M(G,X)$ ($L^p(G,X)$ for $1 < p < \infty$) by

$$\mu_\alpha(a) = L(\phi(a)e_\alpha \quad \text{for any} \quad a \in A.$$

Then $\mu_\alpha$ is a bounded linear operator since L and $\phi$ are bounded linear. It is easy to see that the mapping

$$(f,a) \in L^1(G) \times A \longrightarrow \mu_\alpha(a) * f \in L^1(G,X)$$

is bounded bilinear, it follows from the universal property of tensor product that there exists a bounded linear map

$$T_\alpha:\ L^1(G) \hat{\otimes}_\gamma A \longrightarrow L^1(G,A)$$

given by

$$T_\alpha(f \otimes a) = \mu_\alpha(a) * f.$$

Since $L^1(G) \hat{\otimes}_\gamma A = L^1(G,A)$, $T_\alpha$ is a bounded linear transformation of $L^1(G,A)$ to $L^1(G,X)$ such that

$$\begin{aligned} T_\alpha(f \otimes a) &= T_\alpha(fa) \\ &= \mu_\alpha(a) * f \\ &= L(\phi(a))e_\alpha * f. \end{aligned}$$

Thus

$$\lim_\alpha T_\alpha(T \otimes a) = L(\phi(a))f.$$

We write

$$T(f \otimes a) = T(fa) = L(\phi(a))f.$$

This $T$ is a bounded linear operator from $L^1(G,A)$ to $L^1(G,X)$. If $\tau_t$ is a translation operator for $t \in G$, then it is obvious that

$$\begin{aligned} \tau_t T(fa) &= \lim \{\mu_\alpha(a) * \tau_t f\} \\ &= T(\tau_t fa), \qquad t \in G. \end{aligned}$$

Hence $T$ is invariant. But

$$\begin{aligned} T(fa * gb) &= T((f * g)ab) \\ &= \lim_\alpha T_\alpha((f * g)ab) \\ &= \lim_\alpha \{\mu_\alpha(ab) * (f * g)\} \\ &= L(\phi(ab))(f * g) \\ &= \phi(b)L(\phi(a))f * g \quad \text{since } L \in \mathrm{Hom}_A(A,X) \\ &= T(fa) * \phi(b)g \\ &\neq T(fa) * bg \qquad \text{since } \phi(b) \neq b. \end{aligned}$$

Therefore $T \notin \mathrm{Hom}_{L^1(G,A)}(L^1(G,A), L^1(G,X))$. Q.E.D.

REMARK 3. If $X = A$, $p = 1$, then Theorem 6 is reduced to Theorem A.

In view of Theorem 6, we ask under what conditions

$$(L^1(G,A), L^p(G,X)) \cong \mathrm{Hom}_{L^1(G,A)}(L^1(G,A), L^p(G,X)).$$

The answer is that $A$ must be isometrically isomorphic to the complex field $\mathbb{C}$.

THEOREM 7. Let $A$ be a commutative Banach algebra with identity of norm 1 and $X$ an A-module. Then each invariant operator

$T: L^1(G,A) \to L^p(G,X)$ for $1 \leq p < \infty$ is a multiplier if and only if $A \cong \mathbb{C}$.

Proof. The sufficient condition follows from Theorem 1. Thus we have only to show the condition $A \cong \mathbb{C}$ is necessary. Suppose that

$$(4.1) \quad (L^1(G,A),\ L^p(G,X)) \cong \mathrm{Hom}_{L^1(G,A)}(L^1(G,A),\ L^p(G,X)), \quad \text{for } 1 \leq p < \infty.$$

Applying Theorem 6, if $A \not\cong \mathbb{C}$ and so $\dim A > 1$, then

$$\mathrm{Hom}_{L^1(G,A)}(L^1(G,A),\ L^p(X,G)) \subsetneqq (L^1(G,A),\ L^p(G,X)).$$

This contradicts the assumption (4.1). Hence $A \cong \mathbb{C}$, and the theorem is proved.

REMARK 4. If $p = \infty$, we take $C_0(G,X)$ instead of $L^p(G,X)$ in the above discussions, then we could get the same conclusions above.

## REFERENCES

[ 1] Diestel, J. and Uhl,Jr., J.J., *Vector Measures*, Math. Surveys, Amer. Math. Soc. No.15, 1977.
[ 2] Dinculeanu, N., *Vector Measures*, Pergaman, Oxford, 1967.
[ 3] Dinculeanu, N., *Integration on Locally Compact Spaces*, Noordhoff International Publishing, 1974.
[ 4] Johnson, G.P., Spaces of functions with values in a Banach algebra, Trans. Amer. Math. Soc. 92(1959), 411-429.
[ 5] Khalil, R., Multipliers for some sapces of vector-valued functions, J. Univ. Kuwait (Sci), 8 (1981), 1-7.
[ 6] Lai, H.C., Multipliers of a Banach algebra in the second conjugate algebra as an idealizer, Tohoku Math. J., 26 (1974), 431-452.
[ 7] Lai, H.C., Multipliers for some spaces of Banach algebra valued functions, Rocky Mountain J. Math., 15(1985), 157-166.
[ 8] Lai, H.C., Multipliers of Banach-valued function spaces, J. Austral. Math. Soc., 39(Series A) (1985), 51-62.
[ 9] Lai, H.C., Duality of Banach function spaces and the Radon-Nikodym property, Acta Math. (Hung.), 47(1-2) (1986), 45-52.
[10] Quek, T.S., Multipliers of certain vector valued function spaces, Preprint.
[11] Rieffel, M.A., Multipliers and tensor products on $L^p$-spaces of locally compact group, Studia Math., 33(1969), 71-82.
[12] Tewari, U., Dutta, M. and Vaidya, D.P., Multipliers of group algebras of vector valued functions, Proc. Amer. Math. Soc. 81(1981), 223-229.

Proceedings of the Analysis Conference, Singapore 1986
S.T.L. Choy, J.P. Jesudason, P.Y. Lee (Editors)




# A PROOF OF THE GENERALIZED DOMINATED CONVERGENCE THEOREM FOR DENJOY INTEGRALS

**Lee Peng Yee**

We give an independent proof of the generalized dominated convergence theorem for the Denjoy integral.

Recently, Lee and Chew proved several convergence theorems for the Denjoy integral (see [6], [7] and [8]). In this note, we give an independent proof of the generalized dominated convergence theorem. As a consequence, other convergence theorems follow.

First, we define Denjoy integrable functions. A function F is said to be $AC_*(X)$ if for every $\varepsilon > 0$ there is $\eta > 0$ such that for every finite or infinite sequence of non-overlapping intervals $\{[a_i, b_i]\}$ with $a_i$, $b_i \in X$ satisfying $\sum_i |b_i - a_i| < \eta$ we have

$$\sum_i \omega(F;[a_i,b_i]) < \varepsilon$$

where $\omega$ denotes the oscillation of F over $[a_i,b_i]$. A function F is said to be $ACG_*$ if $[a,b]$ is the union of a sequence of closed sets $X_i$ such that on each $X_i$ the function F is $AC_*(X_i)$. A function f is said to be Denjoy integrable on $[a,b]$ if there exists a function F which is continuous on $[a,b]$ and $ACG_*$ such that its derivative $F'(x) = f(x)$ almost everywhere in $[a,b]$. For further results concerning the Denjoy integral, see [9] and [5]. It is well-known that the Denjoy integral is equivalent to the Perron integral, and also the Henstock-Kurzweil integral [4].

A function H is said to be a major function of f in $[a,b]$ if

$$f(x) \leq \underline{D}\, H(x) \neq -\infty \text{ for every } x$$

where $\underline{D}$ denotes the lower derivative. A function G is said to be a minor function of f in $[a,b]$ if $-G$ is a major function of $-f$ in $[a,b]$.

The following convergence theorem was proved in [8] as a consequence of the controlled convergence theorem. Here we give an independent proof.

**GENERALIZED DOMINATED CONVERGENCE THEOREM.** Let $f_n$ be Denjoy integrable on $[a,b]$ such that $f_n(x) \to f(x)$ almost everywhere in $[a,b]$ as $n \to \infty$, and the primitives $F_n$ of $f_n$ converge uniformly on $[a,b]$. If $f_n$ have at least one common major function and at least one common minor function in $[a,b]$, then f is Denjoy integrable on $[a,b]$ and we have

$$\int_a^b f_n(x)dx \to \int_a^b f(x)dx \quad \text{as } n \to \infty.$$

**Proof.** We say that a point x is regular if the consequence of the theorem holds for some open subinterval containing x. Then the set Q of all points x not regular is closed. Let H and G be respectively the common major and minor functions of f. Then both are $VBG_*$ (see [9, page 234]). Hence, in view of Baire's category theorem, the set of regular points is non-empty.

Let $(a_i,b_i)$, i = 1,2,..., be the subintervals in [a,b] which are contiguous to Q. Then f is Denjoy integrable on any interval $[u,v] \subset (a_i,b_i)$. Since $F_n$ converges uniformly on [a,b], f is Denjoy integrable on $[a_i,b_i]$ for each i and

$$\int_{a_i}^{b_i} f_n(x)dx \to \int_{a_i}^{b_i} f(x)dx \quad \text{as} \quad n \to \infty.$$

Again, by Baire's category theorem, there is a portion $Q_0$ of Q such that H and G are $VB_*$ on $Q_0$. Take $J_0$ to be the smallest interval that contains $Q_0$. Then it follows from Lebesgue's dominated convergence theorem that f is Lebesgue integrable on the closure of $Q_0$. Also, in view of the fact that

$$\omega(F_n;I) \leq \omega(H;I) + \omega(G;I)$$

for any interval I belonging to $J_0 - Q_0$, the series of oscillations of $F_n$ over the intervals in $J_0$ contiguous to $Q_0$ converges uniformly in n. By the fact that the Denjoy integral is closed under Harnack extension (see [9, page 249]), f is Denjoy integrable on $J_0$ which is a contradiction. Hence the proof is complete.

If the common major and minor functions are both continuous, then by Marcinkiewicz's theorem [9, page 253] the functions $\sup\{f_n; n\geq 1\}$ and $\inf\{f_n; n\geq 1\}$ are Denjoy integrable on [a,b]. Hence the theorem reduces to the following corollary.

**COROLLARY 1.** Let $f_n$ be Denjoy integrable on [a,b] such that $f_n(x) \to f(x)$ almost everywhere in [a,b] as $n \to \infty$, and g, h are also Denjoy integrable on [a,b]. If $g(x) \leq f_n(x) \leq h(x)$ almost everywhere in [a,b] for all n, then the consequence of the theorem holds.

For convenience, we write $F_n(u,v) = F_n(v) - F_n(u)$ in the following.

**COROLLARY 2.** Let $f_n$ be Denjoy integrable on [a,b] with primitive $F_n$ such that $f_n(x) \to f(x)$ almost everywhere in [a,b] as $n \to \infty$. If for every $\xi \in [a,b]$ and $\varepsilon > 0$ there exist an integer N and $\delta(\xi) > 0$ such that

$$|F_n(u,v) - F_m(u,v)| < \varepsilon|v-u|$$

whenever m, $n \geq N$ and $\xi - \delta(\xi) < u \leq \xi \leq v < \xi + \delta(\xi)$, then the consequence of the theorem holds.

We remark that the condition in Corollary 2 may hold only for $\xi \in [a,b] - D$ where D is countable and for each $\xi \in D$ the sequence $F_n$ converges uniformly in an open neighbourhood of $\xi$.

**COROLLARY 3.** Let $f_n$ be Denjoy integrable on [a,b] with primitive $F_n$ such that $f_n(x) \to f(x)$ almost everywhere in [a,b] as $n \to \infty$. If $F_n$ are uniformly differentiable in [a,b], i.e., for every $\xi \in [a,b]$ and $\varepsilon > 0$ there exists $\delta(\xi) > 0$ such that

$$|F_n(u,v) - f_n(\xi)(v-u)| < \varepsilon|v-u|$$

whenever $\xi - \delta(\xi) < u \leq \xi \leq v < \xi + \delta(\xi)$ and for all n, then the consequence of the theorem holds.

We remark that other convergence theorems by Djvarsheishvili [2], Lee and Chew [6,8], and Grimshaw [3] will also follow from the above theorem. Furthermore, the theorem can be extended to more general integrals, for example, the Burkill approximately continuous integral [1].

**References**

[1] P. S. Bullen, The Burkill approximately continuous integral, J. Austral. Math. Soc. (series A) 35(1983), 236-253.

[2] A. G. Djvarsheishvili, On a sequence of integrals in the sense of Denjoy, Akad. Nauk Gruzin. SSR Trudy Mat. Inst. Rajmadze 18(1951), 221-236. MR 14-628.

[3] M. E. Grimshaw, A convergence theorem for non-absolutely convergent integrals, J. London Math. Soc. 4(1929), 439-444.

[4] R. Henstock, Theory of integration, London 1963.

[5] R. L. Jeffery, The theory of functions of a real variable, Univ. of Toronto Press 1951.

[6] P. Y. Lee and T. S. Chew, A better convergence theorem for Henstock integrals, Bull. London Math. Soc. 17(1985), 557-564.

[7] P. Y. Lee and T. S. Chew, A Riesz-type definition of the Denjoy integral, Real Analysis Exchange 11(1985/86), 221-227.

[8] P. Y. Lee and T. S. Chew, On convergence theorems for the nonabsolute integrals, Bull. Australian Math. Soc. 34(1986), 133-140.

[9] S. Saks, Theory of the integral, Warsaw 1937.

National University of Singapore
Republic of Singapore

Proceedings of the Analysis Conference, Singapore 1986
S.T.L. Choy, J.P. Jesudason, P.Y. Lee (Editors)



# A Factorization Theorem for the Real Hardy Spaces

Akihiko MIYACHI

Department of Mathematics, Hitotsubashi University
Kunitachi, Tokyo, 186 Japan *)

The purpose of this article is to give a generalization of the factorization theorem for the real Hardy spaces and its application to the majorant property of the real Hardy spaces.

## 1. INTRODUCTION

First we shall recall the factorization theorem for the classical Hardy spaces.

Let $\underline{H}^p$, $0 < p < \infty$, denotes the classical Hardy space over the upper half plane, i.e., this is the set of those functions $F$ which are holomorphic in the upper half plane and for which

$$\sup_{y>0} \left( \int_{-\infty}^{\infty} |F(x+iy)|^p \, dx \right)^{1/p} < \infty.$$

The factorization theorem for these spaces reads as follows: Let $0 < p, q, r < \infty$ and $1/p = 1/q + 1/r$ ; then, a holomorphic function $F$ on the upper half plane belongs to $\underline{H}^p$ if and only if there exist $G \in \underline{H}^q$ and $H \in \underline{H}^r$ such that $F = GH$ (pointwise product of $G$ and $H$ ). As for this theorem, see e.g. Zygmund [20; Chapt. VII, §7], Duren [8; esp. Chapters 2 and 11] , or Koosis [12; esp. Chapters IV and VI].

We can restate this theorem in terms of the Hilbert transform and the boundary values of the holomorphic functions. We define $\mathrm{Re}\, H^p$, $0 < p < \infty$, as follows: $f$ belongs to $\mathrm{Re}\, H^p$ if $f$ is a tempered distribution on $\underline{R}$ (the real line) and if there exists an $F$ in $\underline{H}^p$ such that

$$f = \lim_{y \to 0} \mathrm{Re}\, F(\cdot + iy),$$

where the limit is taken in the sense of tempered distribution. The above function $F$ is uniquely determined by $f$. If $f$ and $F$ have the above relation, then we define $\tilde{f}$ by

$$\tilde{f} = \lim_{y \to 0} \mathrm{Im}\, F(\cdot + iy),$$

where the limit is taken, again, in the sense of tempered distribution. This $\tilde{f}$ is called the Hilbert transform of $f$. Now, taking the boundary values of

*) Partly supported by the Japan Society for the Promotion of Science and by the Grant-in-Aid for Scientific Research (C 61540088), the Ministry of Education, Japan.

the real parts of $F = GH$, we can restate the factorization theorem as follows: Let p, q, and r be the same as before; then, a tempered distribution f on $\underline{R}$ belongs to $\mathrm{Re}\, H^p$ if and only if there exist $g \in \mathrm{Re}\, H^q$ and $h \in \mathrm{Re}\, H^r$ such that

$$f = gh - \tilde{g}\tilde{h}. \tag{1}$$

(This is not a precise statement unless we determine the meaning of the right hand side of (1); the terms $gh$ and $\tilde{g}\tilde{h}$, each by itself, may not have the meanings as tempered distributions if $p < 1$.) If, instead of the formula $F = GH$, we use the formula $F = -\sqrt{-1}\, GH$, then we obtain another restatement of the factorization theorem, i.e., we see that the factorization theorem for the spaces $\mathrm{Re}\, H^p$ holds if we replace (1) by

$$f = g\tilde{h} + \tilde{g}h.$$

Coifman-Rochberg-Weiss [6], Uchiyama [18], [19], Chanillo [5], Komori [11], and the author [14], [15] obtained generalizations, in a weak form, of the above factorization theorem for $\mathrm{Re}\, H^p$ to the case of the $H^p$ spaces over $\underline{R}^n$ (see Remark (iv) in the next section). The purpose of the present article is to give a further generalization of these results and, as an application, to give a proof of the majorant property of the $H^p$ spaces.

## 2. PRELIMINARIES

Hereafter, we fix a Euclidean space $\underline{R}^n$; the letter n always denotes the dimension of this space.

We denote by $\underline{S}$ and $\underline{S}'$ the Schwartz class of rapidly decreasing smooth functions on $\underline{R}^n$ and the space of tempered distributions on $\underline{R}^n$ respectively. The Fourier transform and the inverse Fourier transform are denoted by $\hat{}$ and $F^{-1}$ respectively.

For $p > 0$, we denote by $H^p$ the $H^p$-space given by C. Fefferman and E. M. Stein [9; §11], i.e., this is the set of those f in $\underline{S}'$ for which the functions

$$f^*(x) = \sup_{t > 0} |(t^{-n}\phi(t^{-1}\cdot) * f)(x)|$$

belong to $L^p$, where $\phi$ is a fixed function in $\underline{S}$ such that $\int \phi(x)dx \neq 0$. For f in $H^p$, we set

$$\|f\|_{H^p} = \|f^*\|_{L^p}.$$

If $1 < p < \infty$, then $H^p = L^p$ with equivalent norms. If $n = 1$, the relation between $H^p$ and $\mathrm{Re}\, H^p$ is as follows: A tempered distribution f on $\underline{R}$ belongs to $H^p$ if and only if both its real part and imaginary part belong to $\mathrm{Re}\, H^p$.

For $\lambda$ with $0 \leqq \lambda < n$, we denote by $G(\lambda)$ the set of those smooth functions $f$ on $\underline{R}^n \setminus \{0\}$ such that

$$|\partial_x^\alpha f(x)| \leqq C_\alpha |x|^{-\lambda - |\alpha|}$$

for all multi-indices $\alpha$. We shall regard the elements of $G(\lambda)$ as locally integrable functions on $\underline{R}^n$ and the set $G(\lambda)$ as a subset of $\underline{S}'$.

We denote by $G'(n)$ the set of those smooth functions $f$ on $\underline{R}^n \setminus \{0\}$ such that

$$|\partial_x^\alpha f(x)| \leqq C_\alpha |x|^{-n - |\alpha|} \qquad \text{for all } \alpha$$

and

$$\sup_{0 < a < b < \infty} \left| \int_{a < |x| < b} f(x)\, dx \right| < \infty.$$

If $f \in G'(n)$, then there exists a sequence $\{a_j\}$ such that $a_j > 0$, $\lim_{j \to 0} a_j = 0$ and the limit

$$f' = \lim_{j \to \infty} \chi[\{ x \mid |x| > a_j \}] f$$

exists in $\underline{S}'$, where $\chi[E]$ denotes the characteristic function of the set $E$. We denote by $G(n)$ the set of all those tempered distributions of the form $f' + c\delta$, where $f'$ is the tempered distribution arising from $f \in G'(n)$ in the above way, $c$ is a complex number, and $\delta$ denotes Dirac's distribution.

Proposition 1. Let $0 \leqq \lambda \leqq n$. Then $f \in G(\lambda)$ if and only if $\hat{f} \in G(n-\lambda)$.

We can prove this proposition by elementary calculations (the integration by parts).

If $m \in G(\lambda)$ with $0 \leqq \lambda < n$, then an operator $T$ from $\underline{S}$ to $\underline{S}'$ is defined by

$$Tf = F^{-1}(m\hat{f}), \qquad f \in \underline{S}.$$

We call $T$ the operator associated with $m$, and $m$ the multiplier corresponding to $T$. If we set $k = F^{-1}m$, then the operator $T$ associated with $m$ is given by

$$Tf = k * f, \qquad f \in \underline{S},$$

where $*$ denotes the convolution.

For $\lambda$ with $0 \leqq \lambda < n$, we denote by $K(\lambda)$ the set of the operators associated with $m \in G(\lambda)$.

Let $m \in G(\lambda)$, $0 \leqq \lambda < n$, and let $T$ be the operator associated with $m$. We define $\check{m}$ by $\check{m}(\xi) = m(-\xi)$ and denote by $T'$ the operator associated with $\check{m}$. We call $T'$ the conjugate of $T$. The operator $T$ and its conjugate $T'$ satisfy

$$\langle Tf, g \rangle = \langle f, T'g \rangle \qquad \text{for all } f, g \in \underline{S}.$$

Proposition 2. Suppose $T \in K(\lambda)$, $0 \leqq \lambda < n$, $p > 0$, $q > 0$, and $1/p - 1/q = \lambda/n$. Then there exists a constant $C$ depending only on $T$, $p$, $q$, and $n$ for which the inequality

$$\|Tf\|_{H^q} \leqq C\|f\|_{H^p}$$

holds for all $f$ in $\underline{S} \cap H^p$.

As for a proof of this proposition, see e.g. [4; §4]. If $p \leqq 1$, then we can also easily prove it by using the atomic decomposition for $H^p$. (As for the atomic decomposition, see [13].)

Let $J$ be a finite set. Suppose $T_j \in K(\lambda_j)$, $0 \leqq \lambda_j < n$, where $j \in J$, and $\lambda = \sum_{j \in J} \lambda_j < n$. Then we define the product of the operators $T_j$, $j \in J$, by

$$\left(\prod_{j \in J} T_j\right) f = F^{-1}\left(\hat{f} \prod_{j \in J} m_j\right), \qquad f \in \underline{S},$$

where $m_j$ is the multiplier corresponding to $T_j$. This product is an operator in $K(\lambda)$. If $J$ is the empty set, then $\prod_{j \in J} T_j$ shall be understood as the identity operator.

## 3. MAIN RESULT

The following is the main result of this article.

Theorem 1. Let $N$ be a positive integer and let $\lambda$ and $\lambda_j$ where $j = 1, \cdots, N$, be nonnegative numbers satisfying $\lambda = \sum_{j=1}^{N} \lambda_j < n$. Let $T_j \in K(\lambda_j)$, $j = 1, \cdots, N$. For $g$ and $h$ in $\underline{S}$, we set

$$P(T_1, \cdots, T_N; g, h) = \sum_J (-1)^{|J|} \left\{ \left(\prod_{j \in J} T_j'\right) g \right\} \left\{ \left(\prod_{j \in J^c} T_j\right) h \right\}, \tag{2}$$

where the summation is taken over all subsets $J$ of $\{1, \cdots, N\}$, $|J|$ denotes the cardinality of $J$, and $J^c$ denotes the complement of $J$ with respect to $\{1, \cdots, N\}$. Suppose $0 < p, q, r < \infty$, $1/q > \lambda/n$, $1/r > \lambda/n$, and $1/p = 1/q + 1/r - \lambda/n < 1 + N/n$.

(i) Then, there exists a constant $C_1$ depending only on $T_1, \cdots, T_N$, $p$, $q$, $r$, and $n$ for which the inequality

$$\|P(T_1, \cdots, T_N; g, h)\|_{H^p} \leqq C_1 \|g\|_{H^q} \|h\|_{H^r}$$

holds for all $g \in \underline{S} \cap H^q$ and all $h \in \underline{S} \cap H^r$.

(ii) In addition to the above assumptions, assume further that $p \leqq 1$ and that the multipliers $m_j$ corresponding to $T_j$ have the following properties: (a) $m_j$ is a homogeneous function of degree $-\lambda_j$, i.e.,

$$m_j(t\xi) = t^{-\lambda_j} m_j(\xi), \qquad t > 0, \quad \xi \neq 0,$$

where $j = 1, \cdots, N$; (b) For every $\xi \in \underline{R}^n \setminus \{0\}$, there exists an $\eta \in \underline{R}^n \setminus \{0\}$ such that

$$\begin{cases} \prod_{j=1}^{N} ( m_j(\xi) - m_j(\eta) ) \neq 0 \\ \text{and } m_j(\eta) \neq 0 \quad \text{for } j \text{ with } \lambda_j > 0. \end{cases}$$

Then, every $f$ in $H^p$ can be decomposed as

$$f = \sum_{k=1}^{\infty} a_k \, P(T_1, \cdots, T_N; g_k, h_k),$$

where $a_k$ are complex numbers, $g_k \in \underline{S} \cap H^q$, $h_k \in \underline{S} \cap H^r$,

$$\| g_k \|_{H^q} \| h_k \|_{H^r} \leqq C_2,$$

and

$$\left( \sum_{k=1}^{\infty} |a_k|^p \right)^{1/p} \leqq C_3 \| f \|_{H^p} .$$

Here $C_2$ and $C_3$ are constants depending only on $T_1, \cdots, T_N$, $p$, $q$, $r$, and $n$.

Remark. (i) For $g$ and $h$ in $\underline{S}$, the right hand side of (2) is well defined since, by Proposition 2 and Hölder's inequality, each term in the summation in (2) belongs to $L^s$ for all sufficiently large $s$ and hence, a fortiori, to $\underline{S}'$.

(ii) In terms of the Fourier transform, the product in the theorem can be redefined as follows:

$$P(T_1, \cdots, T_N; g, h)\hat{}(\xi) = \int_{\underline{R}^n} \hat{g}(\eta)\hat{h}(\xi-\eta) \prod_{j=1}^{N} (m_j(\xi-\eta) - m_j(-\eta)) \, d\eta.$$

(iii) If $n = 1$ and $T_1 = \cdots = T_N =$ the Hilbert transform, then $P(T_1, \cdots, T_N; g, h)$ is equal to $g\tilde{h} + \tilde{g}h$ (if $N$ is an odd integer) or $gh - \tilde{g}\tilde{h}$ (if $N$ is an even integer) multiplied by a nonzero constant.

(iv) Theorem 1 for some special cases have been known. Coifman-Rochberg-Weiss [6] gave the theorem for the case $N = 1$, $\lambda = 0$, and $p = 1$ (they gave (ii) for $T_j =$ the Riesz transforms). Uchiyama [18], [19] treated the case $N = 1$, $\lambda = 0$, and $p > n/(n+1)$. Chanillo [5] treated the case $N = 1$, $0 < \lambda < n$, and $p = 1$. Komori [11] treated the case $N = 1$, $0 < \lambda < n$, and $p > n/(n+1)$. The author [14], [15] treated the case $N \geqq 1$ and $\lambda = 0$.

## 4. SKETCH OF THE PROOF

We can prove Theorem 1 by only slightly modifying the arguments in [14] and [15]. So, we shall give only a sketch of the proof.

First we shall sketch the proof of Theorem 1 (i). The basic idea of the proof of this part is due to Uchiyama [19].

Let $0 \leqq \mu < n$, $x \in \underline{R}^n$, and $k$ be a nonnegative integer. We define the set

$T_k^\mu(x)$ as follows: $g$ belongs to $T_k^\mu(x)$ if $g$ is a smooth function on $\underline{R}^n$ and if there exists a positive number $t$ such that $\operatorname{supp} g \subset \{ y \mid |y-x| \leqq t \}$ and $|\partial_x^\alpha g(x)| \leqq t^{\mu-n-|\alpha|}$ for $|\alpha| \leqq k$. For $f$ in $\underline{S}'$, we set

$$M_k^\mu(f)(x) = \sup\{ |\langle f, g\rangle| \mid g \in T_k^\mu(x) \}.$$

Let $0 \leqq \mu < n$ and $s > 0$; for measurable functions $f$ on $\underline{R}^n$, we set

$$f^*_{\mu,s}(x) = \sup_{r>0} \left( r^{\mu s-n} \int_{|x-y| \leqq r} |f(y)|^s \, dy \right)^{1/s}.$$

In order to prove Theorem 1 (i), we use the following lemmas.

Lemma 1. If $0 < p, q < \infty$, $0 \leqq \mu < n$, $1/p - 1/q = \mu/n$, and if $k$ is a nonnegative integer satisfying $k > n/p - n$, then

$$\| M_k^\mu(f) \|_{L^q} \leqq C \, \| f \|_{H^p}.$$

If $p \leqq 1$, then we can prove the above lemma by using the atomic decomposition. If $p > 1$, then the lemma is a corollary to the lemma below.

Lemma 2. If $0 \leqq \mu < n$, $0 < s < p < \infty$, $0 < q < \infty$, and $1/p - 1/q = \mu/n$, then

$$\| f^*_{\mu,s} \|_{L^q} \leqq C \| f \|_{L^p}.$$

This lemma is due to Chanillo [5; Lemma 2].

Now we shall prove Theorem 1 (i). For subsets $J$ of $\{ 1, \cdots, N \}$ we use the following notations:

$$T'^J = \prod_{j \in J} T'_j, \quad T^J = \prod_{j \in J} T_j, \quad \lambda(J) = \sum_{j \in J} \lambda_j.$$

By slightly modifying the arguments in [14], we can prove the following: There exist $C$, $k$, $k'$, $u$, and $v$ such that the inequality

$$\begin{aligned} & M_{k'}^0( P(T_1, \cdots, T_N; g, h) )(x) \\ & \leqq C \sum_J \left\{ M_k^0(T'^J g)(x) + M_k^{\lambda(J)}(g)(x) \right\} \left\{ M_k^0(T^{J^c} h)(x) + M_k^{\lambda(J^c)}(h)(x) \right\} \\ & \quad + C (M_k^0(g))^*_{\lambda,u}(x) \, (M_k^0(h))^*_{0,v}(x) \end{aligned}$$

holds for all $g, h \in \underline{S}$ and all $x \in \underline{R}^n$; here $k$ and $k'$ are sufficiently large positive integers, $0 < u < q$, $0 < v < r$, and $C$ is a constant depending only on $T_1, \cdots, T_N$, $p$, $q$, $r$, $k$, $k'$, $u$, and $v$. From this inequality, we can easily deduce Theorem 1 (i) with the aid of Proposition 2, Lemmas 1 and 2, and Hölder's inequality.

Next we shall sketch the proof of Theorem 1 (ii).

For a positive integer $M$ and for $p$ with $0 < p \leqq 1$, we denote by $A_{p,M}$ the set of those $L^2$-functions $f$ on $\underline{R}^n$ whose Fourier transforms $\hat{f}$ satisfy

$$\hat{f}(\xi) = 0 \quad \text{if} \quad |\xi| \leqq 1/t$$

and

$$\| \partial_\xi^\alpha \hat{f}(\xi) \|_{L_\xi^2} \leqq t^{|\alpha|-n/p+n/2} \quad \text{if} \quad |\alpha| \leqq M$$

for some $t > 0$.

Lemma 3. If $0 < p \leqq 1$ and $M > n/p - n/2$, then every $f$ in $H^p$ can be decomposed as follows:

$$f = \sum_{k=1}^{\infty} a_k f_k(\cdot - x_k),$$

where $a_k$ are complex numbers, $f_k \in A_{p,M}$, $x_k \in \underline{R}^n$, and

$$\left( \sum_{k=1}^{\infty} |a_k|^p \right)^{1/p} \leqq C\|f\|_{H^p}.$$

Here $C$ is a constant depending only on $M$, $n$, and $p$.

This is a modification of the atomic decomposition for $H^p$. As for a proof, see [15].

By slightly modifying the argument in [15], we can prove the following: For every $f$ in $A_{p,M}$ and every $\varepsilon > 0$, there exist $g$ and $h$ in $\underline{S}$ such that

$$\| f - P(T_1, \cdots, T_N; g, h)\|_{H^p} \leqq \varepsilon$$

and

$$\|g\|_{H^q} \|h\|_{H^r} \leqq C_\varepsilon,$$

where $C_\varepsilon$ is a constant depending only on $T_1, \cdots, T_N, p, q, r, M, n$, and $\varepsilon$. Combining this with Lemma 3, we can easily prove Theorem 1 (ii).

## 5. AN APPLICATION

We shall give a proof of the following theorem.

Theorem 2. Let $0 < p \leqq 1$. Then, for every $f$ in $H^p$, there exists a $g$ in $H^p$ such that $\hat{g}(\xi) \geqq |\hat{f}(\xi)|$ for all $\xi \in \underline{R}^n$ and

$$\|g\|_{H^p} \leqq C\|f\|_{H^p}.$$

Here $C$ is a constant depending only on $p$ and $n$.

This theorem has already been proved by several methods. Proof for the case $n = p = 1$ can be found in Zygmund's book [20; Chapt. VII, Proof of Theorem (8.7), p.287]. Coifman and Weiss [7; p.584] used the atomic decomposition to give a new proof (also for the case $n = p = 1$ ). Baernstein and Sawyer [2], [3; §8], Aleksandrov [1], and the author [16] extended the method of Coifman and Weiss to prove the theorem in the general case. The author [16] also gave two other different proofs, one of which is based on the duality between $H^1$ and BMO and is valid for the case $p = 1$, and the other is similar to the one to be given below.

Here we shall give a proof of Theorem 2 using our factorization theorem; this is an extension of one of the proofs given in [16]. We shall introduce a terminology: We say that $H^p$, where $0 < p \leq 2$, has the lower majorant property if for every $f$ in $H^p$, there exists a $g$ in $H^p$ such that $\hat{g}(\xi) \geq |\hat{f}(\xi)|$ for all $\xi \in \mathbb{R}^n$ and

(3) $$\|g\|_{H^p} \leq C \|f\|_{H^p}.$$

(Thus, Theorem 2 asserts that $H^p$ with $0 < p \leq 1$ has the lower majorant property*).)

Proof of Theorem 2. In order to prove the theorem, it is sufficient to prove the two facts: (a) $H^2$ has the lower majorant property; (b) If $0 < p < q \leq 2$, $p \leq 1$, $1/p \leq 1/q + 1/2$, and if $H^q$ has the lower majorant property, then $H^p$ also has the lower majorant property. The fact (a) is obvious by virtue of Plancherel's theorem. We shall prove (b). Let $p$ and $q$ be as mentioned there and suppose $H^q$ has the lower majorant property. Define $\lambda$ by $1/p = 1/q + 1/2 - \lambda/n$. Then $0 \leq \lambda < n/2$. Take a positive integer $N$ satisfying $1/p < 1 + N/n$ and take $\lambda_j$ and $m_j \in G(\lambda_j)$, $j = 1, \cdots, N$, such that $\lambda = \sum_{j=1}^{N} \lambda_j$, $m_j$'s satisfy the conditions in Theorem 1 (ii), and

(4) $$\prod_{j \in 1}^{N} ( m_j(\xi) - m_j(\eta) ) \geq 0$$

for all $\xi$ and $\eta$ in $\mathbb{R}^n \setminus \{0\}$. (This condition (4) is certainly satisfied if $N$ is an even integer, $m_1 = \cdots = m_N$, and $m_j$ is real valued.) Take an arbitrary $f$ in $H^p$ and decompose it as in Theorem 1 (ii) (take $r = 2$). Since $H^q$ and $H^2$ have the lower majorant property, we can find $g_k' \in H^q$ and $h_k' \in H^2$ such that

$$\hat{g}_k'(\xi) \geq |\hat{g}_k(\xi)|, \qquad \hat{h}_k'(\xi) \geq |\hat{h}_k(\xi)|,$$

$$\|g_k'\|_{H^q} \leq C \|g_k\|_{H^q}, \qquad \|h_k'\|_{H^2} \leq C \|h_k\|_{H^2}.$$

Define $g$ by

$$g = \sum_{k=1}^{\infty} |a_k| \, P(T_1, \cdots, T_N; g_k', h_k').$$

Using (4) and the formula in Remark (ii) (§3), we easily see that $\hat{g}(\xi) \geq |\hat{f}(\xi)|$. On the other hand, by virtue of Theorem 1 (i), the inequality (3) holds. This completes the proof. (To be precise, some limiting arguments are necessary

*) As for this property for $H^p = L^p$ with $p > 1$, see [10] and [17].

since $g'_k$ and $h'_k$ may not belong to $\underline{S}$; we omitted the limiting arguments.)

REFERENCES

[1] A. B. Aleksandrov, The majorant property for the multi-dimensional Hardy-Stein-Weiss classes (in Russian), Vestnik Leningrad Univ. 13 (1982), 97-98.
[2] A. Baernstein II and E. T. Sawyer, Fourier transforms of $H^p$ spaces, Abstracts Amer. Math. Soc., Vol. 1, No. 5 (1980), 779-42-8, p. 444.
[3] A. Baernstein II and E. T. Sawyer, Embedding and multiplier theorems for $H^p(\underline{R}^n)$, Mem. Amer. Math. Soc. 53 (1985), no. 318.
[4] A. P. Calderón and A. Torchinsky, Parabolic maximal functions associated with a distribution, II, Advances in Math. 24 (1977), 101-171.
[5] S. Chanillo, A note on commutators, Indiana Univ. Math. J. 31 (1982), 7-16.
[6] R. R. Coifman, R. Rochberg, and G. Weiss, Factorization theorems for Hardy spaces in several variables, Ann. of Math. 103 (1976), 611-635.
[7] R. R. Coifman and G. Weiss, Extensions of Hardy spaces and their use in analysis, Bull. Amer. Math. Soc. 83 (1977), 569-645.
[8] P. L. Duren, Theory of $H^p$ spaces, Academic Press, New York-San Francisco-London, 1970.
[9] C. Fefferman and E. M. Stein, $H^p$ spaces of several variables, Acta Math. 129 (1972), 137-193.
[10] E. T. Y. Lee and G. Sunouchi, On the majorant properties in $L^p(G)$, Tôhoku Math. J. 31 (1979), 41-48.
[11] Y. Komori, The factorization of $H^p$ and the commutators, Tokyo J. Math. 6 (1983), 435-445.
[12] P. Koosis, Introduction to $H_p$ Spaces, London Math. Soc. Lecture Note Series 40, Cambridge Univ. Press, Cambridge, 1980.
[13] R. H. Latter, A characterization of $H^p(\underline{R}^n)$ in terms of atoms, Studia Math. 62 (1978), 93-101.
[14] A. Miyachi, Products of distributions in $H^p$ spaces, Tôhoku Math. J. (2) 35 (1983), 483-498.
[15] A. Miyachi, Weak factorization of distributions in $H^p$ spaces, Pacific J. Math. 115 (1984), 165-175.
[16] A. Miyachi, Majorant properties in Hardy spaces, Research Reports Dept. Math. Hitotsubashi Univ., 1983.
[17] M. Rains, Majorant problems in harmonic analysis, Ph. D. dissertation, Univ. of British Columbia, Vancouver, 1976.
[18] A. Uchiyama, On the compactness of operators of Hankel type, Tôhoku Math. J. 30 (1978), 163-171.
[19] A. Uchiyama, The factorization of $H^p$ on the space of homogeneous type, Pacific J. Math. 92 (1981), 453-468.
[20] A. Zygmund, Trigonometric Series, 2nd ed., Vols I, II, Cambridge Univ. Press, Cambridge, 1959.

Proceedings of the Analysis Conference, Singapore 1986
S.T.L. Choy, J.P. Jesudason, P.Y. Lee (Editors)



# Estimates for Pseudo-differential Operators of class $S^m_{\rho,\delta}$ in $L^p$, $h^p$, and bmo

Akihiko MIYACHI

Department of Mathematics, Hitotsubashi University
Kunitachi, Tokyo, 186 Japan *)

The purpose of this article is to give some estimates for the operator norms of pseudo-differential operators as operators between $h^p$, $L^p$, and bmo by means of certain Lipschitz norms of their symbols and to give also some negative results concerning these estimates. The negative results will show that most of our norm estimates are sharp in a sense. The results are slight generalizations of those given at the author's lecture at the Analysis Conference, Singapore, 1986.

## 1. INTRODUCTION

The notations used in this article will be explained in the next section.

In this article, we shall consider the pseudo-differential operator of the following form:

$$(a(X,D)f)(x) = (2\pi)^{-n}\int_{\underline{R}^n} e^{ix\xi}a(x,\xi)\hat{f}(\xi)d\xi,$$

where ^ denotes the Fourier transform. The function $a(x,\xi)$ is called the symbol of the pseudo-differential operator $a(X,D)$. We shall consider only those symbols $a(x,\xi)$ which satisfy $|a(x,\xi)| \leqq C(1+|\xi|)^m$ for some constants $C$ and $m$. For these symbols, the operators $a(X,D)$ are well defined on $\underline{S}(\underline{R}^n)$. We shall say that $a(X,D)$ is bounded in $L^p$ if there exists a constant $C$ such that the inequality

$$\|a(X,D)f\|_{L^p} \leqq C\|f\|_{L^p}$$

holds for all $f$ in $\underline{S}(\underline{R}^n)$; we shall use the similar expression replacing $L^p$ by other function spaces.

The following theorem is known. (See the remark given at the end of this section.)

Theorem A. If $0 \leqq \delta \leqq \rho \leqq 1$, $\delta < 1$, $0 < p < \infty$, $m \leqq -n(1-\rho)|1/p - 1/2|$, and if

(1.1) $\quad |\partial^\beta_x \partial^\alpha_\xi a(x,\xi)| \leqq C_{\alpha\beta}(1+|\xi|)^{m+\delta|\beta|-\rho|\alpha|}$

for $|\beta| \leqq k$ and $|\alpha| \leqq k'$ with $k$ and $k'$ sufficiently large, then $a(X,D)$ is bounded in $h^p$ (if $p \leqq 1$) or $L^p$ (if $p > 1$).

*)Partly supported by the Japan Society for the Promotion of Science and by the Grant-in-Aid for Scientific Research (C61540088), the Ministry of Education, Japan.

It is also known that the above condition on $m$ cannot be relaxed. That is, if $\delta$, $\rho$, and $p$ are as mentioned in the above theorem and if $m > -n(1-\rho)|1/p - 1/2|$, then there is a symbol $a(x,\xi)$ which satisfies (1.1) for all $\alpha$ and $\beta$ but for which $a(X,D)$ is not bounded in $h^p$ (if $p \leqq 1$) or $L^p$ (if $p > 1$).

The purpose of this article is to find the smallest $k$ and $k'$ for which Theorem A holds. We shall do this in a generalized setting, that is, we shall consider $k$ and $k'$ not necessarily integers. More precisely, we shall do the followings. We shall introduce a class $S^m_{\rho,\delta}(\kappa,\kappa')$ for arbitrary positive numbers $\kappa$ and $\kappa'$. This class can be considered, figuratively, as the class of those symbols which satisfy (1.1) for $|\beta| \leqq \kappa$ and $|\alpha| \leqq \kappa'$. Then we shall find the numbers $\kappa_0$ and $\kappa'_0$ which are critical in the sense that if $\kappa > \kappa_0$ and $\kappa' > \kappa'_0$, then the pseudo-differential operators with symbols in the class $S^m_{\rho,\delta}(\kappa,\kappa')$ are bounded in $L^p$ but if $\kappa < \kappa_0$ or $\kappa' < \kappa'_0$, then there are symbols in the class $S^m_{\rho,\delta}(\kappa,\kappa')$ for which the associated pseudo-differential operators are not bounded in $L^p$; we shall also consider the spaces $h^p$ and bmo in place of $L^p$.

Remark. The condition such as (1.1) was introduced by Hörmander [9]. The negative results mentioned below Theorem A was also pointed out by him. The difficulty of the proof of Theorem A lies in the case $0 \leqq \delta = \rho < 1$. Theorem A for this case is due to Calderón-Vaillancourt [2], [3] (the case $p = 2$), C. Fefferman [6] (the case $1 < p < \infty$; cf. also Wang-Li [16; p.194]), and Päivärinta-Somersalo [13] (the case $0 < p \leqq 1$). In some cases, the critical orders $\kappa_0$ and $\kappa'_0$ mentioned above have already been obtained. They are found in Cordes [5; Theorem D] (the case $p = 2$ and $\delta = \rho = 0$), Miyachi [10], [11] (the case $0 < p < \infty$ and $0 \leqq \delta = \rho < 1$), Muramatu [12] (the case $p = 2$, $0 \leqq \delta \leqq \rho \leqq 1$, and $\delta < 1$), and Sugimoto [15] (the case $0 < p < \infty$ and $\delta = \rho = 0$).

## 2. NOTATIONS AND FUNCTION SPACES

The following notations are used throughout this article.

$\underline{R}$ denotes the set of real numbers.

We fix a Euclidean space $\underline{R}^n$; the letter $n$ always denotes the dimension of this space.

If $x = (x_1, \cdots, x_n)$ and $\xi = (\xi_1, \cdots, \xi_n)$ are elements of $\underline{R}^n$, then $x\xi = \sum_{j=1}^n x_j\xi_j$, $|x| = \sqrt{xx}$, and $\langle x \rangle = (1+|x|^2)^{1/2}$.

A multi-index $\alpha = (\alpha_1, \cdots, \alpha_n)$ is an $n$-tuple of nonnegative integers. If $\alpha = (\alpha_1, \cdots, \alpha_n)$ is a multi-index, then the length $|\alpha|$ is defined by $|\alpha| = \alpha_1 + \cdots + \alpha_n$ and the differential operator $\partial^\alpha$ is defined by

$$\partial_x^\alpha f(x) = (\partial/\partial x_1)^{\alpha_1} \cdots (\partial/\partial x_n)^{\alpha_n} f(x),$$

where $x = (x_1, \cdots, x_n) \in \underline{R}^n$.

If $u \in \underline{R}^n$ and $f$ is a function on $\underline{R}^n$, then

$$\Delta_x^2(u)f(x) = f(x+2u) - 2f(x+u) + f(x).$$

The Fourier transform is defined by

$$\hat{f}(\xi) = \int_{\underline{R}^n} f(x)e^{-ix\xi}dx.$$

We shall explain the function spaces considered in this article.

$\underline{S}(\underline{R}^n)$ and $\underline{S}(\underline{R}^n \times \underline{R}^n)$ are the spaces of rapidly decreasing smooth functions on $\underline{R}^n$ and $\underline{R}^n \times \underline{R}^n$ respectively.

$L^p$, $0 < p \leqq \infty$, denotes the set of those measurable functions $f$ on $\underline{R}^n$ for which the followings are finite:

$$\|f\|_{L^p} = \begin{cases} \left(\int_{\underline{R}^n} |f(x)|^p dx\right)^{1/p} & \text{if } p < \infty \\ \text{essential supremum of } |f(x)| & \text{if } p = \infty. \end{cases}$$

$H^p$, $0 < p \leqq 1$, is defined as follows. Fix a function $\phi$ in $\underline{S}(\underline{R}^n)$ such that $\int \phi(x)dx \neq 0$. For a tempered distribution $f$ on $\underline{R}^n$, we define the function $f^*$ on $\underline{R}^n$ by

$$f^*(x) = \sup_{t>0} |(t^{-n}\phi(t^{-1}\cdot)*f)(x)|.$$

Then $H^p$, $0 < p \leqq 1$, is the set of those tempered distributions $f$ on $\underline{R}^n$ for which $f^*$ belong to $L^p$. For $f$ in $H^p$, we define

$$\|f\|_{H^p} = \|f^*\|_{L^p}.$$

It is known that $H^p$ does not depend on the choice of $\phi$.

$h^p$, $0 < p \leqq 1$, is defined as follows. Let $\phi$ be the same as above. For a tempered distribution $f$ on $\underline{R}^n$, we define the function $f^{*,1}$ on $\underline{R}^n$ by

$$f^{*,1}(x) = \sup_{0<t\leqq 1} |(t^{-n}\phi(t^{-1}\cdot)*f)(x)|.$$

Then $h^p$, $0 < p \leqq 1$, is the set of those tempered distributions $f$ on $\underline{R}^n$ for which $f^{*,1}$ belong to $L^p$. For $f$ in $h^p$, we define

$$\|f\|_{h^p} = \|f^{*,1}\|_{L^p}.$$

It is known that $h^p$ does not depend on the choice of $\phi$.

BMO is the set of those locally integrable functions $f$ on $\underline{R}^n$ for which

$$\|f\|_{BMO} = \sup_Q \frac{1}{|Q|}\int_Q |f(x)-f_Q|dx < \infty,$$

where the supremum is taken over all cubes $Q$ in $R^n$, $|Q|$ denotes the Lebesgue measure of $Q$, and $f_Q = |Q|^{-1}\int_Q f(x)dx$.

bmo is the set of those locally integrable functions $f$ on $\underline{R}^n$ for which

$$\|f\|_{bmo} = \sup_{|Q|\leqq 1} \frac{1}{|Q|}\int_Q |f(x)-f_Q|\,dx + \sup_{|Q|>1} \frac{1}{|Q|}\int_Q |f(x)|\,dx < \infty,$$

where the notations are the same as above.

As for $H^p$ and BMO, see Fefferman-Stein [7]. As for $h^p$ and bmo, see Goldberg [8].

If $a$ is in $\underline{S}(\underline{R}^n \times \underline{R}^n)$, we define the operator $a(X,D)^*$ by

$$(a(X,D)^* g)(y) = \int g(x)K(x,x-y)\,dx,$$

where the function $K$ is defined by

$$K(x,z) = (2\pi)^{-n}\int e^{iz\xi}a(x,\xi)\,d\xi.$$

Note that if $a$ belongs to $\underline{S}(\underline{R}^n \times \underline{R}^n)$, then the operators $a(X,D)$ and $a(X,D)^*$ are well defined on $\underline{S}(\underline{R}^n)$ and the equality

$$\int g(x)(a(X,D)f)(x)\,dx = \int (a(X,D)^* g)(y)f(y)\,dy$$

holds for all $f$ and $g$ in $\underline{S}(\underline{R}^n)$.

## 3. CLASSES OF SYMBOLS

We shall introduce the following class.

Definition. Let $m \in \underline{R}$, $0 \leqq \delta, \rho \leqq 1$, $\kappa > 0$, and $\kappa' > 0$. Let $k$ and $k'$ be the nonnegative integers satisfying $k < \kappa \leqq k+1$ and $k' < \kappa' \leqq k'+1$. Then $S^m_{\rho,\delta}(\kappa,\kappa')$ denotes the set of those functions $a = a(x,\xi)$ on $\underline{R}^n_x \times \underline{R}^n_\xi$ which have the following estimates:

(i) if $|\beta| \leqq k$ and $|\alpha| \leqq k'$, then the derivative $\partial^\beta_x \partial^\alpha_\xi a(x,\xi)$ exists in the classical sense and

$$|\partial^\beta_x \partial^\alpha_\xi a(x,\xi)| \leqq A\langle\xi\rangle^{m+\delta|\beta|-\rho|\alpha|};$$

(ii) if $|\beta| = k$, $|\alpha| \leqq k'$, $u \in \underline{R}^n$, and $|u| \leqq \langle\xi\rangle^{-\delta}$, then

$$|\Delta^2_x(u)\partial^\beta_x \partial^\alpha_\xi a(x,\xi)| \leqq A\langle\xi\rangle^{m+\delta\kappa-\rho|\alpha|}|u|^{\kappa-k};$$

(iii) if $|\beta| \leqq k$, $|\alpha| = k'$, $\eta \in \underline{R}^n$, and $|\eta| \leqq \langle\xi\rangle^\rho/4$, then

$$|\Delta^2_\xi(\eta)\partial^\beta_x \partial^\alpha_\xi a(x,\xi)| \leqq A\langle\xi\rangle^{m+\delta|\beta|-\rho\kappa'}|\eta|^{\kappa'-k'};$$

(iv) if $|\beta| = k$, $|\alpha| = k'$, $u, \eta \in \underline{R}^n$, $|u| \leqq \langle\xi\rangle^{-\delta}$, and $|\eta| \leqq \langle\xi\rangle^\rho/4$, then

$$|\Delta^2_x(u)\Delta^2_\xi(\eta)\partial^\beta_x \partial^\alpha_\xi a(x,\xi)| \leqq A\langle\xi\rangle^{m+\delta\kappa-\rho\kappa'}|u|^{\kappa-k}|\eta|^{\kappa'-k'}.$$

Here $A$ is a constant which does not depend on $\alpha$, $\beta$, $x$, $\xi$, $u$, and $\eta$. The smallest such constant $A$ is denoted by $\|a\|_{m,\rho,\delta,\kappa,\kappa'}$.

It is easy to see that $S^m_{\rho,\delta}(\kappa,\kappa')$ with the norms $\|\ \|_{m,\rho,\delta,\kappa,\kappa'}$ are Banach spaces. These are generalizations of the Lipschitz spaces $B^\alpha_{\infty,\infty}$ over $\underline{R}^n$, which are the special ones of the Besov spaces $B^\alpha_{p,q}$, to the function spaces over $\underline{R}^n \times \underline{R}^n$. Many properties of the Lipschitz spaces $B^{\alpha}_{\infty,\infty}$ can be generalized to our spaces $S^m_{\rho,\delta}(\kappa,\kappa')$. Here we shall mention only one such property.

Proposition. Let $m(0)$, $m(1) \in \underline{R}$, $0 \leqq \rho$, $\delta \leqq 1$, $\kappa_0$, $\kappa_1$, $\kappa_0'$, $\kappa_1' > 0$, $0 < \theta < 1$, $m(\theta) = (1-\theta)m(0) + \theta m(1)$, $\kappa_\theta = (1-\theta)\kappa_0 + \theta\kappa_1$, and $\kappa_\theta' = (1-\theta)\kappa_0' + \theta\kappa_1'$. Then

$$[S^{m(0)}_{\rho,\delta}(\kappa_0,\kappa_0'), S^{m(1)}_{\rho,\delta}(\kappa_1,\kappa_1')]_\theta = S^{m(\theta)}_{\rho,\delta}(\kappa_\theta,\kappa_\theta'),$$

where the left hand side denotes the complex intermediate space.

As for a proof of this proposition, together with other properties of the spaces $S^m_{\rho,\delta}(\kappa,\kappa')$, see [10; §2], [11; §2], and the interpolation argument in [11; §4, Proofs of Theorems 3.1 and 3.2].

The main results of this article, which will be given in the next section, will treat only those classes $S^m_{\rho,\delta}(\kappa,\kappa')$ with $0 \leqq \delta \leqq \rho \leqq 1$ and $\delta < 1$.

4. MAIN RESULTS.

We shall use the following notation. Let $(Y,Z)$ be a couple of function spaces over $\underline{R}^n$. Suppose nonnegative functions $\| \ \|_Y$ and $\| \ \|_Z$ are defined on $Y$ and $Z$ respectively. We shall extend $\| \ \|_Y$ and $\| \ \|_Z$ in such a way that $\|f\|_Y = \infty$ if $f$ does not belong to $Y$ and similarly for $\| \ \|_Z$. Then we shall write as $\Psi^m_{\rho,\delta}(\kappa,\kappa') \subset \underline{L}(Y,Z)$ if there exists a constant $C$ depending only on $n$, $m$, $\rho$, $\delta$, $\kappa$, $\kappa'$, $Y$, and $Z$ for which the inequality

$$(4.1) \qquad \|a(X,D)f\|_Z \leqq C\,\|a\|_{m,\rho,\delta,\kappa,\kappa'}\,\|f\|_Y$$

holds for all $a \in \underline{S}(\underline{R}^n \times \underline{R}^n)$ and all $f \in \underline{S}(\underline{R}^n)$. Similarly, we shall write as $(\Psi^m_{\rho,\delta}(\kappa,\kappa'))^* \subset \underline{L}(Y,Z)$ if the inequality

$$(4.2) \qquad \|a(X,D)^* f\|_Z \leqq C\,\|a\|_{m,\rho,\delta,\kappa,\kappa'}\,\|f\|_Y$$

holds with $C$, $a$, and $f$ being the same as above. We shall use the symbol $\not\subset$ to indicate the negations of the above statements.

Remark. Although the class $\underline{S}(\underline{R}^n \times \underline{R}^n)$ is not dense in $S^m_{\rho,\delta}(\kappa,\kappa')$, the restriction that $a$ is in $\underline{S}(\underline{R}^n \times \underline{R}^n)$ mentioned above is not an essential one. In fact, if the inequality (4.1) holds for all $a \in \underline{S}(\underline{R}^n \times \underline{R}^n)$ and all $f \in \underline{S}(\underline{R}^n)$ and if $Z$ has a certain good property, then (4.1) holds, possibly with a larger constant $C$, for all $a \in S^m_{\rho,\delta}(\kappa,\kappa')$ and all $f \in \underline{S}(\underline{R}^n)$. A property of $Z$ which guarantees this is as follows: If $\{g_j\}$ is a sequence of uniformly bounded smooth functions on $\underline{R}^n$ with $\sup_j \|g_j\|_Z = A < \infty$ and if $g_j$ converges as $j \to \infty$ to a function $g$ uniformly on each compact subset of $\underline{R}^n$, then $\|g\|_Z \leqq C'A$. In fact, for every $a \in S^m_{\rho,\delta}(\kappa,\kappa')$, we can find a sequence $\{a_j\}$ of functions in $\underline{S}(\underline{R}^n \times \underline{R}^n)$ such that $a_j(x,\xi)$ converges as $j \to \infty$ to $a(x,\xi)$ uniformly on each compact subset of $\underline{R}^n_x \times \underline{R}^n_\xi$ and $\|a_j\|_{m,\rho,\delta,\kappa,\kappa'} \leqq$

$C'' \|a\|_{m,\rho,\delta,\kappa,\kappa'}$ with $C''$ depending only on $n$, $m$, $\rho$, $\delta$, $\kappa$, and $\kappa'$. If $\{a_j\}$ is such a sequence, $f$ is in $\underline{S}(\underline{R}^n)$, and $g_j = a_j(X,D)f$, then $\{g_j\}$ is a sequence of uniformly bounded smooth functions,

$$\|g_j\|_Z \leq C''C \|a\|_{m,\rho,\delta,\kappa,\kappa'} \|f\|_Y,$$

and $g_j$ converges to $g = a(X,D)f$ uniformly on each compact subset of $\underline{R}^n$. Hence, if $Z$ has the property mentioned above, then, by a limiting argument involving the above sequence $\{a_j\}$, we see that the inequality (4.1) with $C$ replaced by $C'C''C$ holds for all $a \in S^m_{\rho,\delta}(\kappa,\kappa')$ and all $f \in \underline{S}(\underline{R}^n)$. The spaces $L^p$ $(0 < p \leq \infty)$, $H^p$ $(0 < p \leq 1)$, $h^p$ $(0 < p \leq 1)$, BMO, and bmo have that property for $Z$. If one finds a nice representation for $a(X,D)^*f$ for general $a \in S^m_{\rho,\delta}(\kappa,\kappa')$, then the similar argument may show that inequality (4.2) can also be automatically extended to general $a \in S^m_{\rho,\delta}(\kappa,\kappa')$.

Now we shall give the main theorems of this article. In the following theorems, we assume

$$0 \leq \delta \leq \rho \leq 1 \quad \text{and} \quad \delta < 1.$$

Theorem 1. (1) If $0 < p \leq 1$, $-n(1-\rho)/p \leq m \leq -n(1-\rho)(1/p - 1/2)$, $(1-\delta)\kappa > n(1-\rho)/p + m + n\rho(1/p - 1)$, and $\kappa' > n/p$, then $\Psi^m_{\rho,\delta}(\kappa,\kappa') \subset \underline{L}(h^p,h^p)$.
(2) If $0 < p \leq 2$, $-n(1-\rho)/p \leq m \leq -n(1-\rho)(1/p - 1/2)$, $(1-\delta)\kappa > n(1-\rho)/p + m$, and $\kappa' > n/p$, then $\Psi^m_{\rho,\delta}(\kappa,\kappa') \subset \underline{L}(h^p,L^p)$ (if $0 < p < 1$) or $\subset \underline{L}(h^1,h^1)$ (if $p = 1$)*) or $\subset \underline{L}(L^p,L^p)$ (if $1 < p \leq 2$).
(3) If $2 < p < \infty$, $-n(1-\rho)/2 \leq m \leq -n(1-\rho)(1/2 - 1/p)$, $(1-\delta)\kappa > n(1-\rho)/2 + m$, and $\kappa' > n/2$, then $\Psi^m_{\rho,\delta}(\kappa,\kappa') \subset \underline{L}(L^p,L^p)$.
(4) If $m = -n(1-p)/2$, $\kappa > 0$, and $\kappa' > n/2$, then $\Psi^m_{\rho,\delta}(\kappa,\kappa') \subset \underline{L}(\text{bmo},\text{bmo})$.

Theorem 2. (1) If $0 < p \leq 1$, $m = -n(1-\rho)(1/p - 1/2)$, $\kappa > n(1/p - 1)$, and $\kappa' > n(1/p - 1/2)$, then $(\Psi^m_{\rho,\delta}(\kappa,\kappa'))^* \subset \underline{L}(h^p,h^p)$.
(2) If $1 < p \leq 2$, $-n(1-\rho)/2 \leq m \leq -n(1-\rho)(1/p - 1/2)$, $(1-\delta)\kappa > n(1-\rho)/2 + m$, and $\kappa' > n/2$, then $(\Psi^m_{\rho,\delta}(\kappa,\kappa'))^* \subset \underline{L}(L^p,L^p)$.
(3) If $2 < p < \infty$, $-n(1-\rho)(1 - 1/p) \leq m \leq -n(1-\rho)(1/2 - 1/p)$, $(1-\delta)\kappa > n(1-\rho)(1 - 1/p) + m$, and $\kappa' > n(1 - 1/p)$, then $(\Psi^m_{\rho,\delta}(\kappa,\kappa'))^* \subset \underline{L}(L^p,L^p)$.
(4) If $-n(1-\rho) \leq m \leq -n(1-\rho)/2$, $(1-\delta)\kappa > n(1-\rho) + m$, and $\kappa' > n$, then $(\Psi^m_{\rho,\delta}(\kappa,\kappa'))^* \subset \underline{L}(\text{bmo},\text{bmo})$.

Theorem 3. (1) If $0 < p \leq 2$, $-n(1-\rho)/p < m \leq -n(1-\rho)(1/p - 1/2)$, and $(1-\delta)\kappa < n(1-\rho)/p + m$, then $\Psi^m_{\rho,\delta}(\kappa,\kappa') \not\subset \underline{L}(H^p,L^p)$ (if $0 < p \leq 1$) or $\not\subset \underline{L}(L^p,L^p)$ (if $1 < p \leq 2$) for every $\kappa' > 0$.
(2) If $2 < p < \infty$, $-n(1-\rho)/2 < m \leq -n(1-\rho)(1/2 - 1/p)$, and $(1-\delta)\kappa < n(1-\rho)/2 + m$, then $\Psi^m_{\rho,\delta}(\kappa,\kappa') \not\subset \underline{L}(L^p,L^p)$ for every $\kappa' > 0$.
(3) There exists a symbol $a(x,\xi)$ which satisfies

*) This result for $p = 1$ is included in (1).

$$\sup_x |\partial_\xi^\alpha a(x,\xi)| \leqq C_\alpha \langle\xi\rangle^{-n(1-\rho)/2-\rho|\alpha|}$$

for all multi-indices $\alpha$ and for which $a(X,D)$ is not bounded from $L^\infty$ to BMO.

(4) If $0 < p \leqq 2$, then $\Psi^m_{\rho,\delta}(\kappa,n/p) \not\subset \underline{L}(H^p,L^p)$ (if $0 < p \leqq 1$) or $\not\subset \underline{L}(L^p,L^p)$ (if $1 < p \leqq 2$) for every $m \in \underline{R}$ and every $\kappa > 0$.

(5) If $\kappa' < n/2$, then for every $m \in \underline{R}$ and every $\kappa > 0$ it holds that $\Psi^m_{\rho,\delta}(\kappa,\kappa') \not\subset \underline{L}(L^p,L^p)$ for $2 < p < \infty$ and $\Psi^m_{\rho,\delta}(\kappa,\kappa') \not\subset \underline{L}(L^\infty,\text{BMO})$.

Theorem 4. (1) If $0 < p < 1$ and $\kappa < n(1/p-1)$, then $(\Psi^m_{\rho,\delta}(\kappa,\kappa'))^* \not\subset \underline{L}(H^p,L^p)$ for every $m \in \underline{R}$ and every $\kappa' > 0$.

(2) There exists a symbol $a(x,\xi)$ which satisfies

$$\sup_x |\partial_\xi^\alpha a(x,\xi)| \leqq C_\alpha \langle\xi\rangle^{-n(1-\rho)/2-\rho|\alpha|}$$

for all multi-indices $\alpha$ and for which $a(X,D)^*$ is not bounded from $H^1$ to $L^1$.

(3) If $1 < p \leqq 2$, $-n(1-\rho)/2 < m \leqq -n(1-\rho)(1/p-1/2)$, and $(1-\delta)\kappa < n(1-\rho)/2+m$, then $(\Psi^m_{\rho,\delta}(\kappa,\kappa'))^* \not\subset \underline{L}(L^p,L^p)$ for every $\kappa' > 0$.

(4) If $2 < p < \infty$, $-n(1-\rho)(1-1/p) < m \leqq -n(1-\rho)(1/2-1/p)$, and $(1-\delta)\kappa < n(1-\rho)(1-1/p)+m$, then $(\Psi^m_{\rho,\delta}(\kappa,\kappa'))^* \not\subset \underline{L}(L^p,L^p)$ for every $\kappa' > 0$.

(5) If $-n(1-\rho) < m \leqq -n(1-\rho)/2$ and $(1-\delta)\kappa < n(1-\rho)+m$, then $(\Psi^m_{\rho,\delta}(\kappa,\kappa'))^* \not\subset \underline{L}(L^\infty,\text{BMO})$ for every $\kappa' > 0$.

(6) If $0 < p \leqq 1$ and $\kappa' < n(1/p-1/2)$, then $(\Psi^m_{\rho,\delta}(\kappa,\kappa'))^* \not\subset \underline{L}(H^p,L^p)$ for every $m \in \underline{R}$ and every $\kappa > 0$.

(7) If $1 < p \leqq 2$ and $\kappa' < n/2$, then $(\Psi^m_{\rho,\delta}(\kappa,\kappa'))^* \not\subset \underline{L}(L^p,L^p)$ for every $m \in \underline{R}$ and every $\kappa > 0$.

(8) If $2 < p < \infty$, then $(\Psi^m_{\rho,\delta}(\kappa,n-n/p))^* \not\subset \underline{L}(L^p,L^p)$ for every $m \in \underline{R}$ and every $\kappa > 0$.

(9) We have $(\Psi^m_{\rho,\delta}(\kappa,n))^* \not\subset \underline{L}(L^\infty,\text{BMO})$ for every $m \in \underline{R}$ and every $\kappa > 0$.

Theorems 3 and 4 show that most of the results in Theorems 1 and 2 are sharp in a sense.

## 5. PROBLEMS AND FURTHER RESULTS

First, there is a problem: Can one relax the condition on $\kappa$ in Theorem 1 (1)? If $\rho = 0$ or if $p = 1$, then Theorem 3 (1) shows that it cannot be essentially relaxed; the problem arises in the case $\rho > 0$ and $p < 1$. The present author has checked that if $0 < p < 1$, $0 < \rho \leqq 1$, $m = -n(1-\rho)(1/p-1/2)$, and $\Psi^m_{\rho,0}(\kappa,\kappa') \subset \underline{L}(h^p,h^p)$ for some $\kappa' > 0$, then $\kappa \geqq n\rho(1/p-1)$, but does not know whether one can relax that condition in Theorem 1 (1).

Secondly, there is a problem concerning the condition on $\kappa$ in Theorem 1 (2) in the case $m = -n(1-\rho)/p$. In this case, the condition on $\kappa$ in that assertion reads as $\kappa > 0$; so there arises a problem whether one can discard entirely the condition on the continuity of the symbol with respect to x. More precisely, does the condition

$$\sup_x |\partial_\xi^\alpha a(x,\xi)| \leqq C_\alpha \langle\xi\rangle^{-n(1-\rho)/p-\rho|\alpha|},$$

where $0 \leqq \rho \leqq 1$ and $0 < p \leqq 2$, imply that $a(X,D)$ is bounded from $h^p$ to $L^p$ (if $0 < p \leqq 1$) or $L^p$ to $L^p$ (if $1 < p \leqq 2$)? The followings are partial answers to this question. If $\rho = 1$, the answer is NO; this can be seen from the counter example given by Coifman and Meyer [4; pp.39-40]. If $\rho = 0$ and $1 < p \leqq 2$, the answer is again NO; this can be seen from the following counter example:

$$a(x,\xi) = \langle\xi\rangle^{-n/p}\exp(-|x|^2-ix\xi).$$

On the other hand, the answer is YES if $0 \leqq \rho < 1$ and $0 < p \leqq 1$; this can be shown by arguments similar to those in [10; §4] or [11; §4].

Problems similar to the above one arises in connection with the condition $\kappa > 0$ in Theorem 1 (1) with $m = -n(1-\rho)/p$ and $\rho(1/p-1) = 0$, in Theorem 1 (3) with $m = -n(1-\rho)/2$, in Theorem 2 (2) with $m = -n(1-\rho)/2$, in Theorem 2 (3) with $m = -n(1-\rho)(1-1/p)$, and in Theorem 2 (4) with $m = -n(1-\rho)$.

Thirdly, there is a problem of sharpening Theorems 1 and 2 further. Theorems 1 ~ 4 give those numbers $\kappa_0 = \kappa_0(n,m,\rho,\delta,Y,Z)$ and $\kappa_0' = \kappa_0'(n,m,\rho,\delta,Y,Z)$, where $Y$ and $Z$ are $h^p$ or $L^p$ or bmo, which are critical in the sense that the inclusion $\Psi^m_{\rho,\delta}(\kappa,\kappa') \subset \underline{L}(Y,Z)$ holds if $\kappa > \kappa_0$ and $\kappa' > \kappa_0'$ but does not hold if $\kappa < \kappa_0$ or $\kappa' < \kappa_0'$. The problem is: Try to show $\Psi^m_{\rho,\delta}(\kappa,\kappa') \subset \underline{L}(Y,Z)$ with $\kappa = \kappa_0$ and $\kappa' = \kappa_0'$. In fact, this will not bring many results so far as we consider only the classes $S^m_{\rho,\delta}(\kappa,\kappa')$#). But, if we introduce some classes of symbols which are generalizations of the Besov spaces $B^\alpha_{p,q}$ over $\underline{R}^n$ to the function spaces over $\underline{R}^n \times \underline{R}^n$, then we can expect much. Muramatu [12] and Sugimoto [15] have already obtained some results in this direction.

## 6. SKETCHES OF THE PROOFS

Theorems 1 ~ 4 are generalizations of the theorems given in [10] and [11]. We can prove Theorems 1 ~ 4 by only slightly modifying the arguments in these papers. So we shall omit the details and give only the sketches of the Proofs.

First, we shall give a sketch of the proof of Theorem 1 (2) for the case $p \neq 1$ and $m = -n(1-\rho)(1/p-1/2)$, which is typical of our argument. The proof goes by three steps. The first step is to prove the result for $p = 2$, i.e., the result for the $L^2$-boundedness. The fact is that Theorem 1 (2) for $p = 2$ and $m = 0$ has been proved by Cordes [5; Theorem D] (the case $\rho = \delta = 0$; cf. also [10; §4]) and Muramatu [12; Theorems 4.5 and 4.6] (the general case). (In fact, Muramatu's theorems are sharper than Cordes's theorem and our theorem; Muramatu

#) We might be able to replace $\kappa > n/p - n$ in Theorem 2 (1) in the case $p < 1$ by $\kappa = n/p - n$.

treats the case $\kappa = n(1-\rho)/2(1-\delta)$ and $\kappa' = n/2$, whereas Cordes and we treat the case $\kappa > n(1-\rho)/2(1-\delta)$ and $\kappa' > n/2$.) The second step is to prove the following weak version of the assertion: if $0 < p < 1$, $m = -n(1-\rho)(1/p - 1/2)$, $(1-\delta)\kappa > n(1-\rho)/2$, and $\kappa' > n/p + 1$, then $\Psi^m_{\rho,\delta}(\kappa,\kappa') \subset \underline{L}(h^p,L^p)$. We can prove this as follows: We use Cordes's or Muramatu's $L^2$-boundedness theorem mentioned above to show the weighted $L^2$-estimate, and then we use Hölder's inequality to obtain the desired $L^p$-estimate. The atomic decomposition theorem for $h^p$ spaces (see Goldberg [8; Lemma 5]) is of much help in technical calculations in this step. As for details in this step, cf. [10; Proposition 4.1] or [11; Proposition 4.1]. The third step is to use the complex interpolation. Suppose, for simplicity, $1 < p < 2$. (The following argument holds true also in the case $0 < p < 1$ if we replace $\underline{L}(L^p,L^p)$ by $\underline{L}(h^p,L^p)$.) Take a number $q$ such that $0 < q < 1$, and let $\theta$ be the number such that $1/p = (1-\theta)/q + \theta/2$. The results in the first and the second steps read as

$$\Psi^0_{\rho,\delta}\left(\frac{1-\rho}{1-\delta}\cdot\frac{n}{2} + \varepsilon, \frac{n}{2} + \frac{\varepsilon}{2}\right) \subset \underline{L}(L^2,L^2)$$

and

$$\Psi^{m(q)}_{\rho,\delta}\left(\frac{1-\rho}{1-\delta}\cdot\frac{n}{2} + \varepsilon, \frac{n}{q} + \frac{\varepsilon}{2} + 1\right) \subset \underline{L}(h^q,L^q),$$

where $m(q) = -n(1-\rho)(1/q - 1/2)$ and $\varepsilon$ is an arbitrary positive number. Hence, by interpolation (see the proposition in Section 2 of this article and [14] and [1; §3]), we obtain

$$\Psi^{m(p)}_{\rho,\delta}\left(\frac{1-\rho}{1-\delta}\cdot\frac{n}{2} + \varepsilon, \frac{n}{p} + \frac{\varepsilon}{2} + 1 - \theta\right) \subset \underline{L}(L^p,L^p),$$

where $m(p) = -n(1-\rho)(1/p - 1/2)$. If we let $q$ tends to $0$, then $1 - \theta$ tends to $0$; hence we can take $1 - \theta < \varepsilon/2$ by taking sufficiently small $q$. This proves the desired result since $\varepsilon$ is arbitrary.

Next we shall prove Theorem 1 (2) for the case $p \neq 1$ and $-n(1-\rho)/p \leqq m < -n(1-\rho)(1/p - 1/2)$. Suppose $p$ and $m$ satisfy these inequalities, $(1-\delta)\kappa > n(1-\rho)/p + m$, and $\kappa' > n/p$. We may and shall assume, without loss of generality, $(1-\rho)n/2 > (1-\delta)\kappa > (1-\rho)\kappa' + m$. We can find $\tilde{m}$, $\tilde{\rho}$, and $\tilde{\delta}$ such that $0 \leqq \tilde{\delta} \leqq \tilde{\rho} < 1$, $\tilde{m} = n(1-\tilde{\rho})(1/p - 1/2)$, $(1-\tilde{\delta})\kappa > n(1-\tilde{\rho})/2$, and $S^m_{\rho,\delta}(\kappa,\kappa') \subset S^{\tilde{m}}_{\tilde{\rho},\tilde{\delta}}(\kappa,\kappa')$ with continuous embedding. (The last embedding holds if the inequalities $m + \delta t - \rho s \leqq \tilde{m} + \tilde{\delta} t - \tilde{\rho} s$ hold for $(t,s) = (0,0)$, $(0,\kappa')$, $(\kappa,0)$, and $(\kappa,\kappa')$.) Thus the assertion in Theorem 1 (2) for the case under consideration can be derived from that for the case $p \neq 1$ and $m = -n(1-\rho)(1/p - 1/2)$.

We can prove Theorem 1 (1) and Theorem 2 (1), (2) in the same way as above; the only additional tool we need is the singular integral characterization of $h^p$ (cf. [11; the paragraph just below Proposition 4.2]), by virtue of which we can reduce the $h^p \to h^p$ estimate to the $h^p \to L^p$ estimate. For details, cf. [11; §4].

Theorem 1 (3), (4) and Theorem 2 (3), (4) are derived from Theorem 1 (1), (2) and Theorem 2 (1), (2) by the use of the duality between $L^p$ and $L^q$, where $1/p + 1/q = 1$, or the duality between $h^1$ and bmo. As for the latter duality, see Goldberg [8].

These are the sketches of the proofs of Theorems 1 and 2.

Next we shall proceed to the proofs of Theorems 3 and 4. Note that Theorem 3 (2), (3), (5) and Theorem 4 (4), (5), (8), (9) are derived from the rest of the theorems by the use of duality. Most of the results in these theorems can be proved in almost the same way as in [10; §5] and [11; §5]. In the following, we shall give a proof of Theorem 3 (1); this is the only proof which requires an extra idea which is not in those papers.

Let $p$ and $m$ be as mentioned in Theorem 3 (1). Take a smooth function $\phi$ on $\underline{R}^n$ such that $\operatorname{supp}\phi \subset \{x \mid 1 \leqq |x| \leqq 2\}$ and $\phi \neq 0$. Let $t > 1$ and define $a_t$ and $f_t$ by

$$a_t(x,\xi) = \phi(t^\rho x)e^{-ix\xi}\phi(t^{-1}\xi)$$

and

$$\hat{f}_t(\xi) = \overline{\phi(t^{-1}\xi)}.$$

It is easy to see that

$$\|a_t(X,D)f_t\|_{L^p} / \|f_t\|_{H^p} = Ct^{n(1-\rho)/p}$$

($H^p$ shall be replaced by $L^p$ if $p > 1$), where $C$ is a positive constant depending only on $n$, $\rho$, $p$, and $\phi$. On the other hand, it holds that

$$\|a_t\|_{m,\rho,\delta,\kappa,\kappa'} \leqq C't^{-m+(1-\delta)\kappa},$$

where $C'$ is a constant depending only on $n$, $m$, $\rho$, $\delta$, $\kappa$, $\kappa'$, and $\phi$. (If $\kappa$ and $\kappa'$ are integers, then we can prove the above estimate by elementary calculations; in the general case, we can prove it by using the proposition in Section 3.) Now suppose $\kappa, \kappa' > 0$ and $\Psi^m_{\rho,\delta}(\kappa,\kappa') \subset \underline{L}(H^p,L^p)$ (if $p \leqq 1$) or $\subset \underline{L}(L^p,L^p)$ (if $p > 1$). Then, from the above two inequalities, it follows that $t^{n(1-\rho)/p} \leqq C''t^{-m+(1-\delta)\kappa}$ for all $t > 1$ and hence $n(1-\rho)/p \leqq -m + (1-\delta)\kappa$. This proves Theorem 3 (1).

REFERENCES

[1] A. P. Calderón and A. Torchinsky, Parabolic maximal functions associated with a distribution, II, Advances in Math. 24 (1977), 101-171.

[2] A. P. Calderón and R. Vaillancourt, On the boundedness of pseudo-differential operators, J. Math. Soc. Japan 23 (1971), 374-378.

[3] A. P. Calderón and R. Vaillancourt, A class of bounded pseudo-differential operators, Proc. Nat. Acad. Sci. USA 69 (1972), 1185-1187.

[4] R. R. Coifman and Y. Meyer, Au-delà des opérateurs pseudo-différentiels, 2nd ed., Astérisque 57, Soc. Math. France, Paris, 1978.

[5] H. O. Cordes, On compactness of commutators of multiplications and convolutions, and boundedness of pseudodifferential operators, J. Funct. Anal. 18 (1975), 115-131.

[6] C. Fefferman, $L^p$ bounds for pseudo-differential operators, Israel J. Math. 14 (1973), 413-417.

[7] C. Fefferman and E. M Stein, $H^p$ spaces of several variables, Acta Math. 129(1972), 137-193.

[8] D. Goldberg, A local version of real Hardy spaces, Duke Math. J. 46 (1979), 27-42.

[9] L. Hörmander, Pseudo-differential operators and hypoelliptic equations, Proc. Symp. Pure Math. X, pp.138-183, Amer. Math. Soc., Providence, 1967.

[10] A. Miyachi, Estimates for pseudo-differential operators of class $S_{0,0}$, to appear in Math. Nachr..

[11] A. Miyachi, Estimates for pseudo-differential operators with exotic symbols, preprint.

[12] T. Muramatu, Estimates for the norm of pseudo-differential operators by means of Besov spaces I, $L_2$-theory, preprint.

[13] L. Päivärinta and E. Somersalo, A generalization of Calderón-Vaillancourt theorem to $L^p$ and $h^p$, preprint.

[14] E. M.Stein and G. Weiss, On the interpolation of analytic families of operators acting on $H^p$-spaces, Tôhoku Math. J. (2) 9 (1957), 318-339.

[15] M. Sugimoto, $L^p$-boundedness of pseudo-differential operators satisfying Besov estimates I, II, preprints.

[16] R.-H. Wang and C.-Z. Li, On the $L^p$-boundedness of several classes of pseudo-differential operators, Chin. Ann. of Math. 5B (2) (1984), 193-213.

Proceedings of the Analysis Conference, Singapore 1986
S.T.L. Choy, J.P. Jesudason, P.Y. Lee (Editors)



# A NOTE ON A LIFTING PROPERTY FOR CONVEX PROCESSES

K. F. Ng
Chinese University, HONG KONG.

L.S. Liu
Zhongshen University, CHINA.

Let E and F be locally convex spaces, and T be an open multi-valued map from E to F such that its graph is a closed cone in $E \times F$ and kernel ker T complete metrizable, then we show that each compact subset of F is contained in the image by T of a compact subset of E.

Following Rockafellar [3], by a convex process is meant a map T of points in a locally convex space E into the subsets of another locally convex space F such that $o \in To$, $T(\lambda x) = \lambda Tx$ and $Tx_1 + Tx_2 \subseteq T(x_1+x_2)$ for all $\lambda > o$, $x_1$, $x_2$ and x in E. This is the case if and only if the graph G(T) of T is a convex cone in $E \times F$. T is called a closed convex process if G(T) is a closed convex cone. Recall also that T is said to be opened if it maps open sets to open sets. The following result was proved by Fakhoury in the special case when T was single-valued (see [2] or [1, Proposition VI. 3.5].

THEOREM 1. Let $T: E \to 2^F$ be a closed convex process. Suppose T is open and that its kernel $T^{-1}(o)$ is complete metrizable with respect to the relative uniformity. Then every compact subset of F is contained in the image by T of a compact subset of E.

Remark. If E, F are assumed to be complete metrizable then, by a generalized open mapping theorem (see [5], [6]), the condition "T is open" can be replaced by "T is onto F".

To begin our proof, we write $\phi$ for $T^{-1}$. Then $\phi$ is also a closed convex process (from F to E), and is lower semi-continuous (l.s.c.) in the sense that $\{y \in F : \phi(y) \cap \omega \neq \phi\}$ is open for each open set $\omega$ in E, because T is open. Note that, if $y \in F$, $x_1 \in \phi(y)$ and $x_2 \in \phi(-y)$ then

$$x_1 + \phi(o) \subseteq \phi(y) \subseteq \phi(o) - x_2 \qquad (1)$$

Take countably many circled convex neighbourhoods $\{V_i\}_{i=1}^{\infty}$ of o in E with $V_{i+1} + V_{i+1} \subseteq V_i$ such that $\{V_i^* \cap \phi(o)^2\}$ is a filter base for the relative uniformity in $\phi(o)$, where

$$V_i^* = \{(x_1,x_2) \varepsilon E^2 : x_1-x_2 \varepsilon V_i\}.$$

By (1), the relative uniformity for $\phi(y)$ is also determined by $\{V_i^*\}$ in the above manner for each y. We now adopt an idea of Fakhoury as presented in [1, Proposition VI. 3.5] to apply Michael's selection Theorem : For any given compact subset K of F there exists $V_1$-selection $\psi_1$ of $\phi$ on K, that is $\psi_1$ is a continuous (single-valued) function from K into E such that

$$\psi_1(y) \varepsilon \phi(y) + V_1, \qquad y \varepsilon K.$$

Define a multivalued function $\phi_1$ by

$$\phi_1(y) = \phi(y) \cap [\psi_1(y) + V_1], \quad y \varepsilon K.$$

Then it is easily seen that $\phi_1$ is l.s.c. and $\phi_1(y)$ is convex for all y. Hence one has $V_2$-selection $\psi_2$ of $\phi_1$. Inductively we have $\{\phi_n\}$ and $\{\psi_n\}$ with

$$\begin{aligned} \phi_n(y) &= \phi(y) \cap [\psi_n(y) + V_n] \\ \psi_{n+1}(y) &\varepsilon \phi_n(y) + V_{n+1} \end{aligned} \qquad (2)$$

for all y $\varepsilon$ K and all n. Let $\psi(y)$ denote the limit of the Cauchy sequence $\{\psi_n(y)\}$ on the complete set $\phi(y)$. Do this for all y in K. Then, by (2) and continuity of $\psi_n$, $\psi$ is continuous on K and $\psi(y) \varepsilon \overline{\phi(y)} = \phi(y) = T^{-1}(y)$ for all y in K. Let K' = $\psi$(K). Then K' is compact and $T(K') \supseteq K$. [Thus $K' \cap T^{-1}(K)$ is a compact subset of E, and is mapped under T to the exact image K (note that $T^{-1}(K)$ is certainly closed as G(T) is closed and K is compact), if T is single-valued.]

## REFERENCES

[1] De Vilde, M., Closed graph theorems and webbed spaces (Pitman, 1978).
[2] Fakhoury, M., Sélections continues dans les spaces uniformes, C.R. Acad. Sc. Paris, 280 (1975), 213-216.
[3] Rockafellar, R. T., Monotone processes of convex and concave type, Mem. Amer. Math. Soc. 77 (1967).
[4] Michael, E., Continuous selections, Ann. of Math., 63 (1956), 361-382.
[5] Ng, K. F., An open mapping theorem, Proc. Camb. Phil. Soc., 74 (1973), 61-66.
[6] Ng, K. F., An inequality implicit-function theorem, Preprint.

Proceedings of the Analysis Conference, Singapore 1986
S.T.L. Choy, J.P. Jesudason, P.Y. Lee (Editors)




# WEAK $L_p$-SPACES AND WEIGHTED NORM INEQUALITIES FOR THE FOURIER TRANSFORM ON LOCALLY COMPACT VILENKIN GROUPS

**C. W. Onneweer**

**Department of Mathematics and Statistics**
**University of New Mexico**
**Albuquerque, New Mexico 87131**
**USA**

**In this paper we consider the weighted norm inequality problem for the Fourier transform for functions defined on a certain class of topological groups. We study the case in which the weight functions belong to suitable weak $L_p$-spaces.**

## 1. INTRODUCTION

In his 1978 survey lecture on weighted norm inequalities for certain operators, B. Muckenhoupt posed the so-called weighted norm inequality problem for the Fourier transform [8]. This is the problem of characterizing, for given $p$ and $q$ with $1 < p, q < \infty$, those nonnegative measurable functions $u$ and $v$ on $\mathbf{R}$ or, more general, on $\mathbf{R}^n$ so that the inequality

$$\|\hat{f}u\|_q \leq C\|fv\|_p \tag{1.1}$$

holds for all Lebesgue integrable functions $f$.

Significant progress towards the solution of this problem has been made by, among others, Muckenhoupt himself [9], [10], by W. B. Jurkat and G. Sampson [7] and especially by H. P. Heinig and his co-authors J. J. Benedetto and R. Johnson [1], [2] and [5]. The characterizations obtained by these authors for the weight functions $u$ and $v$ that are equivalent to (1.1) all impose the restrictions that both $u$ and $v$ are radial functions and that they satisfy certain monotonicity conditions. Conditions imposed on $u$ and $v$ that imply (1.1) and that deal with less restricted classes of weight functions are often rather difficult to apply and it seems likely that techniques different from those used in [2], [7] or [9] will be needed to solve Muckenhoupt's problem for nonradial functions $u$ and $v$.

In this paper we consider Muckenhoupt's problem for functions defined on certain groups different from $\mathbf{R}$ or $\mathbf{R}^n$ and we study the case in which the weight functions $u$ and $v$ belong to certain weak $L_p$-spaces.

In the remainder of this section we briefly describe the class of groups considered here. In addition, we introduce most of the notation to be used and we state some of the known results that are needed afterwards. Section 2 will contain the statements and proofs of some sufficient conditions for (1.1) to hold. The final section complements Section 2 by presenting some conditions implied by inequality (1.1).

Throughout this paper $G$ will denote a locally compact Abelian topological group with a suitable family of compact open subgroups, cf. [4, §4.1]. This means that there exists a sequence $(G_n)_{-\infty}^{\infty}$ such that

(i) each $G_n$ is an open compact subgroup of $G$,

(ii) $G_{n+1} \subsetneqq G_n$ and order $(G_n/G_{n+1}) < \infty$,

(iii) $\bigcup_{-\infty}^{\infty} G_n = G$ and $\bigcap_{-\infty}^{\infty} G_n = \{0\}$.

Moreover, we shall assume that $G$ is order-bounded, that is,

(iv) $\sup\{\text{order}(G_n/G_{n+1})\ ;\ n \in \mathbf{Z}\} < \infty$.

Such groups are the locally compact analogue of the Vilenkin groups [12]. Several examples of such groups are given in [4, §4.1.2]. Additional examples are the $p$-adic numbers and, more general, the additive group of a local field [11].

Let $\Gamma$ denote the dual group of $G$ and for each $n \in \mathbf{Z}$ let $\Gamma_n$ denote the annihilator of $G_n$, that is,

$$\Gamma_n = \{\gamma \in \Gamma\ ;\ \gamma(x) = 1 \quad \text{for all} \quad x \in G_n\}\ .$$

Then we have, cf. [4, §4.1.4],

(i)* each $\Gamma_n$ is an open compact subgroup of $\Gamma$,

(ii)* $\Gamma_n \subsetneqq \Gamma_{n+1}$ and $\text{order}(\Gamma_{n+1}/\Gamma_n) = \text{order}(G_n/G_{n+1})$,

(iii)* $\bigcap_{-\infty}^{\infty} \Gamma_n = \{1\}$ and $\bigcup_{-\infty}^{\infty} \Gamma_n = \Gamma$.

If we choose Haar measures $\mu$ on $G$ and $\lambda$ on $\Gamma$ so that $\mu(G_0) = \lambda(\Gamma_0) = 1$ then $\mu(G_n) = (\lambda(\Gamma_n))^{-1}$ for each $n \in \mathbf{Z}$. We set $m_n = \lambda(\Gamma_n)$.

If we define the function $d : G \times G \to \mathbf{R}_+$ by

$$d(x,y) = \begin{cases} 0 & \text{if } x = y, \\ (m_n)^{-1} & \text{if } x - y \in G_n \setminus G_{n+1}, \end{cases}$$

then $d$ defines a metric on $G \times G$ and the topology on $G$ induced by this metric is the same as the original topology on $G$. For $x \in G$ we define $\|x\|$ by $\|x\| = d(x,0)$; then $\|x\| = (m_n)^{-1}$ if and only if $x \in G_n \setminus G_{n+1}$. In a similar way we can define a metric $\tilde{d}$ on $\Gamma \times \Gamma$; if we set $\|\gamma\| = \tilde{d}(\gamma, 1)$, then $\|\gamma\| = m_n$ if and only if $\gamma \in \Gamma_{n+1} \setminus \Gamma_n$. A function $f : G \to \mathbf{C}$ is called radial if $f(x) = f(\|x\|)$; thus, a radial function on $G$ is constant on each subset $G_n \setminus G_{n+1}$ in $G$ ($n \in \mathbf{Z}$). A similar definition can be given for radial functions on $\Gamma$.

For $p$ with $1 \le p \le \infty$ we denote its conjugate by $p'$, thus $p' = p/(p-1)$ if $1 < p < \infty$ and $p' = 1$ or $\infty$ if $p = \infty$ or $1$. For a given set $A$ we denote its characteristic function by $\xi_A$. The symbols ^ and ˇ will be used to denote the Fourier transform and the inverse Fourier transform, respectively. It is easy to see that for each $n \in \mathbf{Z}$ we have

$$(\xi_{G_n})^\wedge = (\lambda(\Gamma_n))^{-1} \xi_{\Gamma_n} . \tag{1.2}$$

As usual, $C$ will denote a constant whose value may change from one occurrence to the next.

We now give the definition of the Lorentz spaces. Let $(X, \mathcal{A}, \mu)$ be a measure space and let $f$ be a measurable complex-valued function on $X$. For $y > 0$ let $f_*(y) = \mu(\{x \in X \; ; \; |f(x)| > y\})$ and define $f^* : \mathbf{R}_+ \to \mathbf{R}_+$, the non-increasing rearrangement of $f$, by $f^*(t) = \inf\{y > 0 \; ; \; f_*(y) \le t\}$. The Lorentz space $L(p,q;X)$ is the set of all measurable functions $f$ on $X$ such that $\|f\|_{p,q} < \infty$, where

$$\|f\|_{p,q} = \begin{cases} \left(\int_0^\infty (f^*(t) t^{1/p})^q t^{-1} dt\right)^{1/q} & \text{if } 0 < p, q < \infty, \\ \sup\{f^*(t) t^{1/p} \; ; \; t > 0\} & \text{if } 0 < p < \infty \, , \; q = \infty. \end{cases}$$

The spaces $L(p, \infty; X)$ are also known as the weak $L^p$-spaces or the Marcinkiewicz spaces on $X$.

For future reference we state here some properties of the Lorentz spaces, see [3, §§1.3 and 5.3] for further details.

If $0 < p < \infty$ then

$$L(p,p;X) = L^p(X)\,. \tag{1.3}$$

If $0 < p < \infty$ and $0 < q_1 \le q_2 \le \infty$ then

$$L(p,q_1;X) \subset L(p,q_2;X) \quad \text{and} \quad \|f\|_{p,q_2} \le C\|f\|_{p,q_1}\,. \tag{1.4}$$

If $0 < p < \infty$ then $f \in L(p,\infty;X)$ if and only if

$$f_*(t) \le Ct^{-p} \quad \text{for all} \quad t > 0\,. \tag{1.5}$$

The Marcinkiewicz interpolation theorem for Lorentz spaces. Let $(X,\mathcal{A},\mu)$ and $Y,\mathcal{B},\nu)$ be two $\sigma$-finite measure spaces. Let $1 \le p_i, q_i, r_i \le \infty (i = 0,1)$ with $p_0 \ne p_1$ and $q_0 \ne q_1$. Assume that

$$T : L(p_i, r_i; X) \to L(q_i, \infty; Y)$$

is a bounded linear operator for $i = 0,1$. Let $1/p = (1-\theta)/p_0 + \theta/p_1$ and $1/q = (1-\theta)/q_0 + \theta/q_1$ for some $\theta$ with $0 < \theta < 1$. Then for each $r$, $1 \le r \le \infty$ the operator

$$T : L(p,r;X) \to L(q,r;Y) \tag{1.6}$$

is a bounded linear operator.

## 2. SUFFICIENT CONDITIONS

**Theorem 1.** Let $1 < p \le 2$ and $1 < p \le q < \infty$. Assume that $u : \Gamma \to \mathbf{R}_+$ belongs to $L(\beta_1,\infty;\Gamma)$ and that $v : G \to \mathbf{R}_+$ is a function such that $v^{-1}$ belongs to $L(\beta_2,\infty;G)$, where $1/\beta_1 = 1/q - 1/r'$ and $1/\beta_2 = 1/r - 1/p$ for some $r$ such that $1 < r < p \le q < r'$. Then there exists a $C > 0$ so that for all $f \in L_1(G)$ we have

$$\|\hat{f}u\|_q \le C\|fv\|_p\,. \tag{2.1}$$

The proof of Theorem 1 will be preceded by a lemma that is essentially the analogue on $G$ and $\Gamma$ of Lemma 1 in [9].

**Lemma 1.** Let $1 < p \le q < p'$ and let $1/\beta = 1/p + 1/q - 1$. If $u : \Gamma \to \mathbf{R}_+$ belongs to $L(\beta,\infty;\Gamma)$ then there exists a $C > 0$ so that for all $f \in L_1(G)$ we have $\|\hat{f}u\|_q \le C\|f\|_{p,q}$.

**Proof.** For $f \in L_1(G)$ define $Tf : \Gamma \to \mathbf{C}$ by $Tf(\gamma) = \hat{f}(\gamma)(u(\gamma))^a$, where $a = -\beta q'/p'$ and define the measure $d\sigma$ on $\Gamma$ by $d\sigma(\gamma) = (u(\gamma))^b d\lambda(\gamma)$, where $b = \beta + \beta q'/p'$. We first prove that $T$ is of weak type (1,1) from $L_1(G, d\mu)$ to $L_1(\Gamma, d\sigma)$. For each $t > 0$ we have

$$\begin{aligned} E_t &= \{\gamma \in \Gamma\,;\ |Tf(\gamma)| > t\} \\ &\subset \{\gamma \in \Gamma\,;\ \|f\|_1 (u(\gamma))^a \geq t\} \\ &= \bigcup_{n=0}^{\infty} \{\gamma \in \Gamma\,;\ 2^{n+1}t > \|f\|_1 (u(\gamma))^a \geq 2^n t\} \\ &= \bigcup_{n=0}^{\infty} E_{t,n}\,. \end{aligned}$$

Therefore,

$$\begin{aligned} \sigma(E_t) &\leq \sum_{n=0}^{\infty} \int_{E_{t,n}} (u(\gamma))^b d\lambda \\ &\leq \sum_{n=0}^{\infty} \int_{\{\gamma \in \Gamma\,;\ \|f\|_1 (u(\gamma))^a < 2^{n+1} t\}} (2^n t / \|f\|_1)^{b/a} d\lambda \\ &\leq C \sum_{n=0}^{\infty} (2^n t / \|f\|_1)^{b/a} (2^{n+1} t / \|f\|_1)^{-\beta/a}\,, \end{aligned}$$

because $u \in L(\beta, \infty; \Gamma)$, cf. (1.5). Thus our choice of $a$ and $b$ implies that

$$\sigma(E_t) \leq C \sum_{n=0}^{\infty} (2^n t / \|f\|_1)^{-1} = C\|f\|_1 / t\,.$$

Next we show that if we define $\alpha$ by $\alpha = (p' + q')/p'$, so that $\alpha' = (p' + q')/q'$, then $T$ is of type $(\alpha, \alpha')$ from $L_\alpha(G, d\mu)$ to $L_{\alpha'}(\Gamma, d\sigma)$.

$$\begin{aligned} \|Tf\|_{L_{\alpha'}(d\sigma)} &= \left(\int_\Gamma |\hat{f}(\gamma)(u(\gamma))^a|^{\alpha'} (u(\gamma))^b d\lambda\right)^{1/\alpha'} \\ &= \left(\int_\Gamma |\hat{f}(\gamma)|^{\alpha'} d\lambda\right)^{1/\alpha'} \\ &\leq \left(\int_G |f(x)|^\alpha d\mu\right)^{1/\alpha}\,, \end{aligned}$$

because our choice of $\alpha$ implies that $a\alpha' + b = 0$ and $1 < \alpha \leq 2$, so that we can apply the Hausdorff-Young inequality. Furthermore, since

$$\frac{1}{p} = \left(1 - \frac{1}{\beta}\right) \cdot \frac{1}{\alpha} + \frac{1}{\beta} \cdot 1 \quad \text{and} \quad \frac{1}{q} = \left(1 - \frac{1}{\beta}\right) \frac{1}{\alpha'} + \frac{1}{\beta} \cdot 1\,,$$

it follows from the Marcinkiewicz interpolation theorem that

$$\|Tf\|_{L_q(d\sigma)} \leq C\|f\|_{p,q}\,,$$

that is, since $aq + b = q$,

$$\|\hat{f}u\|_q \leq C\|f\|_{p,q}\,.$$

Lemma 1 easily yields the following sufficient conditions for the one weight norm inequality for the Fourier transform.

**Corollary 1.** Let $1 < p \leq q < p'$ and $1/\beta = 1/p + 1/q - 1$.

(a) If $u : \Gamma \to \mathbf{R}_+$ belongs to $L(\beta, \infty; \Gamma)$ then there exists a $C > 0$ so that for all $f \in L_1(G)$ we have

$$\|\hat{f}u\|_q \leq C\|f\|_p . \tag{2.2}$$

(b) If $v : G \to \mathbf{R}_+$ and if $v^{-1} \in L(\beta, \infty; G)$ then there exists a $C > 0$ so that for all $f \in L_1(G)$ we have

$$\|\hat{f}\|_{p'} \leq C\|fv\|_{q'} . \tag{2.3}$$

**Proof (a).** Since $p \leq q$ the first part of the corollary follows immediately from Lemma 1, (1.3) and (1.4).

(b). For any $f \in L_1(G) \cap L_p(G)$ and any $\varphi \in S(\Gamma)$, where $S(\Gamma)$ is the set of all functions on $\Gamma$ with support in some $\Gamma_k$ and constant on the cosets of some $\Gamma_\ell$ in $\Gamma$, we have, cf. [6, (31.48)],

$$\begin{aligned} \left|\int_\Gamma \hat{f}(\gamma)\varphi(\gamma)d\lambda\right| &= \left|\int_G f(x)\check{\varphi}(x)d\mu\right| \\ &\leq \|fv\|_{q'}\|\check{\varphi}v^{-1}\|_q . \end{aligned}$$

Since $v^{-1} \in L(\beta, \infty; G)$ it follows from part (a), after interchanging the rôle of $G$ and $\Gamma$, that

$$\|\check{\varphi}v^{-1}\|_q \leq C\|\varphi\|_p .$$

Thus, since $S(\Gamma)$ is a dense subset of $L_p(\Gamma)$, cf. [11, Chapter II, Proposition (1.3)], we see that

$$\begin{aligned} \|\hat{f}\|_{p'} &= \sup\left|\int_\Gamma \hat{f}(\gamma)\varphi(\gamma)d\lambda\right| ; \|\varphi\|_p \leq 1\} \\ &\leq C\|fv\|_{q'} . \end{aligned}$$

**Remark.** If in Corollary 1(a) we replace the assumption that $u \in L(\beta, \infty; \Gamma)$ by $u \in L_\infty(\Gamma)$ and choose $q = p'$ then it follows immediately from the Hausdorff-Young inequality that $\|\hat{f}u\|_q \leq C\|f\|_p$. Therefore, Corollary 1(a), or Lemma 1, can be considered as a generalization of the Hausdorff-Young inequality.

**Proof of Theorem 1.** Since $1 < r < q < r'$ and $1/\beta_1 = 1/q - 1/r'$, Lemma 1 implies that

$$\|\hat{f}u\|_q \leq C\|f\|_{r,q} \,. \tag{2.4}$$

Next, choose $\epsilon$ so that $0 < \epsilon < 1/r'$ and define $r_1$ and $p_1$ by $1/r_1 = 1/r + \epsilon$ and $1/p_1 = 1/p + \epsilon$. Then

$$\begin{aligned} \|f\|_{r_1}^{r_1} &= \int_G |f(x)v(x)|^{r_1}(v(x))^{-r_1}\,d\mu \\ &\leq \int_0^\infty ((fv)^{r_1})^*(t)(v^{-r_1})^*(t)\,dt\,. \end{aligned}$$

Since $v^{-1} \in L(\beta_2, \infty; G)$ we have $v^{-r_1} \in L(\beta_2/r_1, \infty; G)$. Therefore,

$$\begin{aligned} (v^{-r_1})^*(t) &\leq Ct^{-r_1/\beta_2} \\ &= Ct^{r_1/p_1 - 1}\,. \end{aligned}$$

Thus,

$$\begin{aligned} \|f\|_{r_1,\infty} &\leq \|f\|_{r_1} \\ &\leq C\Big(\int_0^\infty ((fv)^*)(t))^{r_1}t^{r_1/p_1 - 1}\,dt\Big)^{1/r_1} \\ &= C\|fv\|_{p_1,r_1}\,. \end{aligned}$$

If we define $r_2$ and $p_2$ by $1/r_2 = 1/r - \epsilon$ and $1/p_2 = 1/p - \epsilon$, then a similar argument shows that

$$\|f\|_{r_2,\infty} \leq C\|fv\|_{p_2,r_2}\,.$$

Thus it follows from (1.6), the Marcinkiewicz interpolation theorem for Lorentz spaces, that

$$\|f\|_{r,q} \leq C\|fv\|_{p,q} \leq C\|fv\|_p\,,$$

because $p \leq q$. Combining this inequality with (2.4) we may conclude that

$$\|\hat{f}u\|_q \leq C\|fv\|_p\,,$$

which concludes the proof of Theorem 1.

We now show that Theorem 1 implies a version of Pitt's Theorem for functions on $G$. This is an easy consequence of the following simple facts that will also be used in Section 3.

**Lemma 2.** (a) If $\varphi(x) \leq C\|x\|^{-1/\alpha}$ on $G$ for some $\alpha > 0$ then $\varphi \in L(\alpha, \infty; G)$.

(b) If $\psi(\gamma) \leq C\|\gamma\|^{-1/\alpha}$ on $\Gamma$ for some $\alpha > 0$ then $\psi \in L(\alpha, \infty; \Gamma)$.

**Proof.** (a) Fix $t > 0$ and choose $n_0 \in Z$ so that $C(m_{n_0-1})^{1/\alpha} < t \leq C(m_{n_0})^{1/\alpha}$, where $C$ is the same constant as in the statement of (a). Then

$$\begin{aligned}\{x \in G\,;\, \varphi(x) > t\} &\subset \{x \in G\,;\, \varphi(x) > C(m_{n_0-1})^{1/\alpha}\} \\ &\subset G_{n_0}\,,\end{aligned}$$

because $\varphi(x) \leq C(m_n)^{1/\alpha}$ for $x \in G_n \setminus G_{n+1} (n \in \mathbb{Z})$ and the sequence $(C(m_n)^{1/\alpha})_{-\infty}^{\infty}$ is monotone increasing. Therefore,

$$\varphi_*(t) = \mu(\{x \in G\,;\, \varphi(x) > t\}) \leq (m_{n_0})^{-1} < \tilde{C}t^{-\alpha}\,,$$

due to our choice of $n_0$. Thus (1.5) implies that $\varphi \in L(\alpha, \infty; G)$.

The proof of (b) is similar and will be omitted.

**Corollary 2.** (Pitt's Theorem on $G$) Let $1 < p \leq q < p'$, let $0 \leq \alpha < 1/p'$ and let $\eta = \alpha + 1/p + 1/q - 1$. Then there exists a $C > 0$ so that for all $f \in L_1(G)$ we have

$$\left(\int_\Gamma (|\hat{f}(\gamma)|\|\gamma\|^{-\eta})^q d\lambda\right)^{1/q} \leq C\left(\int_G (|f(x)|\|x\|^{\alpha})^p d\mu\right)^{1/p}.$$

**Proof.** Let $r = (\alpha + 1/p)^{-1}$. Then $1 < r < p \leq q < r'$. According to Lemma 2, if $u(\gamma) = \|\gamma\|^{-\eta}$ then $u \in L(\eta^{-1}, \infty; \Gamma)$ with $\eta = 1/q - 1/r'$. Also, if $v(x) = \|x\|^\alpha$ then $v^{-1} \in L(\alpha^{-1}, \infty; G)$ with $\alpha = 1/r - 1/p$. Thus an application of Theorem 1 completes the proof of this corollary.

## 3. NECESSARY CONDITIONS

For the one weight norm inequalities for the Fourier transform that were proved in Corollary 1 we have the following complementary result.

**Theorem 2.** Let $1 < p \leq q < p'$ and let $1/\beta = 1/p + 1/q - 1$.

(a) If $u : \Gamma \to \mathbf{R}_+$ is radial and if (2.2) holds then $u \in L(\beta, \infty; \Gamma)$.

(b) If $v : G \to \mathbf{R}_+$ is radial and if (2.3) holds then $v^{-1} \in L(\beta, \infty; G)$.

**Proof.** (a) Since $u$ is a radial function $u$ can be represented as

$$u(\gamma) = \sum_{k=-\infty}^{\infty} \alpha_k \xi_{\Gamma_{k+1} \setminus \Gamma_k}(\gamma)\,.$$

Let $f(x) = \xi_{G_{k+1}}(x)$ so that $\hat{f}(\gamma) = (m_{k+1})^{-1}\xi_{\Gamma_{k+1}}(\gamma)$, cf. (1.2). The order-boundedness of $G$ implies that

$$
\begin{aligned}
\left(\int_{\Gamma_{k+1}\setminus\Gamma_k} |\hat{f}(\gamma)u(\gamma)|^q d\lambda\right)^{1/q} &= \alpha_k(m_{k+1})^{-1}(m_{k+1}-m_k)^{1/q} \\
&\geq C\alpha_k(m_{k+1})^{1/q-1} .
\end{aligned}
$$

We also have

$$
\begin{aligned}
\left(\int_{\Gamma_{k+1}\setminus\Gamma_k} |\hat{f}(\gamma)u(\gamma)|^q d\lambda\right)^{1/q} &\leq \|\hat{f}u\|_q \\
&\leq C\|f\|_p = C(m_{k+1})^{-1/p} .
\end{aligned}
$$

Therefore, for each $k \in \mathbf{Z}$ we have

$$
\alpha_k \leq C(m_{k+1})^{1-1/p-1/q} = C(m_k)^{-1/\beta} ,
$$

that is, $u(\gamma) \leq C\|\gamma\|^{-1/\beta}$. Thus it follows from Lemma 2(b) that $u \in L(\beta, \infty; \Gamma)$.

(b) Since $v$ is radial, $v$ can be represented as

$$
v(x) = \sum_{n=-\infty}^{\infty} a_n \xi_{G_n\setminus G_{n+1}}(x) .
$$

Let $f(x) = \xi_{G_n\setminus G_{n+1}}(x)$; then

$$
\hat{f}(\gamma) = (m_n)^{-1}\xi_{\Gamma_n}(\gamma) - (m_{n+1})^{-1}\xi_{\Gamma_{n+1}}(\gamma) .
$$

Therefore,

$$
\begin{aligned}
\|\hat{f}\|_{p'} &\geq \left(\int_{\Gamma_{n+1}\setminus\Gamma_n} |\hat{f}(\gamma)|^{p'} d\lambda\right)^{1/p'} \\
&= (m_{n+1})^{-1}(m_{n+1}-m_n)^{1/p'} \\
&\geq C(m_n)^{-1/p} .
\end{aligned}
$$

Also

$$
\begin{aligned}
\|fv\|_{q'} &= \left(\int_{G_n\setminus G_{n+1}} (a_n)^{q'} d\mu\right)^{1/q'} \\
&\leq a_n(m_n)^{-1/q'} .
\end{aligned}
$$

Thus we see that $a_n \geq C(m_n)^{-1/\beta}$ for each $n \in \mathbf{Z}$, that is, $v(x) \geq C\|x\|^{1/\beta}$. Thus Lemma 2(a) implies that $v^{-1} \in L(\beta, \infty; G)$.

As a partial converse to Theorem 1 we present the following necessary conditions for the two weights inequality (2.1).

**Theorem 3.** Let $1 < p \leq 2$ and $1 < r < p \leq q < r'$. Let $1/\beta_1 = 1/q - 1/r'$ and $1/\beta_0 = 1/r - 1/p$. Assume $u : \Gamma \to \mathbf{R}_+$ and $v : G \to \mathbf{R}_+$ are radial functions for which (2.4) holds.

(a) If $(v(x))^{-1} \geq C\|x\|^{-1/\beta_2}$ then $u \in L(\beta_1, \infty; \Gamma)$.

(b) If $u(\gamma) \geq C\|\gamma\|^{-1/\beta_1}$ then $v^{-1} \in L(\beta_2, \infty; G)$.

**Proof.** Let $u$ and $v$ have the same representations as in the proof of Theorem 2. If we take $f(x) = \xi_{G_n \setminus G_{n+1}}(x)$ then we have

$$\begin{aligned} \alpha_n (m_{n+1})^{-1} (m_{n+1} - m_n)^{1/q} &\leq \|\hat{f}u\|_q \leq C\|fv\|_p \\ &\leq Ca_n (m_n)^{-1/p}. \end{aligned}$$

Thus for all $n \in \mathbf{Z}$ we have $\alpha_n \leq Ca_n (m_n)^{1-1/p-1/q}$. If $v(x) \leq C\|x\|^{1/\beta_2}$ then for each $n \in \mathbf{Z}$, $a_n \leq C(m_n)^{-1/\beta_2}$. Therefore, in this case $\alpha_n \leq C(m_n)^{-1/\beta_1}$, that is, $u(\gamma) \leq C\|\gamma\|^{-1/\beta_1}$, which implies that $u \in L(\beta_1, \infty; \Gamma)$.

On the other hand, if $u(\gamma) \geq C\|\gamma\|^{-1/\beta_1}$ then $\alpha_n \geq C(m_n)^{-1/\beta_1}$ for each $n \in \mathbf{Z}$ and we easily see that $(a_n)^{-1} \leq C(m_n)^{-1/\beta_2}$, that is, $(v(x))^{-1} \leq C\|x\|^{-1/\beta_2}$, which implies that $v^{-1} \in L(\beta_2, \infty; G)$.

**Remark.** After this paper had been written the author learned of a paper on the same subject by T. S. Quek in which he gives conditions for radial weight functions $u$ and $v$ that are equivalent to inequality (2.1), see [13].

## REFERENCES

[1] Benedetto, J. J. and Heinig, H. P., Weighted Hardy spaces and the Laplace transform (Springer Lecture Notes in Math. 992, 240-277, Springer Verlag, 1982).

[2] Benedetto, J. J., Heinig, H. P. and Johnson, R., Boundary values of functions in weighted Hardy spaces (preprint, 1984).

[3] Bergh, J. and Löfström, J., Interpolation Spaces (Springer Verlag, 1976).

[4] Edwards, R. E. and Gaudry, G. I., Littlewood-Paley and Multiplier Theory (Ergeb. Math. 90, Springer Verlag, 1977).

[5] Heinig, H. P., Weighted norm inequalities for classes of operators, Indiana Univ. Math. J. 33 (1984), 573-582.

[6] Hewitt, E. and Ross, K. A., Abstract Harmonic Analysis, Vol. II (Grundl. der Math 152, Springer Verlag, 1970).

[7] Jurkat, W. B. and Sampson, G., On rearrangement and weight inequalities for the Fourier transform, Indiana Univ. Math. J. 33 (1984), 257-270.

[8] Muckenhoupt, B., Weighted norm inequalities for classical operators (Amer. Math. Soc., Proc. Symp. Pure Math. 35(I) (1979), 69-83).

[9] Muckenhoupt, B., Weighted norm inequalities for the Fourier transform, Trans. Amer. Math. Soc. 276 (1983), 729-742.

[10] Muckenhoupt, B., A note on two weight function conditions for a Fourier transform norm inequality, Proc. Amer. Math. Soc. 88 (1983), 97-100.

[11] Taibleson, M. H., Fourier Analysis on Local Fields (Math. Notes, Princeton Univ. Press, 1975).

[12] Vilenkin, N. Ja., On a class of complete orthonormal systems, Amer. Math. Soc. Transl. (2) 28 (1963), 1-35.

[13] Quek, T. S., Weighted Norm Inequalities for the Fourier Transform on Certain Totally Disconnected Groups (Preprint).

Proceedings of the Analysis Conference, Singapore 1986
S.T.L. Choy, J.P. Jesudason, P.Y. Lee (Editors)



# Multipliers of Segal Algebras

Ouyang Guangzhong
Department of Mathematics, Fudan University, Shanghai, China.

The main result of this article is that:

$M(L_1(G), S(G)) \cong M(G)$ and some applications of this result.

$M(A_p(G), L_1(G)) \cong M(G)$,

$M(A_p(G), L_1(G)) \cong A_p(G)$ when G is discrete.

$M(A_{1,p}(G), L_1(G)) \cong M(G)$, $M(A_{1,p}(G), A_{1,p}(G)) \cong M(G)$,

$M(A_{1,p}(G), L_1(G)) \cong A_{1,p}(G)$, $M(A_{1,p}(G), A_{1,p}(G)) \cong A_{1,p}(G)$ when G is discrete.

Let $S_1(G)$ and $S_2(G)$ be two Segal algebras on G, where G is a locally compact abelian group; $S_1(G)$ and $S_2(G)$ may be the same. The bounded linear operator $T : S_1(G) \to S_2(G)$ is called a multiplier from $S_1(G)$ to $S_2(G)$ iff T commutes with every translation operator $\tau_x$, that is $T \circ \tau_x = \tau_x \circ T$. We denote the whole of these multipliers by $M(S_1(G), S_2(G))$. It is a Banach algebra. In this paper, we will discuss the character of $M(S_1(G), S_2(G))$ for some $S_1(G)$ and $S_2(G)$.

1. Multipliers from $L_1(G)$ to Segal algebra $S(G)$.
2. Multipliers of Segal algebra $A_p(G)$.
3. Segal algebra $A_{1,p}(G)$ and its multipliers.

## §1. Multipliers from $L_1(G)$ to Segal algebra $S(G)$

We have known that for every $T \in M(L_1(G), S(G))$ there exists a unique measure $\mu \in M(G)$ ($M(G)$ is a Banach algebra which consists of all bounded regular Borel measures on G), such that $Tf = \mu * f$ for every $f \in L_1(G)$. Denote the whole of these measures $\mu$ by $M_S(G)$. It is obvious that $M_S(G)$ is dependent on the given Segal algebra $S(G)$. Now the problem we face is that $M_S(G)$ may not be the whole $M(G)$, and it may only be a proper subalgebra of $M(G)$. In this case, what character does $M_S(G)$ have, and how is it determined by $S(G)$ ?

R. Goldberg and S. Seltzer [1] have given a necessary and sufficient condition on $M_S(G)$. But the condition is so complex that no one can use it to do further research. Here we will give a much simpler necessary and sufficient condition.

Suppose that

(i) $\{u_\alpha\}$ is an approximate identity of $L_1(G)$ and $\|u_\alpha\|_1 = 1$ for every $\alpha$, $\hat{u}_\alpha$ has a compact support.

(ii) $M_s(G) = \{\mu \in M(G) : \|u_\alpha * \mu\|_s \leq C_u\}$, where $C_u$ is a constant only dependent on u. In $M_s(G)$, define the norm by

$$\|\mu\|_{M_s} = \overline{\lim_\alpha}\, \|u_\alpha * \mu\|_s.$$

**Theorem 1 [6].** The following are equivalent:

(1) $T \in M(L_1(G), S(G))$,

(2) There exists a unique $\mu \in M_s(G)$ such that $Tf = \mu * f$ for every $f \in L_1(G)$.

Moreover, the correspondence between T and $\mu$ defines an isometric algebraic isomorphism from $M(L_1(G), S(G))$ onto $M_s(G)$.

**Proof.** Suppose $\mu \in M_s(G)$, then $u_\alpha * \mu \in S(G)$. Let $Tf = \mu * f$ for every $f \in L_1(G)$. Now we are going to prove $Tf = \mu * f \in S(G)$.

For any given $f \in L_1(G)$, denote

$$g_\alpha = \mu * u_\alpha * u_\alpha * f \ ;$$

$g_\alpha \in S(G)$ because $u_\alpha \in S(G)$ and $S(G)$ is an ideal in $L_1(G)$. Consider

$$\begin{aligned} g_\alpha - g_{\alpha^1} &= (\mu * u_\alpha * u_\alpha * f) - (\mu * u_{\alpha^1} * u_{\alpha^1} * f) \\ &= (\mu * u_\alpha + \mu * u_{\alpha^1}) * (u_\alpha * f - u_{\alpha^1} * f). \end{aligned}$$

Notice that $u_\alpha * f - u_{\alpha^1} * f \in L_1(G)$, so we obtain

$$\begin{aligned} \|g_\alpha - g_{\alpha^1}\|_s &\leq \|\mu * u_\alpha + \mu * u_{\alpha^1}\|_s \cdot \|u_\alpha * f - u_{\alpha^1} * f\|_1 \\ &\leq 2C\|u_\alpha * f - u_{\alpha^1} * f\|_1 \to 0, \qquad (\alpha, \alpha^1 \to \infty); \end{aligned}$$

therefore, there exists $g \in S(G)$ such that

$$\lim_\alpha g_\alpha = g \ \text{ in } \ (S(G), \|\ \|_s).$$

Of course, the above limit is also tenable in $L_1(G)$. Then

$$\lim_{\alpha} \hat{g}_\alpha = \hat{g}.$$

But $\hat{g}_\alpha = \hat{\mu}(\hat{u}_\alpha)^2\hat{f}$ and $\{u_\alpha\}$ is an approximate identity, therefore

$$\lim_{\alpha} \hat{g}_\alpha = \hat{\mu}\hat{f}.$$

Thus we have proved that $\hat{\mu}\hat{f} = \hat{g}$ i.e. $Tf = \mu * f \in S(G)$. Besides,

$$\|Tf\|_S = \|\mu * f\|_S = \|g\|_S = \lim_{\alpha} \|g_\alpha\|_S$$

$$\leqslant \overline{\lim_{\alpha}} \|\mu * u_\alpha\|_S \|u_\alpha * f\|_1 \leqslant \|\mu\|_{M_S} \cdot \|f\|_1,$$

we obtain $\|T\| \leqslant \|\mu\|_{M_S}$.

Conversely, suppose $T \in M(L_1(G), S(G))$. Now we will prove that there certainly exists a unique $\mu \in M_S(G)$ such that $Tf = \mu * f$ for every $f \in L_1(G)$.

We have known, for every above-mentioned T, there exists a unique measure $\mu \in M(G)$ such that $Tf = \mu * f$ for every $f \in L_1(G)$. We will prove $\mu \in M_S(G)$.

Since T is bounded,

$$\|T\| = \sup_{\|f\|_1 = 1} \{\|\mu * f\|_S\}.$$

Take $f = u_\alpha$, notice that $\|u_\alpha\|_1 = 1$, thus

$$\|u_\alpha * \mu\|_S \leqslant \|T\|.$$

So we have proved $\mu \in M_S(G)$ and

$$\|\mu\|_{M_S} = \overline{\lim_{\alpha}} \|u_\alpha * \mu\|_S \leqslant \|T\|.$$

The proof of Theorem 1 has been accomplished.

Note 1. It is not difficult to prove that $M_S(G)$ is independent of the approximate identity $\{u_\alpha\}$ of $L_1(G)$, i.e. if $\{v_\alpha\}$ is also an approximate identity of $L_1(G)$ and $\mu \in M(G)$, then $\|u_\alpha * \mu\|_S \leqslant C$ if and only if $\|v_\alpha * \mu\|_S \leqslant C$.

Note 2. Theorem 1 can be denoted by

$$M(L_1(G), S(G)) \cong M_S(G).$$

Now, we will use Theorem 1 to study $M(L_1(G), S(G))$ for some concrete Segal algebras.

**Corollary 1.** Suppose the Segal algebra $A_p(G)$ is the following:

$$A_p(G) = \{f \in L_1(G) : \hat{f} \in L_p(G)\},$$

$$\|f\|_{A_p} = \|f\|_1 + \|\hat{f}\|_p,$$

and suppose

$$B(G) = \{\mu \in M(G) : \hat{\mu} \in L_p(\hat{G})\}.$$

Then $B(G) = M_{A_p}(G)$, that is $B(G) \cong M(L_1(G), A_p(G))$, where "$\cong$" means isometric and algebra isomorphic. Refer to the above note 2.

**Proof.** It follows from Theorem 1 that $\mu \in M_{A_p}(G)$ if and only if $\|u_\alpha * \mu\|_{A_p} \leq C_\mu$. But

$$\|u_\alpha * \mu\|_{A_p} = \|u_\alpha * \mu\|_1 + \|(u_\alpha * \mu)^\wedge\|_p$$

$$= \|u_\alpha * \mu\|_1 + \|\hat{u}_\alpha . \hat{\mu}\|_p$$

and $\|\hat{u}_\alpha . \hat{\mu}\|_p$ is bounded iff $\|\hat{\mu}\|_p < \infty$. The proof of Corollary 1 is complete.

**Corollary 2.** Suppose the Segal algebra $F(R)$ is the following:

$$F(R) = \{f \in L_1(R) : \lim_{n\to\infty} \hat{f}(n) \ln n = 0\}$$

$$\|f\|_F = \|f\|_1 + \sup\{|\hat{f}(n)| \ln n\},$$

and suppose

$$C(R) = \{\mu \in M(G) : \{|\hat{\mu}(n)| \ln n\} \text{ is bounded}\}.$$

Then $C(R) = M_F(R)$, that is $C(R) \cong M(L_1(R), F(R))$.

**Proof.** It follows from Theorem 1 that $\mu \in M_F(R)$ if and only if

$$\|u_m * \mu\|_F = \|u_m * \mu\|_1 + \sup_n\{|\hat{u}_m(n)\hat{\mu}(n)| \ln n\} \leq C_\mu$$

$$(m = 1,2,3,\ldots).$$

Let $u_m$ be a Fejér kernel $(m = 1,2,3,\ldots)$, thus $\|u_m * \mu\|_1 \leq \|\mu\|$ and

$$\hat{u}_m(t) = \begin{cases} 1 - \frac{|t|}{m}, & |t| \leq m, \\ 0, & |t| > m. \end{cases}$$

Then $\sup_n \{|\hat{u}_m(n)\hat{\mu}(n)| \ln n\} \leq C_\mu$ iff $\{|\hat{\mu}(n)| \ln n\}$ is bounded. The proof is complete.

**Corollary 3.** Suppose the Segal algebra $S_f(R)$ is the following:

$$S_f(R) = \{g \in L_1(R) : f * g \in C_0(R)\},$$

$$\|g\|_{S_f} = \|g\|_1 + \|f*g\|_\infty,$$

where $f \in L_1(R)$ and $f \neq 0$, $C_0(R)$ is a linear space consisting of all the continuous functions on R which approach to zero at infinity with the norm $\| \ \|_\infty$, and suppose

$$D(R) = \{\mu \in M(R) : \mu * f \in L_\infty(R)\}.$$

Then $D(R) = M_{S_f}(R)$, that is $D(R) = M(L_1(R), S_f(R))$.

**Proof.** It follows from Theorem 1 that $\mu \in M_{S_f}(R)$ if and only if

$\|u_n*\mu\|_{S_f} = \|u_n*\mu\|_1 + \|u_n*\mu*f\|_\infty \leq C_\mu$ $(n = 1,2,3,\ldots)$. But

$$\|u_n*\mu\|_1 + \|u_n*\mu*f\|_\infty \leq \|\mu\| + \|u_n*\mu*f\|_\infty,$$

Let $u_n$ be a Fejér kernel $(n = 1,2,3,\ldots)$; notice that $\mu * f \in L_1(R)$, $u_n * \mu * f$ is the (C,1) sum of the Fourier integral of $\mu * f$. Then $\|u_n*\mu*f\|_\infty$ is bounded iff $\|\mu*f\|_\infty < \infty$. The proof is complete.

## §2. Multipliers of Segal algebra $A_p(G)$

Let G be a locally compact but noncompact abelian group, and

$$A_p(G) = \{f \in L_1(G) : \hat{f} \in L_p(\hat{G})\}, \quad 1 \leq p < \infty,$$

$$\|f\|_{A_p} = \|f\|_1 + \|\hat{f}\|_p.$$

It has been known that $A_p(G)$ is a Segal algebra.

We have known that $M(A_p(G), A_p(G)) \cong M(G)$ [5].

In § 1 of this paper, we have proved $M(L_1(G), A_p(G)) \cong M_{A_p}(G)$. In this section, $M(A_p(G), L_1(G))$ will be considered. Suppose $f \in P(L_1(G))$, where

$$P(L_1(G)) = \{f \in L_1(G) : \hat{f} \in C_c(\hat{G})\},$$

$C_c(\hat{G})$ is a linear space consisting of all the continuous functions on $\hat{G}$ with the compact support. $P(L_1(G))$ is dense in every Segal algebra $S(G)$.

Suppose $\tau_{x_0}, \tau_{x_1}, \ldots, \tau_{x_n}, \ldots$ are translation operators, $x_0 = e$ the identity of G. Define

$$f_n = \frac{\tau_{x_0} f + \tau_{x_1} f + \ldots + \tau_{x_n} f}{n+1}.$$

We have

$$\hat{f}_n \in C_c(\hat{G}) \text{ and } f_n \in P(L_1(G)).$$

For any $\varepsilon > 0$, there exists $g \in C_c(G)$ such that

$$\|f - g\|_1 < \varepsilon.$$

Let

$$g_n = \frac{\tau_{x_0} g + \tau_{x_1} g + \ldots + \tau_{x_n} g}{n+1}.$$

By the homogeneous structure of $L_1(G)$, it is easy to see that

$$\|f_n - g_n\|_1 < \varepsilon.$$

**Lemma.** Suppose G is a locally compact but noncompact abelian group. There exists a sequence $x_0, x_1, \ldots, x_n, \ldots$ in G such that

(a) $\|g_n\|_2 \to 0, \quad (n \to \infty)$.

(b) $\|f_n\|_2 \to 0, \quad (n \to \infty)$.

(c) There is a subsequence $x_{n_1}, x_{n_2}, \ldots, x_{n_k}, \ldots$, such that $\|\hat{f}_{n_k}\|_p \to 0$, $(k \to \infty)$, $1 \leqslant p < \infty$.

**Proof.** Suppose supp $g = K$, K is a compact set in G. There is a sequence $x_0, x_1, \ldots, x_n, \ldots$ . in G such that

$$(K_{x_i}) \cap (K_{x_j}) = \emptyset, \quad i \neq j.$$

Hence we obtain

$$\text{supp } \tau_{x_i} g \cap \text{supp } \tau_{x_j} g = \emptyset, \ i \neq j,$$

$$\|g_n\|_2^2 = \frac{1}{(n+1)^2} \int_G \{ \sum_{i=0}^{n} [\tau_{x_i} g]^2 + 2 \sum_{0 \leqslant i < j \leqslant n} \tau_{x_i} g \cdot \tau_{x_j} g \} d\mu(x).$$

But

$$\int_G [\tau_i g(x)]^2 d\mu(x) = \|g\|_2^2,$$

$$\int_G \tau_{x_i} g(x).\tau_{x_j} g(x)\, d\mu(x) = 0, \ i \neq j.$$

Therefore

$$\|g_n\|_2^2 = \frac{1}{(n+1)^2}(n+1)\,\|g\|_2^2 \to 0 \quad (n \to \infty).$$

Then (a) has been proved. For (b), we have

$$\|f_n\|_2 \leq \|f_n - g_n\|_2 + \|g_n\|_2.$$

Notice that $f - g$ is a bounded function on $G$ since $g \in C_c(G)$ and $f \in C_0(G)$. Suppose

$$\|f(x) - g(x)\|_\infty \leq C, \quad C \text{ is a constant.}$$

Therefore

$$\|f_n - g_n\|_2^2 \leq C\,\|f_n - g_n\|_1 \to 0, \quad (n \to \infty).$$

(b) has been proved.

At last we will prove (c). $f_n \in P(L_1(G))$ and $\|f_n\|_2 \to 0$ $(n \to \infty)$, implying that $\|\hat{f}_n\|_2 \to 0$ $(n \to \infty)$. Hence there exists a subsequence $\{f_{n_k}\}$ of $\{f_n\}$, $\hat{f}_{n_k}(\gamma) \to 0$ $(k \to \infty)$ almost everywhere on $\hat{G}$. But every $\hat{f}_{n_k}$ is continuous, so $\hat{f}_{n_k}(\gamma) \to 0$ $(k \to \infty)$ for every point on $\hat{G}$.

Notice

$$\hat{f}_n(\gamma) = \frac{1+(x_1,\gamma)+\ldots+(x_n,\gamma)}{n+1}\,\hat{f}(\gamma),$$

$|(x_i,\gamma)| \leq 1$, $i = 1, 2, \ldots$, obviously,

$$|\hat{f}_{n_k}(\gamma)| \leq |\hat{f}(\gamma)|, \ \gamma \in \hat{G}.$$

But $\hat{f}_{n_k} \in C_c(\hat{G})$ and $\hat{f} \in C_c(\hat{G})$, Lebesgue Dominated Convergence Theorem indicates

$$\|\hat{f}_{n_k}\|_p \to 0 \quad (k \to \infty).$$

The lemma has been proved.

**Theorem 2.** Suppose G is a locally compact but noncompact abelian group, then

$$M(A_p(G),\ L_1(G)) \cong M(G).$$

where $\cong$ is of the same meaning as in §1.

**Proof.** Suppose $T \in M(A_p(G), L_1(G))$. For every $f \in P(L_1(G))$, $Tf \in L_1(G)$. Hence for every $\varepsilon > 0$ there exists $s \in G$ such that

$$\|Tf + \tau_s(Tf)\|_1 \geq 2\|Tf\|_1 - \varepsilon,$$

or we can say there exists a sequence $x_0, x_1, \ldots, x_n, \ldots, \{x_n\} \subset G$, such that

$$\|\tau_{x_0}(Tf) + \tau_{x_1}(Tf) + \ldots + \tau_{x_n}(Tf)\|_1 \geq (n+1)\|Tf\|_1 - n\varepsilon.$$

Besides, we can elect $\{x_n\}$ satisfying the condition of the lemma. Notice that T is a multiplier, so we have

$$\|\tau_{x_0}(Tf) + \tau_{x_1}(Tf) + \ldots + \tau_{x_n}(Tf)\|_1 = (n+1)\|Tf_n\|_1$$

$$\|Tf\|_1 \leq \|Tf_n\|_1 + \varepsilon \leq \|T\| . \|f_n\|_{A_p} + \varepsilon$$

$$\leq \|T\|(\|f_n\|_1 + \|\hat{f}_n\|_p) + \varepsilon.$$

Now we elect a subsequence $\{f_{n_k}\}$ of $\{f_n\}$ making (c) of the lemma tenable. Therefore, according to the following inequality:

$$\|Tf\|_1 \leq \|T\| (\|f_{n_k}\|_1 + \|\hat{f}_{n_k}\|_p) + \varepsilon$$

and as $L_1(G)$ is a homogeneous Banach algebra, we obtain

$$\|Tf\|_1 \leq \|T\| \ \|f\|_1 \quad \text{for every } f \in P(L_1(G)).$$

It indicates that T defines a bounded linear operator from $P(L_1(G))$ to $L_1(G)$ and T commutes with any translation. T defines a unique bounded linear operator T (we still use the same sign to denote it) from $L_1(G)$ to $L_1(G)$ because $P(L_1(G))$ is dense in $L_1(G)$. T commutes with translation, and the norm doesn't change. According to a known theorem about the multipliers of $L_1(G)$, there exists a unique measure $\mu \in M(G)$ such that

$$Tf = \mu * f \quad \text{for every } f \in L_1(G).$$

Conversely, for $\mu \in M(G)$, define an operator T by

$$Tf = \mu * f \quad \text{for every } f \in A_p(G).$$

Obviously, $\mu * f \in L_1(G)$ because $L_1(G)$ is an ideal in $M(G)$. Then, according to the property of convolution, we can verify that T is a bounded linear operator from $A_p(G)$ to $L_1(G)$ and T commutes with any translation, i.e. $T \in M(A_p(G), L_1(G))$.

So much for the proof of Theorem 2.

If G is a discrete group, Theorem 2 will become the following theorem.

**Theorem 3.** Suppose G is discrete abelian group, then

$$M(A_p(G), L_1(G)) \cong A_p(G).$$

**Proof.** When G is discrete, we have

$$A_p(G) = L_1(G) = M(G),$$

$L_1(G)$-norm and $M(G)$-norm are the same. Because

$$\|f\|_{A_p} = \|f\|_1 + \|\hat{f}\|_p, \quad \|\hat{f}\|_\infty \leqslant \|f\|_1,$$

and $\hat{G}$ is compact, we have $\|\hat{f}\|_p \leqslant C\|f\|_1$, where C is a constant. Therefore we obtain that $L_1(G)$-norm and $A_p(G)$-norm are equivalent.

Suppose $T \in M(A_p(G), L_1(G))$ and e is an identity of $A_p(G) = L_1(G)$.

Obviously,

$$Tf = T(e * f) = Te * f \text{ for every } f \in A_p(G),$$

and $\|T\| = \|Te\|_M = \|Te\|_1$, $Te \in A_p(G)$.

Conversely, let $g \in A_p(G)$, define

$$Tf = g * f \quad \text{for every } f \in A_p(G).$$

It is easy to verify that $T \in M(A_p(G), L_1(G))$.

## §3. Segal algebra $A_{1,p}(G)$ and its multipliers

Denote the space which consists of the following functions by $\overline{A}_p(G)$:

$$f = \sum_{i=1}^{\infty} u_i * \check{v}_i, \quad u_i \in L_p(G), \quad v_i \in L_q(G),$$

where $1 < p, q < \infty$, $\frac{1}{p} + \frac{1}{q} = 1$, $\check{v}_i(x) = v_i(x^{-1})$ and

$$\sum_{i=1}^{\infty} \|u_i\|_p \|v_i\|_q < \infty.$$

The norm on $\overline{A}_p(G)$ is defined by

$$\|f\|_{\overline{A}_p} = \inf\{\sum_{i=1}^{\infty} \|u_i\|_p \|v_i\|_q\}.$$

Now we will discuss an important subalgebra $A_{1,p}(G)$ of $\overline{A}_p(G)$:

$$A_{1,p}(G) = \overline{A}_p(G) \cap L_1(G) ;$$

its norm is defined by

$$\|f\|_{A_{1,p}} = \|f\|_1 + \|f\|_{\overline{A}_p}.$$

G. Herz [2] has proved that $\overline{A}_p(G)$ is a Banach algebra under the pointwise multiplication. Then Herz [3] has proved that $\overline{A}_p(G)$ is a regular Tauberian algebra. Lai and Chen [4] have proved that $A_{1,p}(G)$ is a commutative Banach algebra under the pointwise multiplication. Now we will discuss the space $A_{1,p}(G)$ under convolution.

**Theorem 4.** $A_{1,p}(G)$ is a character Segal algebra with convolution product.

**Proof.** Only the following facts need to be proved.

(a) For any $f, g \in A_{1,p}(G)$, $f * g \in A_{1,p}(G)$ and $\|f*g\|_{A_{1,p}} \leqslant \|f\|_{A_{1,p}} \cdot \|g\|_{A_{1,p}}$.

(b) $A_{1,p}(G)$ is dense in $L_1(G)$.

(c) For every $f \in A_{1,p}(G)$ and every $\gamma \in \hat{G}$, $\gamma f \in A_{1,p}(G)$ and $\|\gamma f\|_{A_{1,p}} = \|f\|_{A_{1,p}}$.

Now, we will prove these facts.

(a) It is obvious that $f * g \in L_1(G)$ for every $f, g \in A_{1,p}(G)$. Let

$$f = \sum_{i=1}^{\infty} u_i * \check{v}_i, \quad g = \sum_{i=1}^{\infty} s_i * \check{t}_i,$$

$$u_i, s_i \in L_p(G), \quad v_i, t_i \in L_q(G), \quad \frac{1}{p} + \frac{1}{q} = 1.$$

$$\sum_{i=1}^{\infty} \|u_i\|_p \|v_i\|_q < \infty, \qquad \sum_{i=1}^{\infty} \|s_i\|_p \|t_i\|_q < \infty,$$

then

$$f * g = \sum_{i=1}^{\infty} (f * s_i) * t_i ;$$

$f \varepsilon L_1(G)$ and $s_i \varepsilon L_p(G)$, then $f * s_i \varepsilon L_p(G)$ and $\|f*s_i\|_p \leqslant \|f\|_1 \|s_i\|_p$. But $t_i \varepsilon L_q(G)$, hence $f * g \varepsilon A_{1,p}(G)$.

Besides,

$$\begin{aligned} \|f*g\|_{A_{1,p}} &= \|f*g\|_1 + \|f*g\|_{\overline{A}_p} \\ &\leqslant \|1\|_1 \|g\|_1 + \|f\|_1 \sum_{i=1}^{\infty} \|s_i\|_p \cdot \|t_i\|_q \\ &\leqslant (\|f\|_1 + \sum_{i=1}^{\infty} \|u_i\|_p \|v_i\|_q)(\|g\|_1 + \sum_{i=1}^{\infty} \|s_i\|_p \|t_i\|_q). \end{aligned}$$

For every $\varepsilon > 0$, there exist $u_i'$, $v_i'$ $s_i'$, $t_i'$, such that

$$f = \sum_{i=1}^{\infty} u_i' * \check{v}_i' , \qquad g = \sum_{i=1}^{\infty} s_i' * \check{t}_i' ,$$

$u_i'$, $s_i' \varepsilon L_p(G)$, $v_i'$, $t_i' \varepsilon L_q(G)$ , $\frac{1}{p} + \frac{1}{q} = 1$, and above inequalities hold, and

$$\sum_{i=1}^{\infty} \|u_i'\|_p \|v_i'\|_q \leqslant \|f\|_{A_p} + \varepsilon,$$

$$\sum_{i=1}^{\infty} \|s_i'\|_p \|v_i'\|_q \leqslant \|g\|_{\overline{A}_p} + \varepsilon.$$

Thus we obtain

$$\|f*g\|_{A_{1,p}} \leqslant \|f\|_{A_{1,p}} \cdot \|g\|_{A_{1,p}} .$$

(b) Notice that $L_1(G)$ has the factorization property, so there exist $g$, $h \varepsilon L_1(G)$ such that $f = g * h$ for $f \varepsilon L_1(G)$. Besides, $C_c(G)$ is dense in $L_1(G)$, so for any $\varepsilon > 0$ there exist $\tilde{g}$, $\tilde{h} \varepsilon C_c(G)$ such that

$$\|g - \tilde{g}\|_1 < \varepsilon, \qquad \|h - \tilde{h}\|_1 < \varepsilon.$$

Hence we obtain

$$\|f - \tilde{g} * \tilde{h}\|_1 = \|g * h - \tilde{g} * \tilde{h}\|_1 < c\varepsilon.$$

But $\tilde{g} * \tilde{h} \varepsilon C_c(G) * C_c(G)$, it indicates that $C_c(G) * C_c(G)$ is dense in $L_1(G)$.

Since

$$C_c(G) * C_c(G) \subset A_{1,p}(G) \subset L_1(G),$$

(b) has been proved.

(c) Let $\gamma \in \hat{G}$, $f = \sum_{i=1}^{\infty} u_i * \check{v}_i$,

$$\sum_{i=1}^{\infty} \|u_i\|_p \|v_i\|_q \leq \|f\|_{\overline{A}_p} + \varepsilon.$$

$$\gamma f = \sum_{i=1}^{\infty} \gamma u_i * \check{v}_i.$$

But $\gamma u_i \in L_p(G)$ and $\|\gamma u_i\|_p = \|u_i\|_p$, thus $\gamma f \in A_{1,p}(G)$. Besides, by $\|\gamma f\|_1 = \|f\|_1$, we have

$$\|\gamma f\|_{A_{1,p}} \leq \|f\|_{A_{1,p}} + \varepsilon, \quad \text{i.e.} \quad \|\gamma f\|_{A_{1,p}} \leq \|f\|_{A_{1,p}}.$$

Repeat the above process and notice $\gamma^{-1}\gamma = 1$, so we obtain

$$\|f\|_{A_{1,p}} = \|\gamma^{-1}(\gamma f)\|_{A_{1,p}} \leq \|\gamma f\|_{A_{1,p}} < \|f\|_{A_{1,p}},$$

a contradiction.

Thus Theorem 4 has been proved.

**Theorem 5.** Let G be a locally compact but noncompact abelian group. Then

$$M(A_{1,p}(G), L_1(G)) \cong M(G).$$

**Proof.** Let $T \in M(A_{1,p}(G), L_1(G))$. For each $f \in A_{1,p}(G)$ and any $\varepsilon > 0$, there exist $x_0, x_1, \ldots, x_n, \ldots$ in G, $x_0 = e$ the identity of G, such that

$$\|Tf\|_1 \leq \|Tf_n\|_1 + \varepsilon \leq \|Tf_n\|_{A_{1,p}} + \varepsilon$$

$$\leq \|T\| \left(\|f_n\|_1 + \|f_n\|_{\overline{A}_p}\right) + \varepsilon$$

where $f_n = \dfrac{\tau_{x_0} f + \tau_{x_1} f + \ldots + \tau_{x_n} f}{n+1}$. The proof is similar to the proof of Theorem 2.

Let

$$f = \sum_{i=1}^{\infty} u_i * \check{v}_i, \quad u_i \in L_p(G), \quad v_i \in L_q(G),$$

$$\sum_{i=1}^{\infty} \|u_i\|_p \|v_i\|_p < \infty, \qquad \frac{1}{p} + \frac{1}{q} = 1.$$

For $\varepsilon > 0$, there exists an integer N such that

$$\sum_{i=N+1}^{\infty} \|u_i\|_p \|v_i\|_p < \varepsilon.$$

Let

$$g_N = \sum_{i=1}^{N} u_i * \check{v}_i,$$

we have $\|f-g_N\|_{\overline{A}_p} < \varepsilon$. Since $C_c(G)$ is dense in $L_p(G)$ and in $L_q(G)$, there exists $\alpha_i$, $\beta_i \in C_c(G)$, $i = 1, 2, \ldots, N$, such that

$$\|u_i-\alpha_i\|_p < \frac{\varepsilon}{M}, \quad \|v_i-\beta_i\|_q < \frac{\varepsilon}{M},$$

$$M = \max\left(\sum_{i=1}^{N} \|u_i\|_p, \quad \sum_{i=1}^{N} \|v_i\|_q, N\right).$$

Let

$$\psi = \sum_{i=1}^{N} c_i * \check{\beta}_i,$$

we have

$$g_N-\psi = \sum_{i=1}^{N} u_i * (\check{v}_i-\check{\beta}_i) - \sum_{i=1}^{N}(u_i-\alpha_i) * (\check{v}_i-\check{\beta}_i)$$

$$+ \sum_{i=1}^{N}(u_i-\alpha_i) * \check{v}_i,$$

$$\|g_N-\psi\|_{\overline{A}_p} \leqslant \sum_{i=1}^{N} \|u_i\|_p \cdot \|v_i-\beta_i\|_q + \sum_{i=1}^{N} \|u_i-\alpha_i\|_p \cdot \|v_i-\beta_i\|_q$$

$$+ \sum_{i=1}^{N} \|u_i-\alpha_i\|_p \|v_i\|_q$$

$$< 3\varepsilon.$$

Thus

$$\|f-\psi\|_{\overline{A}_p} \leqslant \| f-g_N\|_{\overline{A}_p} + \|g_N-\psi\|_{\overline{A}_p} < 4\varepsilon.$$

Therefore

$$\|f_n\|_{\overline{A}_p} \leqslant \frac{1}{n+1} [\|\tau_{x_o} f+\tau_{x_1} f+\ldots+\tau_{x_n} f-(\tau_{x_o}\psi+\ldots+\tau_{x_n}\psi)\|_{\overline{A}_p} +$$

$$+ \|\tau_{x_o}\psi+\ldots+\tau_{x_n}\psi\|_{\overline{A}_p}]$$

$$\leqslant \frac{1}{n+1}[4(n+1)\varepsilon + \|\tau_{x_o}+\ldots+\tau_{x_n}\psi\|_{\overline{A}_p}].$$

Therefore

$$\|Tf\|_1 \leq \|T\| \, [\|f_n\|_1 + 4\varepsilon + (n+1)^{\frac{1}{p}-1} \sum_{i=1}^{N} \|\alpha_i\|_p \cdot \|\beta_i\|_q].$$

Notice that $p > 1$, also $\varepsilon$ and $n$ are arbitrary, and $\|f_n\|_1 \leq \|f\|_1$, so we obtain

$$\|Tf\|_1 \leq \|T\| \; \|f\|_1.$$

Then it is similar to the proof of Theorem 2, the difference is that: we have used the density of $P(L_1(G))$ in $L_1(G)$ in the proof of Theorem 2, but here we use the density of the Segal algebra $A_{1,p}(G)$ in $L_1(G)$. We can obtain that there exists a unique measure $\mu \in M(G)$ such that

$$Tf = \mu * f \quad \text{for every } f \in A_{1,p}(G).$$

Conversely, let $\mu \in M(G)$. Define an operator T by

$$Tf = \mu * f \quad \text{for every } f \in A_{1,p}(G).$$

It is easy to verify that $T \in M(A_{1,p}(G), L_1(G))$.

Theorem 5 has been proved.

**Theorem 6.** Let G be a locally compact but noncompact abelian group. Then

$$M(A_{1,p}(G), A_{1,p}(G)) \cong M(G).$$

**Proof.** Since

$$\|f\|_1 \leq \|f\|_{A_{1,p}} \quad \text{for every } f \in A_{1,p}(G),$$

each $T \in M(A_{1,p}(G), A_{1,p}(G))$ defines a unique element in $M(A_{1,p}(G), L_1(G))$. Then according to Theorem 5, there exists a unique measure $\mu \in M(S)$ such that

$$Tf = \mu * f \quad \text{for every } f \in A_{1,p}(G)$$

and $\|\mu\| = \|T\|$.

Conversely, if $\mu \in M(G)$, then it is obvious that $\mu * f \in A_{1,p}(G)$ for every $f \in A_{1,p}(G)$. So we can see that $A_{1,p}(G)$ is an ideal in the measure algebra $M(G)$ and

$$\|\mu*f\|_{A_{1,p}} \leq \|\mu\| . \|f\|_{A_{1,p}} \quad \text{for every } f \in A_{1,p}(G).$$

Define an operator T by

$$Tf = \mu * f \quad \text{for every } f \in A_{1,p}(G).$$

It is easy to see that $T \in M(A_{1,p}(G), A_{1,p}(G))$.

The proof of Theorem 6 has been accomplished.

When G is a discrete group, as in the proof of Theorem 3, we have:

**Theorem 7.** If G is a discrete abelian group, then

$$M(A_{1,p}(G), L_1(G)) \cong A_{1,p}(G),$$

$$M(A_{1,p}(G), A_{1,p}(G)) \cong A_{1,p}(G).$$

Some open problems:

1. For each Segal algebra S(G), what is the character of multipliers from S(G) to $L_1(G)$ ?

2. Unni [7] has proved that: For each $T \varepsilon M(S(G), S(G))$, there exists a unique pseudomeasure $\sigma \varepsilon P(G)$, such that $Tf = \sigma * f$ for every $f \varepsilon S(G)$. Denote the whole of the above pseudomeasures $\sigma$ by $P_s(G)$. It is probable that $P_s(G)$ is only a proper subset of P(G). For example, $P_s(G) = M(G)$ for some Segal algebras. Now the problem is that: What is a necessary and sufficient condition describing $P_s(G)$ ?

3. More generally, let $S_1(G)$ and $S_2(G)$ be two Segal algebras. What is the character of $M(S_1(G), S_2(G))$ ?

**References**


[1] Goldberg, R. R. and Seltzer, S. E., Uniformly concentrated sequences and multipliers of Segal algebras, J. Math. Anal. and Appl., 59(1977), 488-497.

[2] Herz, C., Theory of p-space with an application to convolution operators, Trans. Amer. Math. Soc. 154(1971), 69-82.

[3] Herz, C., Harmonic synthesis for subgroups, Ann. Inst. Fourier (Grenoble), 23(1973), 91-123.

[4] Lai, H. and Chen, I., Harmonic analysis on the Fourier algebra $A_{1,p}(G)$, J. Austral. Math. Soc. (Series A) 30(1981), 438-452.

[5] Larsen, R., An introduction to the theory of multipliers, Springer-Verlag (1971).

[6] Ouyang, G., Multiplier operators from $L_1(G)$ to a Segal algebra, Chinese Annals of Math. Vol.5 (Ser. A), 2(1984), 247-252 (Chinese).

[7] Unni, L. R., A note on multipliers of a Segal algebra, Studia Math. 99(1974), 125-127.

Proceedings of the Analysis Conference, Singapore 1986
S.T.L. Choy, J.P. Jesudason, P.Y. Lee (Editors)



# Differentiation in Banach spaces

Minos Petrakis and J. J. Uhl, Jr.

## 1. Part 1. Norm differentiation.

This section is a commentary on that part of mathematics having its origin with the following question of Tamarkin's:

Tamarkin's Question: What are the Banach spaces $X$ such that each Lipschitz function $f : [0,1] \to X$ is norm differentiable almost everywhere? To get some familiarity with the question, we'll look at some examples.

Example 1. The function $f : [0,1] \to L_1[0,1]$ defined by $f(t) = \chi_{[0,t]}$, $0 \leq t \leq 1$ is Lipschitz but is nowhere differentiable.

To check this, note that if $0 \leq s \leq t \leq 1$, then $\|f(t) - f(s)\|_1 = |t - s|$; therefore $f$ is Lipschitz. Also if $0 < h$ and $0 \leq t < 1$, then

$$\lim_{h \to 0} \frac{f(t+h) - f(t)}{h} = \lim_{h \to 0} \frac{\chi_{[t,t+h]}}{h} = 0$$

in measure but not in $L_1$-norm because $\left\|\frac{\chi_{[t,t+h]}}{h}\right\|_1 = 1$ for small $h$.

Example 2. The function $f : [0,1] \to c_0$ defined by $f(t) = \left(\int_0^t \sin(n\pi x)dx\right)_{n=1}^{\infty}$ $0 \leq t \leq 1$, is Lipschitz but not differentiable a.e..

Note that evidently for $0 \leq s \leq t \leq 1$

$$\|f(t) - f(s)\|_{c_0} \leq \int_0^t 1dx = |t - s|;$$

therefore $f$ is Lipschitz. But if $f'(\bar{t})$ exists at some $\bar{t}$ in $[0,1]$, then

$$f'(\bar{t}) = (\sin(n\pi\bar{t})) \in c_0$$

which is impossible except at certain selected $\bar{t}$ from a null set.

Example 3. If $f : [0,1] \to \ell_1$ is Lipschitz, then $f$ is differentiable almost everywhere.

To see why, write

$$f = (f_1, f_2, \ldots, f_n, \ldots)$$

and compute variation of $f = \sum_{n=1}^{\infty}$ variation of $f_n = \sum_{n=1}^{\infty} \int_0^1 |f'_n| = \int_0^1 \sum_{n=1}^{\infty} |f'_n|$. This shows $(f'_n(t))_{n=1}^{\infty} \in \ell_1$ for almost all $t$ in $[0,1]$ and it is not much more work to show

$$f'(t) = f'_n(t))$$

for almost all $t$.

Upon seeing this little argument, Dunford and Morse [6] in 1936 came up with the following definition.

Definition 4. A Banach space $X$ has a boundedly complete basis $(x_n)$ if

(a) Each $x \in X$ has the unique expansion

$$x = \sum_{n=1}^{\infty} \alpha_n x_n$$

where the $\alpha_n$'s are real numbers uniquely determined by $x$, and

(b) If $(\beta_n)$ is a sequence of real numbers and

$$\sup_m \| \sum_{n=1}^{m} \beta_n x_n \| < \infty,$$

then the series

$$\sum_{n=1}^{\infty} \beta_n x_n$$

is convergent in the norm of $X$.

Now suppose $f : [0,1] \to X$ is Lipschitz, and write

$$f(t) = \sum_{n=1}^{\infty} f_n(t)x_n.$$

Use (b) to show

$$\sum_{n=1}^{\infty} f_n'(t)x_n$$

is convergent in $X$ for almost all $t$ and then directly compute

$$f'(t) = \sum_{n=1}^{\infty} f_n'(t)x_n \quad \text{a.e.}$$

This proves Dunford's and Morse's fifty year old theorem:

Theorem 5. *If $X$ has a boundedly complete basis, then every Lipschitz $f : [0,1] \to X$ is differentiable almost everywhere.*

Most familiar Banach spaces (e.g. $\ell_p$ $1 \leq p < \infty$) have boundedly complete bases. Putting (1) and (2) together with (4) proves that *no basis of $c_0$ or $L_1$ is boundedly complete*. This shows that differentiation can be used to study individual Banach spaces. This is correct and much of the work of the last twenty years is motivated by this little observation.

The first response ever given to Tamarkin's question was by Tamarkin's student Clarkson. Recall that a Banach space $X$ is *uniformly convex* if for each $\varepsilon > 0$ there is a $\delta > 0$ such that

$$\|x - y\| < \varepsilon \quad \text{provided} \quad \|x\| = \|y\| = 1$$

and

$$\|x + y\| > 1 - \delta.$$

In 1936, Clarkson [3] introduced the notion of uniform convexity to prove the following theorem

<u>Example</u> 5. <u>If</u> $X$ <u>is uniformly convex</u>, <u>then every Lipschitz</u> $f : [0,1] \to X$ <u>is differentiable a.e.</u>

Clarkson's theorem marked one of the very first investigations into the geometry of Banach spaces. It provides a concrete link between differentiation and geometry and shows, among other things, that neither $c_0$ nor $L_1$ have equivalent unifomrly convex norms.

In case it is not obvious, what we have been talking about is a property of Banach spaces now well known as the Radon-Nikodym property.

<u>Definition</u> 7. A Banach space has the <u>Radon-Nikodym property</u> (RNP) if every Lipschitz $f : [0,1] \to X$ is differentiable a.e.

To get more information on RNP, we'll set up some machinery, the idea of which is present in the classic 1940 paper of Dunform and Pettis [8].

Take a Lipschitz function $f : [0,1] \to X$ and for each partition $\pi$ of $[0,1]$ into intervals, set

$$f_\pi = \sum_{[a,b]\in\pi} \frac{f(b) - f(a)}{b - a} \chi_{[a,b)}$$

where $\chi_{[a,b)}$ is the characteristic or indicator function of $[a,b)$. Note that $\|f_\pi\|_\infty$ is no greater than the Lipschitz constant for $f$.

<u>Definition</u> 8. Let $(\pi_n)$ be a sequence of partitions of $[0,1]$ such that for all $n$ the partition $\pi_{n+1}$ refines $\pi_n$. If

$$\lim_{n\to\infty} \max_{[a,b]\in\pi_n} |b - a| = 0,$$

we call $(f_{\pi_n})$ a <u>difference quotient martingale</u>. It is not hard to prove the following theorem.

<u>Theorem</u> 9. <u>A Lipschitz function</u> $f : [0,1] \to X$ <u>is differentiable everywhere if and only if there is a difference quotient martingale</u> $(f_{\pi_n})$

$$\lim_{m,n\to\infty} \int_{[0,1]} \|f_{\pi_n} - f_{\pi_m}\|_X = 0.$$

<u>In which case</u> $\lim_n f_{\pi_n} = f'$ a.e. <u>and all other difference quotient martingales converge to</u> $f'$ <u>as well</u>.

Some might recognize Theorem 9 is a primitive version of the martingale convergence theorem but this sets us off our track.

Suffice it to say that once Theorem 9 is understood, then the Chatterji-Ionescu-Tuleca (see [3,13]) characterization of RNP via martingales is an immediate corollary.

Not so easy but not terribly hard is Pettis's theorem:

<u>Theorem</u> 10. <u>Suppose</u> $X$ <u>is separable</u>. <u>A Lipschitz function</u> $f : [0,1] \to X$ <u>is differentiable almost everywhere if and only if there is a function</u> $g : [0,1] \to X$ <u>such that</u>

$$x^*(f(t) - f(0)) = \int_0^t x^* g$$

<u>for all</u> $x^*$ <u>in a separating subset of the dual space</u> $X^*$. <u>In this case</u>, $f' = g$ a.e.

Pettis used this theorem to prove that an absolutely continuous $f : [0,1] \to X$ is differentiable almost everywhere in norm if and only if $f$ is differentiable weakly at each point of a subset of $[0,1]$ whose complement is Lebesgue null. Here are some related consequences. The first is due to Dunford and Pettis [8].

<u>Corollary</u> 12. <u>Separable dual spaces have</u> RNP.

<u>Proof</u>. Let $f : [0,1] \to X$ be Lipschitz and let $(f_{\pi_n})$ be a difference quotient martingale. Since $\|f_{\pi_n}\|_\infty$ is no greater than the Lipschitz constant for $f$, we see that for each $t$ in $[0,1]$, the sequence $(f_{\pi_n}(t))$ is bounded. Let $g(t)$ be an arbitrary weak $^*$-cluster point of $(f_{\pi_n}(t))$. From real variable theory,

$$(f(t) - f(0))x = \int_0^t g \cdot x$$

for all $x$ in $X$. Apply Pettis's theorem to learn $f' = g$ a.e.

The next corollary is due to Dunford-Pettis-Phillips.

Corollary 13. Reflexive spaces have RNP.

Since Lipschitz functions on $[0,1]$ have separable ranges, it is clear that a Banach space has RNP if and only if each of its separable subspaces has RNP. But each separable subspace of a reflexive space is a separable dual space and hence has the RNP.

Proved by a similar argument is the next corollary which has been described by some as the Uhl Tool.

Corollary 14. If $X$ is a Banach space such that each separable subspace of $X$ has a separable dual, then the dual $X^*$ has RNP.

A famous theorem of Stegall's [19] shows that the converge of Corollary 14 is true and hence the statement of Corollary 14 characterizes the RNP in dual spaces.

It is well known that neither $L_1$ nor $c_0$ are dual spaces. The usual argument proceeds by showing that the unit ball of each contains no extreme points. But all this shows is that in their given norms they are not isometric to dual spaces. The above theorems can be used to give more information, a fact first noted by Gelfand.

Corollary 15. Neither $c_0$ nor $L_1$ are isomorphic to subspaces of a separable dual space.

Now let's take up the question of RNP and geometry. One way to block the RNP for a Banach space $X$ is to find a Lipschitz function $f : [0,1] \to X$ that has a difference quotient martingale $(f_{\pi_n})$ such that there is a $\delta > 0$ with the property that

$$\|f_{\pi_{n+1}}(t) - f_{\pi_n}(t)\|_X \geq \delta$$

for all $t \in [0,1)$ and all $n = 1,2,\ldots$ . Let's see what this means for the dyadic partitions:

$$f_{\pi_1} = (f(1) - f(0))\chi_{[0,1]} = x_1\chi_{[0,1]}$$

$$f_{\pi_2} = \frac{f(1/2) - f(0)}{1/2 - 0}\chi_{[0,1/2)} + \frac{f(1) - f(1/2)}{1 - 1/2}\chi_{[1/2,1)}$$

$$= x_2\chi_{[0,1/2)} + x_3\chi_{[1/2,1)}$$

$$f_{\pi_3} = x_4\chi_{[0,1/4)} + x_4\chi_{[1/4,1/2)} + x_5\chi_{[1/2,3/4)} + x_6\chi_{[3/4,1)},$$

etc. Now note

$$x_1 = \frac{x_2 + x_3}{2}$$

and in general

$$x_n = \frac{x_{2n} + x_{2n+1}}{2} ;$$

i.e. the sequence $(x_n)$ is a bounded tree in X. Also for the dyadic partitions $\pi_n$,

$$\|f_{\pi_{n+1}}(t) - f_{\pi_n}(t)\| = \begin{cases} \|x_k - x_{2k}\| \\ \text{or } \|x_k - x_{2k+1}\| \end{cases}$$

depending on whether $t$ is in the left hand side or the right hand side of the interval in $\pi_n$ in which $t$ is found.

<u>Definition</u> 16. A bounded tree $(x_n)$ in a Banach space is called a $\delta$-tree if there exists a $\delta > 0$ such that

$$\|x_n - x_{2n}\| \geq \delta$$

and

$$\|x_n - x_{2n+1}\| \geq \delta.$$

Here are two easy examples of $\delta$-trees:

<u>Example</u> 17. <u>A $\delta$-tree in</u> $L_1[0,1]$

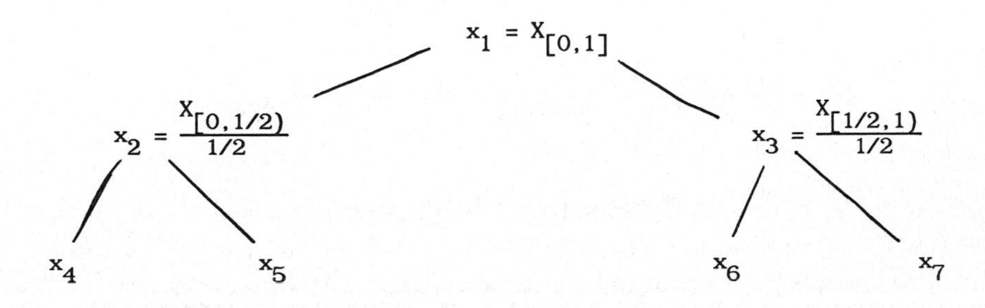

Example 18. A $\delta$-tree in $c_0$

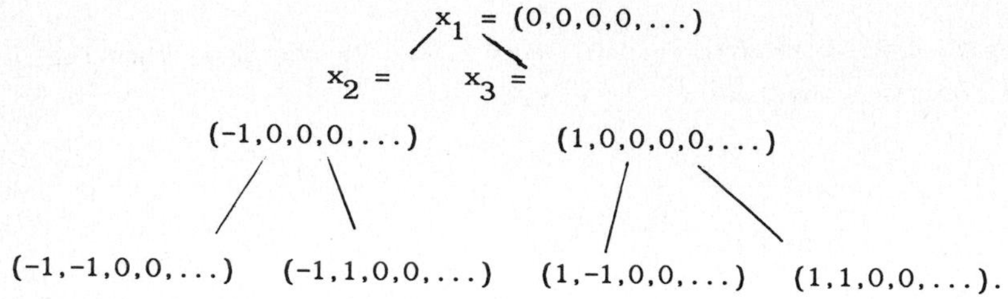

$x_1 = (0,0,0,0,\dots)$

$x_2 =$ $(-1,0,0,0,\dots)$ $x_3 =$ $(1,0,0,0,0,\dots)$

$(-1,-1,0,0,\dots)$ $(-1,1,0,0,\dots)$ $(1,-1,0,0,\dots)$ $(1,1,0,0,\dots)$.

Note that if $X$ contains a $\delta$-tree $(x_n)$, then reading the above backwards and supplying a small continuity argument produces a Lipschitz $f : [0,1]$ such that

$$\|f_{\pi_n}(t) - f_{\pi_{n+1}}(t)\|_X \geq \delta$$

for all $t \in [0,1)$ and $n = 1,2,\dots$ . This proves

Theorem 19. If $X$ contains a bounded $\delta$-tree, then $X$ fails RNP.

Stegall [19] proved the converse dual Banach spaces. Attempts to block non-dyadic difference quotient martingales give rise to the idea of a $\delta$-bush.

One way to help a Banach space to have RNP is to eliminate the possibility of $\delta$-trees or $\delta$-bushes. The idea is due to Rieffel [16].

Definition 20. A subset $D$ of a Banach space is not dentable if there exists an $\varepsilon > 0$ such that $x \in D$ implies

$$x \in \overline{c_0}(D \setminus B_\varepsilon(x)).$$

Here "$\overline{c_0}$" stands for closed convex hull and $B_\varepsilon(x) = \{y \in X : \|y - x\| < \varepsilon\}$. A slice of a set $D$ is any set of the form

$$\{x \in D : x^*(x) \geq \alpha\}$$

for a fixed $x^* \in X^*$ and any real $\alpha$ strictly less than $\sup|x^*(D)|$. The separation theorem guarantees that a subset $D$ of $X$ is dentable if and only if it has slices of arbitrarily small diameter. The next theorem amalgamates theorems of Rieffel [16], Huff [11] and Davis-Phelps [5].

<u>Theorem</u> 21. <u>A Banach space has</u> RNP <u>if and only if every bounded subset of</u> X <u>is dentable</u>. <u>A Banach space</u> X <u>fails</u> RNP <u>if and only if it contains a</u> $\delta$-<u>bush</u>.

The proof boils down to showing that if every subset is dentable, then for some difference quotient martingale $(f_{\pi_n})$ satisfies

$$\lim_{n,m} \int_{[0,1]} \|f_{\pi_n} - f_{\pi_m}\| = 0.$$

Conversely if X contains a non-dentable subset, it is possible to define a Lipschitz $f : [0,1] \to X$ such that

$$\int_{[0,1]} \|f_{\pi_{n+1}} - f_{\pi_n}\|$$

is bounded away from 0 for all $n$. It should be noted here that there are (non-dual) Banach spaces that fail RNP yet have no $\delta$-trees. The first known example of this is the Bourgain-Rosenthal space BR [2].

Related to dentability is the Krein-Milman property.

<u>Definition</u> 22. A Banach space has the Krein-Milman property (KMP) if every closed bounded convex set in X has an extreme point

A separation theorem argument together with the Bishop-Phelps theorem guarantees that if X has KMP, then every closed bounded convex subset of X is the norm-closed convex hull of its extreme points.

A simple and beautiful fact due to Lindenstrauss follows:

<u>Theorem</u> 23. <u>If</u> X <u>has</u> RNP <u>then</u> X <u>has</u> KMP.

<u>Idea of Proof</u>. Take a closed bounded convex set D.

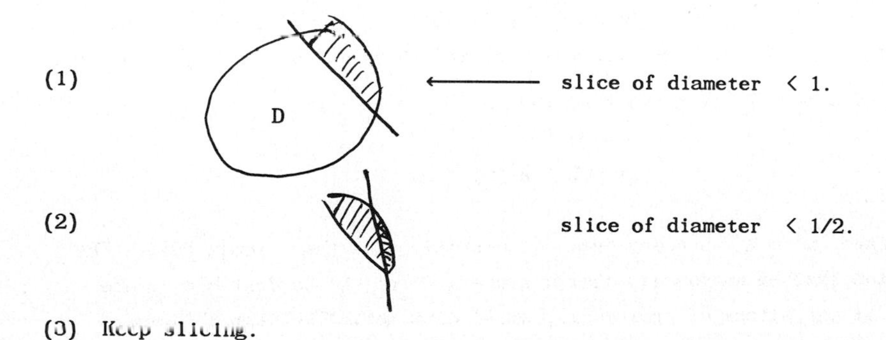


(3) Keep slicing.

The slices are nested and their diameters tend to zero. By completeness, they all contain a common point; this point is an extreme point of $D$.

One outstanding question is

Question 24. Are KMP and RNP equivalent properties?

Huff and Morris [12] proved that for dual spaces, the two properties are equivalent. For non-dual spaces this problem has remained highly resistant to a solution. And it is not from lack of attention that this problem is not yet solved. The implication KMP $\Rightarrow$ RNP is vexing because there is no *a priori* reason to believe KMP is separably determined (see the proof of Corollary 13 to see why RNP is separably determined). Nevertheless Schachermayer [18] has obtained the following tantalizing facts.

(1) *If $X$ is isomorphic to its square, then KMP and RNP are equivalent for $X$.*

(2) *The KMP in dual spaces is separably determined.*

(3) *If $\ell_2(X)$ has KMP, then $X$ has RNP!*

Another way of studying RNP is via operators on $L_1[0,1]$. Let $T : L_1[0,1] \to X$ be a bounded linear operator .ne by

$$\phi(t) = T(\chi_{[0,t]}).$$

Note that if $0 \leq \leq \leq t \leq 1$, then

$$\|\phi(s) - \phi(t)\|_X = \|T(\chi_{(s,t]})\|_X$$

$$\leq \|T\| \, \|\chi_{(s,t]}\|_1 = \|T\| \, |t - s|.$$

Therefore $\phi$ is Lipschitz with constant $\|T\|$. Similarly, Lebesgue-Steiltjes integration with respect to a Lipschitz function $\phi : [0,1] \to X$ defines a bounded linear operator $T : L_1[0,1] \to X$ for which

$$\phi(t) = T(\chi_{[0,t]})$$

for all $t$ in $[0,1]$.

Definition 25. A bounded linear operator $T : L_1[0,1] \to X$ is called *representable* if the norm derivative

$$\frac{d}{dt} T(X_{[0,t]})$$

exists almost everywhere. In view of the discussion above, we have

Corollary 26. A Banach space has RNP if and only if all bounded linear operators from $L_1[0,1]$ into $X$ are representable.

Let's take a short look at the meaning of representability. Take an operator $T : L_1[0,1] \to X$ and let

$$\phi(t) = T(X_{[0,t]}) \quad 0 \leq t \leq 1.$$

If $T$ is representable and $f \in L_1[0,1]$ is a step function, then

$$Tf = \int_{[0,1]} f d\phi = \int_{[0,1]} f\phi' .$$

The upshot of this is that

$$Tf = \int_{[0,1]} f\phi' \quad \text{(Bochner integral)}$$

for all $f \in L_1[0,1]$. In other words, a bounded liner operator $T : L_1[0,1] \to X$ is representable if and only if there is a bounded measurable $g : [0,1] \to X$ (all derivatives are measurable) such that

$$Tf = \int_{[0,1]} fg$$

for all $f \in L_1[0,1]$. Weakly compact operators on $L_1[0,1]$ are representable for precisely the same reason that reflexive spaces have RNP. Here are some facts.

Fact 27. Representable operators are almost compact.

Proof. Let $T : L_1[0,1] \to X$ have the action

$$Tf = \int_{[0,1]} fg$$

for a bounded measurable $g : [0,1] \to X$. Let $\varepsilon > 0$ and choose a sequence $(g_n)$ of simple functions converging to $g$ a.e. Use Egorov's theorem to find a measurable set $E \subseteq [0,1]$ whose comploement has measure less than $\varepsilon$ such that $\lim_{n\to\infty} g_n = g$ uniformly on $E$. Note that if $f \in L_1[0,1]$, then

$$Tf = \int_E fg + \int_{[0,1]\setminus E} fg$$

and that

$$f \to \int_E fg$$

is compact because it is the operator limit of finite rank operators. Thus $T$ is almost compact in this sense.

The following fact is not well known, but it has been around for many years. Our proof follows that of Gretsky-Uhl [10].

Corollary 28. Weakly compact subsets of $L_\infty[0,1]$ are almost compact.

Proof. Let $W \subseteq L_\infty[0,1]$ be weakly compact. Given $\varepsilon > 0$ we want to find a measurable set $E \subseteq [0,1]$ such that $[0,1] \setminus E$ has measure less than $\varepsilon$ and such that

$$\{f\chi_E : f \in W\}$$

is compact in $L_\infty[0,1]$.

To this end, find a reflexive Banach space $R$ and an operator $T : R \to L_\infty$ such that

$$T(B_R) \supseteq W.$$

(Here $B_R$ is the closed unit ball of $R$.) Let $S = T^*$ restricted to $L_1[0,1]$. Since $R^*$ has RNP, the operator

$$S : L_1 \to R^*$$

is representable. A short computation shows $S^* = T$. By Fact 27, there is a measurable set $E \subseteq [0,1]$ such that $[0,1] \setminus E$ has measure less than $\varepsilon$ and such

$$f \to S(fX_E)$$

is compact from $L_1[0,1]$ to $R^*$. Therefore

$$(S(\cdot\ X_E))^*\ X_E T : R \to L_\infty[0,1]$$

is compact. Thus

$$\{fX_E : f \in w\} = X_E \cdot w \subseteq X_E T(B_R)$$

is compact in $L_\infty[0,1]$. This completes the proof.

Here is a quick corollary.

Corollary 29. Weakly convergent sequences in $L_\infty[0,1]$ are almost everywhere convergent.

The last theorem we'll look at in this section is an old chesnut due to Dunford and Pettis. Let us agree that a bounded linear operator $T : L_1[0,1] \to X$ is a Dunford-Pettis operator if $T$ sends weakly compact sets onto norm compact sets.

Corollary 30. Representable operators on $L_1[0,1]$ are Dunford-Pettis operators.

Proof. Let $W \subseteq L_1[0,1]$ be weakly compact and recall that this ensures that

$$\lim_{\mu(E)\to 0} \sup_{f \in w} \int_E |f| = 0.$$

Let $\varepsilon > 0$ and use Fact 27 and the weak compactness of $W$ to find a measurable subset $E$ of $[0,1]$ such that $f \to T(fX_E)$ is compact on $L_1[0,1]$ and

$$\sup_{f \in E} \int_{[0,1] \setminus E} |f| < \frac{\varepsilon}{2\|T\|}.$$

Now $T(X_E W)$ is compact and hence it can be covered by finitely many balls of radius $\varepsilon/2$. Since every point in $T(W)$ is within $\varepsilon/2$ of a point in $T(X_E W)$, we see that $T(W)$ can be covered by finitely many balls of radius $\varepsilon$. Therefore $T(W)$ is totally bounded.

Part II. Dunford-Pettis operators on $L_1[0,1]$. This section is devoted to a study of Dunford-Pettis operators on $L_1[0,1]$. The main purpose of this section is to make a case for the position that maybe Dunford-Pettis operators on $L_1[0,1]$ can be profitably studied by a program modeled on the RNP study of Part I.

The main fact about Dunford-Pettis operators is the following elementary observation:

Fact 1. A bounded linear operator $T : L_1[0,1] \to X$ is a Dunford-Pettis operator if and only if the restriction $T : L_\infty[0,1]$ is compact.

Proof. If $T : L_1[0,1] \to X$ is Dunford-Pettis, then $T(B_{L_\infty})$ is compact because $B_{L_\infty}$ is a weakly compact subset of $L_1[0,1]$.

For the converse, suppose $T : L_1[0,1]$ is a bounded linear operator such that $T : L_\infty[0,1] \to X$ is compact. Let $W$ be a weakly compact subset of $L_1[0,1]$ and recall that $W$ is uniformly integrable. Hence

$$\lim_n \int_{[|f|\geq n]} |f| = 0$$

uniformly in $f \in W$. let $\varepsilon > 0$ and pick $n_0$ such that

$$\sup_{f\in W} \int_{[|f|\geq n_1]} |f| < \frac{\varepsilon}{2\|T\|}.$$

Now

$$T(w) \subseteq \{T(fX_{[|f|\leq n_0]})$$

$$+ T(fX_{[|f|>n_0]}) : f \in W\}.$$

Since $T : L_\infty[0,1] \to X$ is compact it follows that

$$\{T(f\chi_{[|f|\leq n_0]}) : f \in W\}$$

is relatively compact and therefore can be covered by finitely many balls of radius $\varepsilon/2$. From the discussion above, we see that $T(W)$ can be covered by finitely many balls of radius $\varepsilon$. This proves $T(W)$ is relatively compact and hence compact.

The possibility of a strong analogy between representable operators and Dunford-Pettis operators is contained in the next result.

Theorem 2. *Let* $T : L_1[0,1] \to X$ *be a bounded linear operator. For* $0 \leq t \leq 1$, *set*

$$f(t) = T(\chi_{[0,t]})$$

*and let* $(f_{\pi_n})$ *be a difference quotient martingale. Then*

(a) *the operator* $T$ *is representable if and only if* $(f_{\pi_n})$ *is* $L_1$-*Cauchy*; i.e.

$$\lim_{m,n} \int_{[0,1]} \|f_{\pi_n} - f_{\pi_m}\| = 0,$$

and

(b) *the operator* $T$ *is Dunford-Pettis if and only if* $(f_{\pi_n})$ *is Pettis Cauchy*; i.e.

$$\lim_{m,n} \sup_{\|x^*\|\leq 1} \int_{[0,1]} |x^* f_{\pi_m} - x^* f_{\pi_n}| = 0.$$

Proof. The equivalence (a) is just Theorem 1.9. To prove (b), define the operators $E_{\pi_n}$ on $L_\infty[0,1]$ by

$$E_{\pi_n}(\phi) = \sum_{I\in\pi_n} \frac{\int_I f\,d\mu}{\mu(I)} \chi_I$$

where $\mu(I)$ is the Lebesgue measure of the interval $I$. Recall that a bounded subset $K$ of $L_1[0,1]$ is relatively norm compact if and only if $\lim_n E_{\pi_n}(f) = f$ uniformly in $f \in k$. Now $T : L_\infty[0,1] \to X$ is compact if and only if $T^* : X^* \to L_1[0,1]$ is compact which, in turn, is equivalent to saying

$$\lim_n E_{\pi_n} T^* x^* = T^* x^*$$

uniformly in $\|x^*\| \leq 1$. This is the same as saying

$$\lim_{m,n} \|E_{\pi_m} T^* - E_{\pi_n} T^*\| = 0$$

for the operator norm. But a quick calculation shows that for fixed $m$ and $n$

$$\|E_{\pi_m} T^* - E_{\pi_n} T^*\| = \sup_{\|x^*\| \leq 1} \int_{[0,1]} |x^* f_{\pi_m} - x^* f_{\pi_n}|.$$

Putting it all together proves that $T : L_\infty[0,1] \to X$ is compact if and only if

$$\lim_{n,m} \sup_{\|x^*\| \leq 1} \int_{[0,1]} |x^* f_{\pi_n} - x^* f_{\pi_m}| = 0.$$

A glance at Fact 1 completes the proof.

Definition 3. A Banach space $X$ has the complete continuity property (CCP) if each $T : L_1[0,1] \to X$ is a Dunford-Pettis operator.

The property CCP is also known in the literature as CRP (compact range property).

In terms of differentiation, a Banach space $X$ has CCP if and only if every Lipschitz $f : [0,1] \to X$ is scalarly uniformly differentiable in one sense that for all difference quotient martingales

$$\lim_n \int_{[0,1]} |x^* f_{\pi_n} - \frac{d}{dt}(x^* f)| = 0$$

uniformly in $\|x^*\| \leq 1$.

Evidently any Banach space with RNP has CCP, but the converse is false as can be seen from the following theorem, which is an easy consequence of Rosenthal's fundamental $\ell_1$ theorem.

Theorem 4. Let $X$ be a separable Banach space. The dual $X^*$ has CCP if and only if $X$ contains no copy of the space $\ell_1$.

Proof. Suppose $X$ contains no copy of $\ell_1$. Let $T : L_1[0,1] \to X$ be a bounded linear operator. By a compactness argument on the difference quotient martingales $(f_{\pi_n})$ (for $f(t) = T(X_{[0,t]})$ in the style the proof of Corollary 12 produces a bounded function $g : [0,1] \to X^*$ (satisfying $\frac{d}{dt} f(t) \cdot x = x \cdot g$ for all $x \in X$) such that for each $\phi \in L_1[0,1]$

$$T\phi x = \int_{[0,1]} \phi g \cdot x \quad \text{for all } x \text{ in } X.$$

Define an operator $S : X \to L_1[0,1]$ by $Sx = g \cdot x$. Let $(x_n)$ be a bounded sequence in $X$. Since $X$ contains no copy of $\ell_1$, Rosenthal's theorem tells us that $(x_n)$ has a weakly Cauchy subsequence $(x_{n_j})$. Observe that since $g : [0,1] \to X^*$, the sequence $(g \cdot x_{n_j})$ is a pointwise conversent subsequence of the $L_\infty$-bounded sequence $(g \cdot x_n)$. The bounded conversence theorem guarantees that $(g \cdot x_{n_j})$ is conversent in $L_1$-norm. Hence the operator $S : X \to L_1[0,1]$ is compact and $S^* : L_\infty[0,1] \to X^*$ is also compact. But a brief computation shows $S^*$ is exactly the restriction of $T$ to $L_\infty[0,1]$. Hence $T : L_\infty[0,1] \to X$ is compact. This proves $T : L_1[0,1] \to X$ is Dunford-Pettis. For the rest of the proof, see Riddle-Uhl [15].

Using this theorem, we can see that if $X$ is the James Tree space, then $X^*$ fails RNP, but $X^*$ has CCP.

Next we shall take a look at a tree-structure that plays the same role for CCP as the $\delta$-tree structure plays for RNP.

Definition 5. Let $\delta > 0$. A $\delta$-Rademacher tree is a bounded sequence $(x_n)$ in $X$ such that

(a) $x_n = \dfrac{x_{2n} + x_{2n+1}}{2}$

and

(b) $\|x_1\| \geq \delta$

$\|x_2 - x_3\| \geq 2\delta$

$\|x_4 - x_5 + x_6 - x_7\| \geq 4\delta,$ etc.

Both the trees given in Examples 1.17 and 1.18 are 1-Rademacher trees. Using a procedure similar to the discussion preceeding Definition 1.16, we can see that a Banach space X contains a $\delta$-Rademacher tree if and only if there is a Lipschitz function $f : [0,1] \to X$ such that the dyadic difference quotient martingale $(f_{\pi_n})$ satisfies

$$\sup_{\|x^*\| \geq 1} \int_{[0,1]} |x^* f_{\pi_{n+1}} - x^* f_{\pi_n}|$$

$$\geq \left\| \int_{[0,1]} f_{\pi_{n+1}} - f_{\pi_n} \right\| \geq \delta \quad \text{for all } n.$$

This proves:

<u>Fact</u> 5. <u>If a Banach space contains a $\delta$-Rademacher tree, then</u> X <u>fails</u> CCP.

The converse is open but is known to be true in dual spaces, see Riddle-Uhl [15]. We feel that the converse is unlikely to be true in non-dual spaces but have no concrete evidence to offer.

With regard to dentability, we can offer the following definition.

<u>Definition</u> 6. A subset D of a Banach space X is not <u>weak-norm-one dentable</u> if there exists an $\varepsilon > 0$ such that for all finite subsets F of D there is a norm one $x_F^*$ in $X^*$ such that if $x \in F$, then

$$x \in \overline{co}(D \cap \{z \in X : |x_F^*(z - x)| \geq \varepsilon).$$

<u>Theorem</u> 7. <u>If the Banach space</u> X <u>contains a bounded set that is not weak-norm-one dentable, then</u> X <u>fails</u> CCP.

<u>Proof</u>. Refer to the proof of Huff [11] and note that a suitable modification of Huff's proof gives a difference quotient martingale $(f_{\pi_n})$ and a sequence $(x_n^*)$ of norm-one members of $X^*$ such that for some $\varepsilon > 0$

$$|x_n^* f_{\pi_{n+1}} - x_n^* f_{\pi_n}| \geq \varepsilon$$

for all $n = 1,2,\ldots$

Hence for all $n = 1,2,\ldots$

$$\sup_{\|x^*\|\leq 1} \int_{[0,1]} |x^* f_{\pi_{n+1}} - x^* f_{\pi_n}| \geq \varepsilon$$

and this combined with Theorem 2 shows that $X$ fails CCP. The converse is open. Another (possibly related) dentability condition is given in the next definition.

Definition 8. Let $D$ be a bounded subset of a Banach space $X$. Let $\varepsilon$ and $\lambda$ be positive real numbers. A point $x_{\varepsilon,\lambda}$ in $D$ is called an $\varepsilon - \lambda$ denting point if whenever

$$x_{\varepsilon,\lambda} = \sum_{i=1}^{\infty} \sigma_i x_i$$

with $x_i \in D$, $\sigma_i \geq 0$ and $\sum_{ii=1}^{\infty} \sigma_i = 1$, then for each norm-one $x^* \in X^*$

$$\sum_{i\in A_{x^*}} \sigma_i < \varepsilon$$

where the index set $A_{x^*}$ is defined by

$$A_{x^*} = \{i : |x^*(x_i) - x^*(x)| > \lambda\}$$

with a little care, one can prove the following fact.

Fact 9. If every bounded subset of $X$ has an $\varepsilon - \lambda$ denting point for all $\varepsilon,\lambda > 0$, then $X$ has CCP.

The proof is not terribly difficult. It involves taking a Lipschitz $f : [0,1] \to X$, letting $\varepsilon \to 0$ and $\lambda \to 0$ and using the $\lambda$'s to define partitions $\pi_n$ such that

$$\lim_{m,n} \sup_{\|x^*\|\leq 1} \int_{[0,1]} |x^* f_{\pi_n} - x^* f_{\pi_n}| = 0.$$

We'll give an alternate proof of Fact 9 a bit later.

Next we come to a theorem of Bourgain's [1] which we do not believe is yet understood to the full.

Theorem 10. If a Banach space $X$ fails CCP, then $X$ contains a $\delta$-tree.

The $\delta$-tree produced by Bourgain's argument is not the tree corresponding to a non-Dunford-Pettis operator from $L_1[0,1]$ into $X$ but rather the operator corresponding to Bourgain's tree is tree related to a Dunford-Pettis operator. We feel that Bourgain's tree has another unnoticed property that may in fact characterize CCP. After many hours of sweat and grief we are still unable to isolate this property.

Bourgain's theorem has the following immediate corollary.

Corollary 11. If a Banach space $X$ contains no $\delta$-trees, then $X$ has CCP.

The converse of Corollary 11 is false. The dual $JT^*$ of the James Tree space has CCP and does contain $\delta$-trees.

Still there is another condition due to Kunen and Rosenthal [17] that eliminates $\delta$-trees.

Definition 12. Let $\varepsilon > 0$ and let $D$ be a bounded subset of a Banach space $X$. A point $x$ in $D$ is an $\varepsilon$-strongly extreme point of $D$ if there is a $\delta > 0$ such that if $x_1$, $x_2$ are in $D$ and there is a point $u$ on the line segment from $x_1$ to $x_2$ with the property that $\|u - x\| < \delta$, then $\|u - x_1\| < \varepsilon$ or $\|u - x_2\| < \varepsilon$.

A Banach space $X$ has the approximate Krein-Milman property (AKMP) if every non-empty bounded subset of $X$ has an $\varepsilon$-strongly extreme point for every $\varepsilon > 0$.

Rosenthal and Kunen [17] proved that if a Banach space $X$ has the approximate Krein-Milman property, then $X$ contains no $\delta$-trees. Combining this with Bourgain's theorem proves the following observation.

Theorem 13. If a Banach space $X$ has the approximate Krein-Milman property, then $X$ has CCP.

Since the Bourgain-Rosenthal space BR has AKMP, this space also has CCP. On the other hand, the space BR fails KMP and hence the AKMP does not imply KMP. In addition the following question is wide-open.

Question 14. Does a Banach space have CCP if it has KMP?

This question is more modest than question 24, although a negative answer to it would also give a negative answer to question 24.

Theorem 13 can also be used to give an alternate proof of Fact 9 because the hypothesis of Fact 9 rules out $\delta$-trees. One surviving question is: If every closed bounded convex subset of X has an $\varepsilon - \lambda$ denting point for all $\varepsilon, \lambda > 0$, then X does X have CCP? How about the converse?

Finally we mention recent work of Ghoussoub, Godefroy, Maurey and Schachermayer [18]. Say that a Banach space X is strongly regular if for every bounded subset D of X and for every $\varepsilon > 0$, there exist slices $S_1, S_2, \ldots, S_n$ of D and positive real numbers $\sigma_1, \sigma_2, \ldots, \sigma_n$ whose sum is one such that

$$\mathrm{diam}\left(\sum_{i=1}^{n} \alpha_i S_i\right) < \varepsilon.$$

They proved

Theorem 15. If X is a strongly regular Banach space, then X has CCP.

Our point in this section is simple: Our point is that the evidence suggests the strong possibility that CCP is an internal geometric property in the same way that RNP is. Still we must ask

Question 16. What is this internal geometric property?

## REFERENCES

[ 1] J. Bourgain, Dunford-Pettis operators on $L_1$ and the Radon-Nikodym property, Israel J. Math. 37(1980).

[ 2] J. Bourgain and H. P. Rosenthal, Applications of the theory of semi-embeddings to Banach space theory, J. Funct. Anal. 52(1983).

[ 3] S. D. Chatterji, Martingale conversence and the Radon-Nikodym theorem in Banach spaces, Math. Scand. 22(1968), 21-41.

[ 4] J. A. Clarkson, Uniformly convex Banach spaces, Trans. Amer. Math. Soc. 40(1936), 396-414.

[ 5] W. J. Davis and R. R. Phelps, The Radon-Nikodym property and dentable sets in Banach spaces, Proc. Amer. Math. Soc. 45(1974), 119-122.

[ 6] J. Diestel and J. J. Uhl, Jr. Vector measures, Math Surveys no. 15, Amer. Math. Soc. Providence 1977.

[ 7] N. Dunford and A. P. Morse, Remarks on the preceding paper of James A. Clarkson, Trans. Amer. Math. Soc. 40(1936), 415-420.

[ 8] N. Dunford and B. J. Pettis, Linear operations on summable functions, Trans. Amer. Math. Soc. 47(1940), 323-392.

[ 9] N. Ghoussoub, G. Godefroy and B. Maurey, First class functions around sets and geometrically regular Banach spaces (preprint).

[10] N. E. Gretsky and J. J. Uhl, Jr. Korotkov and Carleman operators on Banach spaces, Acta Sci. Math 43(1981), 207-219.

[11] R. E. Huff, Dentability and the Radon-Nikodym property in Banach spaces, Duke Math. J. 41(1976), 111-114.

[12] R. E. Huff and P. D. Morris, Dual spaces with the Krein-Milman property have the Radon-Nikodym property, Proc. Amer. Math. Soc. 49(1975), 104-108.

[13] A. and C. Ionescua-Tulcea, Abstract ergodic theorems, Trans. Amer. Math. Soc. 107(1963), 107-124.

[14] B. J. Pettis, Differentiation in Banach spaces, Duke Math. J. 5(1939), 254-269.

[15] L. H. Riddle and J. J. Uhl, Jr., Martingales and the fine line between Asplund spaces and spaces not containing a copy of $\ell_1$. Martingale theory in Harmonic Analysis and Banach spaces, Springer-Verlag Lecture Notes v. 939, 1982.

[16] M. A. Rieffel, Dentable subsets of Banach spaces with application to a Radon-Nikodym theorem, Proc. Conf. Irvine, Calif. 1966(B. R. Gelbaum, editor) Thompson, Washington D. C., 71-77.

[17] K. Kunen and H. Rosenthal, Martingale proofs of some geometric results in Banach space theory, Pacific J. Math 100(1982).

[18] W. Schachermayer, The Radon-Nikodym property and Krein-Milman property, (preprint) 1986.

[19] C. Stegall, The Radon-Nikodym property in conjugate Banach spaces, Trans. Amer. Math. Soc. 206(1975), 213-223.

Proceedings of the Analysis Conference, Singapore 1986
S.T.L. Choy, J.P. Jesudason, P.Y. Lee (Editors)




# "SPECTRAL SUBSETS" OF $\mathbb{R}^m$ ASSOCIATED WITH COMMUTING FAMILIES OF LINEAR OPERATORS

Werner RICKER *

School of Mathematics and Physics
Macquarie University
North Ryde, 2113
Australia

In some recent papers [5,6,7] McIntosh and Pryde introduced and gave some applications of a notion of "spectral set", $\gamma(\underset{\sim}{T})$, associated with each finite, commuting family of bounded linear operators $\underset{\sim}{T}$ in a Banach space. Unlike most concepts of joint spectrum, the set $\gamma(\underset{\sim}{T})$ is part of *real* Euclidean space. As such, there arises the question of the non-emptiness of $\gamma(\underset{\sim}{T})$. It is known that if there is an even number of operators involved or if each operator $T_j \in \underset{\sim}{T}$ has real spectrum, then necessarily $\gamma(\underset{\sim}{T})$ is non-empty. In this note it is shown that $\gamma(\underset{\sim}{T})$ is non-empty for an arbitrary finite, commuting family of operators $\underset{\sim}{T}$ ( consisting of at least two operators ) whenever the underlying Banach space is finite dimensional. It is still unresolved whether $\gamma(\underset{\sim}{T})$ is non-empty in general.

Let X be a complex Banach space and $\underset{\sim}{T} = (T_1,\ldots,T_m)$ be a commuting m-tuple of continuous linear operators in X; the space of all such operators in X is denoted by L(X). A joint spectral set $\gamma(\underset{\sim}{T})$, of $\underset{\sim}{T}$, is defined by

$$\gamma(\underset{\sim}{T}) = \{(\lambda_1,\ldots,\lambda_m) \in \mathbb{R}^m ;\ \Sigma_{j=1}^m (T_j-\lambda_j)^2 \text{ is not invertible in } L(X)\},$$

where $\mathbb{R}$ denotes the real numbers [5]. This notion has proved to be useful in determining functional calculi for certain m-tuples $\underset{\sim}{T}$ with applications to finding estimates for the solution of linear systems of operator equations [5,6,7].

It is known that $\gamma(\underset{\sim}{T})$ is always a compact subset of $\mathbb{R}^m$ [7; Theorem 4.1]. The question arises of whether or not $\gamma(\underset{\sim}{T})$ is empty ? If $m = 1$, then it is easy to verify that $\gamma(T) = \sigma(T) \cap \mathbb{R}$ and hence, it may happen that $\gamma(T)$ is empty in this case. For $m \geq 2$, it is known that $\gamma(\underset{\sim}{T}) \neq \phi$ if each operator $T_j$, $1 \leq j \leq m$, has *real* spectrum. Indeed, in this case $\gamma(\underset{\sim}{T})$ coincides with the Taylor spectrum of $\underset{\sim}{T}$ [8; Theorem 1] and so is certainly non-empty. Furthermore, McIntosh and Pryde have shown, by a judicious use of the theory of Clifford analysis and monogenic functions, that $\gamma(\underset{\sim}{T}) \neq \phi$ for arbitrary commuting m-tuples $\underset{\sim}{T}$ whenever m is an *even* integer [7; §3]. The aim of this talk is to show that $\gamma(\underset{\sim}{T}) \neq \phi$ for arbitrary commuting m-tuples $\underset{\sim}{T}$ and integers $m \geq 2$ whenever X is finite dimensional. It is still unresolved whether $\gamma(\underset{\sim}{T}) \neq \phi$ in general ( for $m \geq 2$ ).

* The assistance provided by a research travel grant under the Queen Elizabeth II Fellowship Scheme is gratefully acknowledged.

Let X be a complex Banach space, not necessarily finite dimensional. Let $\mathcal{B}$ denote the σ-algebra of Borel subsets of the complex plane $\mathbb{C}$. An operator S in L(X) is a *scalar-type spectral operator* if there exists a spectral measure $P: \mathcal{B} \to L(X)$, supported by $\sigma(S)$, such that

$$S = \int_{\mathbb{C}} z dP(z) = \int_{\sigma(S)} z dP(z) ,$$

where the integral exists in the usual sense of integration with respect to a σ-additive vector measure [3; p.1938]. To say that P is a spectral measure means that P is σ-additive when L(X) is equipped with the strong operator topology, $P(E \cap F) = P(E)P(F)$ for each $E,F \in \mathcal{B}$ and $P(\mathbb{C}) = P(\sigma(S))$ is the identity operator in X. The measure P, necessarily unique [3; XV Corollary 3.8], is called the *resolution of the identity of* S.

An operator $T \in L(X)$ is a *spectral operator* if there exists a spectral measure $P: \mathcal{B} \to L(X)$, necessarily unique, such that T belongs to the commutant of $\{P(E);\ E \in \mathcal{B}\}$ in L(X) and the spectrum of the restriction of T to each closed invariant subspace $P(E)X$, $E \in \mathcal{B}$, is contained in the closure of E in $\mathbb{C}$ [3; p.1930]. This is equivalent to the existence of a scalar-type spectral operator S, whose resolution of the identity is necessarily P, and a quasi-nilpotent operator Q commuting with S such that $T = S + Q$ [3; XV Theorem 4.5]. Furthermore, this decomposition is unique; S is called the *scalar part* of T and Q the *radical part* of T.

Let $\tilde{T} = (T_1,\dots,T_m)$ be a commuting m-tuple in $L(X)^m$ consisting of spectral operators. Let $P_j$ be the resolution of the identity of $T_j$, $1 \le j \le m$. Then define

$$\tilde{P}(E_1 \times \dots \times E_m) = P_1(E_1)P_2(E_2)\dots P_m(E_m) , \tag{1}$$

for each measurable rectangle $E_1 \times \dots \times E_m$ in $\mathbb{C}^m$, where each $E_j \in \mathcal{B}$, $j = 1,\dots,m$. Noting that the measures $P_j$, $1 \le j \le m$, all commute [3; XV Corollary 3.7], it follows that $\tilde{P}$ has an extension to a multiplicative, additive set function on the algebra of sets, $\mathcal{A}$, generated by the measurable rectangles [4; §4]. The extension of $\tilde{P}$ to $\mathcal{A}$ is again denoted by $\tilde{P}$. Observing that the σ-algebra generated by $\mathcal{A}$ is the Borel σ-algebra, $\mathcal{B}_m$, of $\mathbb{C}^m$ it follows that $\tilde{P}$ has an extension to an *unique* spectral measure $\tilde{P}: \mathcal{B}_m \to L(X)$ satisfying (1), in which case it is called the *joint resolution of the identity of* $\tilde{T}$, if and only if, for each $x \in X$ the set $\{\tilde{P}(E)x;\ E \in \mathcal{A}\}$ is a relatively weakly compact subset of X; see Theorem 8 of [4], for example. If the Banach space X is weakly sequentially complete, then the product measure $\tilde{P}$ exists whenever $\{\tilde{P}(E);\ E \in \mathcal{A}\}$ is a bounded subset of L(X) [4; Theorem 9]. In certain Banach spaces, $\{\tilde{P}(E);\ E \in \mathcal{A}\}$ is automatically bounded for *every* finite family of commuting spectral measures $P_1,\dots,P_m$. For example, this is known to be the case if X is finite dimensional or a Hilbert space or an $L^p$-space: see pp.2098-2101 of [3] for a more comprehensive list of such spaces. We remark that if the joint resolution of the identity of $\underset{\sim}{T}$, say $\underset{\sim}{P}$, exists, then there is available a multiplicative functional calculus based on the algebra of $\underset{\sim}{P}$-essentially bounded functions $f:\mathbb{C}^m \to \mathbb{C}$ and defined via integration, namely $f \to \int_{\mathbb{C}^m} f d\underset{\sim}{P}$, [3; XVII §2]. In particular, $S_j = \int_{\mathbb{C}^m} \lambda_j d\underset{\sim}{P}(\underset{\sim}{\lambda})$ where $S_j$ is the scalar part of $T_j$, $1 \le j \le m$. The support of $\underset{\sim}{P}$, denoted by $\mathrm{Supp}(\underset{\sim}{P})$, is the complement of the largest open set $U \subseteq \mathbb{C}^m$ such that $\underset{\sim}{P}(U) = 0$.

THEOREM 1. *Let X be a Banach space and $\underset{\sim}{T}$ be a commuting m-tuple in $L(X)^m$ consisting of spectral operators. Suppose that the joint resolution of the identity of $\underset{\sim}{T}$ exists. If $m \ge 2$, then $\gamma(\underset{\sim}{T})$ is non-empty.*

REMARKS. (1) Since every linear operator in a finite dimensional space is a spectral operator and the joint resolution of the identity always exists in finite dimensional spaces it follows from Theorem 1 that $\gamma(\underset{\sim}{T})$ is non-empty for every commuting m-tuple $\underset{\sim}{T}$ in such spaces ( $m \ge 2$ ).

(2) If X is a Hilbert space and $m \ge 2$, then $\gamma(\underset{\sim}{T}) \ne \phi$ for every commuting m-tuple of normal operators.

(3) If $X = L^p(\mu)$, $1 \le p < \infty$, where $\mu$ is a $\sigma$-finite measure and $T_j$ is the operator in X of multiplication by $\psi_j \in L^\infty(\mu)$, $j = 1,\dots,m$, then the joint resolution of the identity of $\underset{\sim}{T}$ exists and hence $\gamma(\underset{\sim}{T}) \ne \phi$.

The proof of Theorem 1 is via a series of lemmas.

LEMMA 1. *Let $\underset{\sim}{T}$ be a commuting m-tuple in $L(X)^m$ consisting of spectral operators. If $S_j$ is the scalar part of $T_j$, $1 \le j \le m$, then $\gamma(\underset{\sim}{T}) = \gamma(\underset{\sim}{S})$.*

*Proof.* Let $Q_j$ be the radical part of $T_j$, $1 \le j \le m$. It follows from [3; XV Corollary 3.7], the definition of spectral operator and the commutativity of the family $\{T_j;\ 1 \le j \le m\}$ that $\underset{\sim}{S}$ and $\underset{\sim}{Q}$ are commuting m-tuples and $S_j Q_k = Q_k S_j$ for all j and k. If $\underset{\sim}{\lambda} \in \mathbb{R}^m$, then it is easily seen that $\Sigma_{j=1}^m (T_j - \lambda_j)^2$ is of the form $Q_{\underset{\sim}{\lambda}} + \Sigma_{j=1}^m (S_j - \lambda_j)^2$ where $Q_{\underset{\sim}{\lambda}}$ commutes with $\Sigma_{j=1}^m (S_j - \lambda_j)^2$ and, being a finite linear combination of products of operators from $\{S_j, Q_k;\ 1 \le j,k \le m\}$ with each product containing at least one element $Q_j$ ( for some j ), $Q_{\underset{\sim}{\lambda}}$ is quasinilpotent ( cf. proof of XV Theorem 5.6 of [3]). So, $\sigma(\Sigma_{j=1}^m (T_j - \lambda_j)^2) = \sigma(\Sigma_{j=1}^m (S_j - \lambda_j)^2)$ [3; XV Lemma 4.4]. Since $\underset{\sim}{\lambda} \in \gamma(\underset{\sim}{T})$ if and only if zero belongs to the spectrum of $\Sigma_{j=1}^m (T_j - \lambda_j)^2$ it follows that $\gamma(\underset{\sim}{T}) = \gamma(\underset{\sim}{S})$. □

Lemma 1 shows that to establish Theorem 1 it suffices to do so for commuting m-tuples of scalar-type spectral operators.

For each $\underset{\sim}{\mu} \in \mathbb{R}^m$ define a subset $\Sigma_{\underset{\sim}{\mu}}$ of $\mathbb{C}^m$ by

$$\Sigma_{\underset{\sim}{\mu}} = \{\underset{\sim}{\lambda} \in \mathbb{C}^m;\ \Sigma_{j=1}^m (\lambda_j - \mu_j)^2 = 0\}.$$

LEMMA 2. *Let $\underset{\sim}{S}$ be a commuting m-tuple in $L(X)^m$ consisting of scalar-type spectral operators. If the joint resolution of the identity of $\underset{\sim}{S}$ exists, say $\underset{\sim}{P}$, then*

$$\gamma(\underset{\sim}{S}) = \{\underset{\sim}{\mu} \in \mathbb{R}^m;\ \Sigma_{\underset{\sim}{\mu}} \cap \mathrm{Supp}(\underset{\sim}{P}) \ne \phi\}.$$

*Proof.* If $P_j$ is the resolution of the identity of $S_j$, $1 \le j \le m$, then the definition of $\underset{\sim}{P}$ implies that $\underset{\sim}{P}(\Pi_{j=1}^{m}\sigma(S_j)) = \Pi_{j=1}^{m}P_j(\sigma(S_j)) = I$ and hence $\mathrm{Supp}(\underset{\sim}{P}) \subseteq \Pi_{j=1}^{m}\sigma(S_j)$ showing that $\underset{\sim}{P}$ is compactly supported. Accordingly, for each $\underset{\sim}{\mu} \in \mathbb{R}^m$ the function $\psi_{\underset{\sim}{\mu}}$ defined by $\psi_{\underset{\sim}{\mu}}(\underset{\sim}{\lambda}) = \Sigma_{j=1}^{m}(\mu_j-\lambda_j)^2$, $\underset{\sim}{\lambda} \in \mathbb{C}^m$, is $\underset{\sim}{P}$-essentially bounded. It follows from the functional calculus for $\underset{\sim}{S}$ that the operator

$$\int_{\mathbb{C}^m} \psi_{\underset{\sim}{\mu}} d\underset{\sim}{P} = \Sigma_{j=1}^{m}\int_{\mathbb{C}^m} (\mu_j-\lambda_j)^2 d\underset{\sim}{P}(\underset{\sim}{\lambda}) = \Sigma_{j=1}^{m}(\mu_j-S_j)^2$$

is invertible in $L(X)$ if and only if the function $1/\psi_{\underset{\sim}{\mu}}$ is $\underset{\sim}{P}$-essentially bounded [3; XVII Corollary 2.11]. But, $\psi_{\underset{\sim}{\mu}}(\underset{\sim}{\lambda}) = 0$ if and only if $\underset{\sim}{\lambda} \in \Sigma_{\underset{\sim}{\mu}}$. Since $\Sigma_{\underset{\sim}{\mu}}$ is a closed set and $\mathrm{Supp}(\underset{\sim}{P})$ is compact, either $\Sigma_{\underset{\sim}{\mu}} \cap \mathrm{Supp}(\underset{\sim}{P}) \neq \phi$ or the distance between $\Sigma_{\underset{\sim}{\mu}}$ and $\mathrm{Supp}(\underset{\sim}{P})$ is positive. In the latter case it is clear that $\xi_{\underset{\sim}{\mu}} = 1/\psi_{\underset{\sim}{\mu}}$ is $\underset{\sim}{P}$-essentially bounded whereas in the former case $\xi_{\underset{\sim}{\mu}}$ is not $\underset{\sim}{P}$-essentially bounded. Indeed, if it were, then there would exist $E \in \mathcal{B}_m$ satisfying $\underset{\sim}{P}(E) = I$ and $\alpha > 0$ such that $\xi_{\underset{\sim}{\mu}}(s) \leq \alpha$ for every $s \in E$. Let $w \in \Sigma_{\underset{\sim}{\mu}} \cap \mathrm{Supp}(\underset{\sim}{P})$. Since $\psi_{\underset{\sim}{\mu}}$ is continuous and $\psi_{\underset{\sim}{\mu}}(w) = 0$ it follows that $\sup\{|\xi_{\underset{\sim}{\mu}}(v)|;\ v \in B_r(w)\} \geq r$, for every $r > 0$ where $B_r(w)$ is the ball in $\mathbb{C}^m$ with centre $w$ and radius $r$. In particular, if $r > \alpha$, then $|\xi_{\underset{\sim}{\mu}}(v)| > \alpha$ for all $v \in B_r(w)$ and hence $E \cap B_r(w)$ is empty. It follows that $\underset{\sim}{P}(B_r(w)) = \underset{\sim}{P}(E)\underset{\sim}{P}(B_r(w)) = \underset{\sim}{P}(E \cap B_r(w)) = 0$. Since $B_r(w)$ is an open set the definition of $\mathrm{Supp}(\underset{\sim}{P})$ implies that $w \notin \mathrm{Supp}(\underset{\sim}{P})$ which is a contradiction. Accordingly, $1/\psi_{\underset{\sim}{\mu}}$ is not $\underset{\sim}{P}$-essentially bounded if and only if $\Sigma_{\underset{\sim}{\mu}} \cap \mathrm{Supp}(\underset{\sim}{P}) \neq \phi$. That is, $\Sigma_{j=1}^{m}(\mu_j-S_j)^2$ is not invertible in $L(X)$ if and only if $\Sigma_{\underset{\sim}{\mu}} \cap \mathrm{Supp}(\underset{\sim}{P}) \neq \phi$ which is what was to be shown. □

So, to complete the proof of Theorem 1 it suffices to show that there exists $\underset{\sim}{\mu} \in \mathbb{R}^m$ such that $\Sigma_{\underset{\sim}{\mu}} \cap \mathrm{Supp}(\underset{\sim}{P}) \neq \phi$, where $\underset{\sim}{P}$ is the joint resolution of the identity of $\underset{\sim}{T}$ ( hence, also of its scalar part $\underset{\sim}{S}$ ). Let $\underset{\sim}{\lambda} \in \mathrm{Supp}(\underset{\sim}{P})$ and write $\lambda_j = a_j + ib_j$, $1 \leq j \leq m$, with $a_j$ and $b_j$ being real numbers. Then $\underset{\sim}{\mu} \in \mathbb{R}^m$ satisfies $\Sigma_{j=1}^{m}(\mu_j-\lambda_j)^2 = 0$ if and only if

$$\Sigma_{j=1}^{m}(\mu_j-a_j)^2 = \Sigma_{j=1}^{m} b_j^2 \tag{2.1}$$

and simultaneously

$$\Sigma_{j=1}^{m}(\mu_j-a_j)b_j = 0. \tag{2.2}$$

But, considering $\underset{\sim}{\mu}$ as a variable, (2.1) is the equation of a sphere in $\mathbb{R}^m$ with centre $\underset{\sim}{a} = (a_1,\dots,a_m)$ and radius $\|\underset{\sim}{b}\| = (\Sigma_{j=1}^{m} b_j^2)^{\frac{1}{2}}$ and (2.2) is a hyperplane in $\mathbb{R}^m$ with normal $\underset{\sim}{b}$ and passing through $\underset{\sim}{a}$. So, if $m \geq 2$, then there certainly exist simultaneous solutions $\underset{\sim}{\mu}$ of (2.1) and (2.2). Actually, this is the case for *every* $\underset{\sim}{\lambda} \in \mathrm{Supp}(\underset{\sim}{P})$. The proof of Theorem 1 is thereby complete. □

REMARK. The idea of the proof of Theorem 1 applies to other commuting m-tuples which have an "adequate" functional calculus. For example, let $\underset{\sim}{T}$ be a commuting m-tuple of *regular generalized scalar operators* in a Banach space X in the sense of C. Foiaş [2]. Then each $T_j$ has a regular spectral distribution $\Phi_j: C^{\infty}(\mathbb{C}) \to L(X)$ such that $\Phi_j(\lambda) = T_j$, $1 \leq j \leq m$, where $\lambda$ denotes the identity function in $\mathbb{C}$. By regularity, the tensor product $\Phi = \Phi_1 \otimes \ldots \otimes \Phi_m : C^{\infty}(\mathbb{C}^m) \to L(X)$ of the $\Phi_j$, $1 \le j \le m$, exists, is again a spectral distribution [2; Ch.4, Proposition 3.1] and satisfies $\Phi(\pi_j) = T_j$ where $\pi_j : \mathbb{C}^m \to \mathbb{C}$, $1 \le j \le m$, is the projection onto the j-th co-ordinate. Furthermore, $\Phi$ is a $C^{\infty}(\mathbb{C}^m)$-functional calculus for $\underset{\sim}{T}$ ( in the sense of [1] ) with compact support, Supp($\Phi$). Here Supp($\Phi$) is the smallest closed set $K \subseteq \mathbb{C}^m$ such that $\Phi(f) = 0$ whenever $f \in C^{\infty}(\mathbb{C}^m)$ has support disjoint from K. If $\underset{\sim}{\mu} \in \mathbb{R}^m$ and $\psi_{\underset{\sim}{\mu}}$ is the function defined in the proof of Lemma 2, then it follows from the multiplicativity of $\Phi$ that the operator $\Phi(\psi_{\underset{\sim}{\mu}}) = \Sigma_{j=1}^m (\mu_j - T_j)^2$ is invertible in L(X) if and only if there exists $\xi \in C^{\infty}(\mathbb{C}^m)$ such that $\xi$ coincides with $1/\psi_{\underset{\sim}{\mu}}$ in a neighbourhood of Supp($\Phi$). But, this is the case if and only if $\psi_{\underset{\sim}{\mu}} \neq 0$ in a neighbourhood of Supp($\Phi$) which, by definition of $\Sigma_{\underset{\sim}{\mu}}$ and the compactness of Supp($\Phi$), is equivalent to $\Sigma_{\underset{\sim}{\mu}} \cap \text{Supp}(\Phi) = \phi$. So, it follows that

$$\gamma(\underset{\sim}{T}) = \{\underset{\sim}{\mu} \in \mathbb{R}^m ; \Sigma_{\underset{\sim}{\mu}} \cap \text{Supp}(\Phi) \neq \phi\}$$

and hence, arguing as before, $\gamma(\underset{\sim}{T})$ is non-empty.

Examples of regular generalized scalar operators are spectral operators of finite-type, prespectral operators of scalar-type, well bounded operators of type (B), polar operators and some multiplication operators;see [8; §3], for example.

ACKNOWLEDGEMENT

Discussions with Alan McIntosh and Alan Pryde are gratefully acknowledged.

REFERENCES

[1] Albrecht, E. and Frunză, Şt., Non-analytic functional calculi in several variables, Manuscripta Math. 18 (1976), 327-336.
[2] Colojoară, I. and Foiaş, C., Theory of Generalized Spectral Operators (Math. & Applications No.9, Gordon & Breach, New York-London-Paris, 1968).
[3] Dunford, N. and Schwartz, J.T., Linear Operators III: Spectral Operators (Wiley-Interscience, New York, 1971).
[4] Kluvánek, I. and Kovářiková, M., Product of spectral measures, Czechoslovak Math. J. 17(92) (1967), 248-255.
[5] McIntosh, A. and Pryde, A., The solution of systems of operator equations using Clifford algebras, Proc. Centre Mathematical Analysis, Canberra, 9 (1985), 212-222.
[6] McIntosh, A., Clifford algebras and applications in analysis, Lectures at the University of N.S.W.-Sydney University joint analysis seminar, 1985.
[7] McIntosh, A. and Pryde, A., A functional calculus for several commuting operators, Indiana Univ. Math. J. (to appear).
[8] McIntosh, A., Pryde, A. and Ricker, W., Comparison of joint spectra for certain classes of commuting operators, Studia Math. (to appear in Vol. 88).

Current address: Centre for Mathematical Analysis, Australian National University, Canberra 2600, Australia.

Proceedings of the Analysis Conference, Singapore 1986
S.T.L. Choy, J.P. Jesudason, P.Y. Lee (Editors)




# THE CLASS OF MÖBIUS TRANSFORMATIONS OF CONVEX MAPPINGS

Rosihan Mohamed Ali

School of Mathematical and Computer Sciences
Universiti Sains Malaysia, Penang, Malaysia.

## 1. INTRODUCTION

Let S denote the class of analytic univalent functions f defined in the unit disk $D = \{z : |z| < 1\}$ and normalized so that $f(0) = f'(0) - 1 = 0$.

If $f \in S$ and $w \notin f(D)$, then the function

$$\hat{f} = wf \,/\, (w - f)$$

belongs again to S. This Möbius transformation $f \to \hat{f}$ is important in the analysis of the class S and other related classes.

If F is subset of S, let

$$\hat{F} = \{\hat{f} : f \in F,\ w \in C^* \setminus f(D)\}.$$

Here $C^*$ is the extended complex plane which is $C \cup \{\infty\}$. Since we allow $w = \infty$, it follows that $F \subset \hat{F} \subset S$, and since the composition of normalized Mobius transformations is again a normalized Möbius transformation, it follows that $\hat{\hat{F}} = \hat{F}$.

In this article, we shall consider the subclass K of S consisting of those functions f in S which map the unit disk D conformally onto convex domains. Simple examples show that K is a proper subset of $\hat{K}$, and so one expects that some interesting properties of K are not inherited by $\hat{K}$.

We determine the radius of convexity for $\hat{K}$, that is, the radius of the largest disk centered at the origin which is mapped onto a convex domain by each function in $\hat{K}$. We also find the largest disk centered at the origin which is contained in the range of each function in $\hat{K}$. This disk is called the Koebe disk for $\hat{K}$. The size of the Koebe disk and the radius of convexity for $\hat{K}$ have also been independently determined by Barnard and Schober [2]. Finally we derive sharp upper and lower bounds for $|\hat{f}(z)|$, $\hat{f} \in \hat{K}$. In all cases, the results obtained are different from those of the class K.

We will need the results obtained by Barnard and Schober [1], and the following theorems summarized their results.

<u>Theorem A.</u> If $\lambda : \hat{K} \to R$ is an admissible continuous functional, then $\lambda$ assumes its maximum over $\hat{K}$ at a function $\hat{f} = wf \,/\, (w - f)$ where either $\hat{f}$ is a half-plane mapping or else f is a strip mapping and w is a finite point of its boundary $\partial f(D)$.

Barnard and Schober [1] also observed the following application of Theorem A:

Let $\lambda$ be defined by

$$\lambda(\hat{f}) = \text{Re}\,\{\Phi(\log[\hat{f}(z)/z])\},$$

where $\Phi$ is a nonconstant entire function, and $z \in D \setminus \{0\}$ is fixed. By a result of Kirwan [5], $\lambda$ is a continuous functional as defined in [1]. So by choosing $\Phi(w) = \pm w$, Theorem A implies that an extremal function to the problems of maximum and minimum modulus is either a half-plane mapping or is generated by a strip mapping. Notice that the extremal strip domains f(D) need not be symmetric about the origin.

Theorem B. If $\hat{f}(z) = z + a_2z^2 + \cdots$ belongs to $\hat{K}$, then

$$|a_2| \leqslant 2\,x_o^{-1}\sin x_o - \cos x_o \approx 1.3270,$$

where $x_o \simeq 2.0816$ is the unique solution of the equation

$$\cot x = 1/x - x/2$$

in the interval $(0, \pi)$. Equality occurs for the functions $e^{-i\alpha}\hat{f}(e^{i\alpha}z)$, $\alpha \in R$, where $\hat{f}(z) = f(1)f(z)/[f(1) - f(z)]$ and f is the vertical strip mapping defined by

$$f(z) = (1/2i\,\sin x_o)\log[(1+e^{ix_o}z)/(1+e^{-ix_o}z)].$$

## 2. THE RADIUS OF CONVEXITY

Since we shall be concerned with the family $\hat{K}$, it will be convenient to drop the ^ in reference to functions in $\hat{K}$.

Observe that f maps $|z| = r$ onto a convex curve $\gamma(r)$ if the tangent to $\gamma(r)$ at the point $f(re^{i\theta})$ turns continuously in an anticlockwise direction as $\theta$ increases. Since the tangent vector to $\gamma(r)$ at $w = f(re^{i\theta})$ is given by $izf'(z)$, $z = re^{i\theta}$, analytically, the circle $|z| = r$ is mapped onto a convex curve if and only if

$$\frac{\partial}{\partial\theta}\arg\{izf'(z)\} \geqslant 0, \quad z = re^{i\theta},$$

or, equivalently, if and only if

$$1 + \text{Re}\{zf''(z)/f'(z)\} \geqslant 0 \tag{2.1}$$

on that circle. Since a convex curve bounds a convex domain, the radius of convexity is the least upper bound of r for which (2.1) holds.

Let us denote the sharp bound for the second coefficient in $\hat{K}$ by $A_2$, that is,

$$A_2 = \max\{|a_2| : f(z) = z + a_2z^2 + \cdots \in \hat{K}\}.$$

Theorem B gives the explicit value of $A_2$.

Lemma 2.1. For each $f \in \hat{K}$,

$$\text{Re}\{zf''(z)/f'(z)\} \geqslant 2r(r - A_2)/(1 - r^2), \quad |z| = r < 1.$$

Proof. Given $f \in \hat{K}$, fix $\zeta$ in $D$. Let

$$g(z) = [f((z+\zeta)/(1+\bar{\zeta}z)) - f(\zeta)]/f'(\zeta)(1-|\zeta|^2).$$

Since $g(D)$ is an affine transformation of $f(D)$, it follows that $g$ also belongs to $\hat{K}$.

Suppose $g(z) = z + b_2z^2 + \cdots$. Straightforward computations show that

$$b_2 = (\tfrac{1}{2})(1 - |\zeta|^2)f''(\zeta)/f'(\zeta) - \bar{\zeta}.$$

From Theorem B, we see that $|b_2| \leqslant A_2$, and so

$$|\zeta f''(\zeta)/f'(\zeta) - 2|\zeta|^2/(1 - |\zeta|^2)| \leqslant 2A_2|\zeta|/(1 - |\zeta|^2).$$

Since $|w| \leqslant c$ clearly implies $\mathrm{Re}\{w\} \geqslant -c$, the above inequality yields

$$\mathrm{Re}\{\zeta f''(\zeta)/f'(\zeta)\} \geqslant 2r(r - A_2)/(1 - r^2), \quad |\zeta| = r.$$

Replacing $\zeta$ by $z$, we obtain the desired inequality.

Theorem 2.2. The radius of convexity for the class $\hat{K}$ is

$$R_c = A_2 - (A_2^2 - 1)^{\frac{1}{2}} \cong 0.4547.$$

Proof. Let $f \in \hat{K}$. From (2.1) and Lemma 2.1, $f$ maps $|z| = r$ onto a convex curve if $r^2 - 2A_2r + 1 \geqslant 0$, and this is clearly nonnegative whenever $r \leqslant A_2 - (A_2^2 - 1)^{\frac{1}{2}}$. Thus the radius of convexity, $R_c$, satisfies $R_c \geqslant A_2 - (A_2^2 - 1)^{\frac{1}{2}}$.

On the other hand, let

$$g(z) = [f((z+r)/(1 + rz)) - f(r)]/f'(r)(1 - r^2),$$

where $f$ is the extremal function for the second coefficient as given in Theorem B. Formal computations show that at $z = -r$,

$$1 + zg''(z)/g'(z) = (r^2 - 2A_2r + 1)/(1 - r^2).$$

From (2.1), the circles $|z| = r$, for $r > A_2 - (A_2^2 - 1)^{\frac{1}{2}}$, are not mapped by every function $f \in \hat{K}$ onto convex curves. Therefore $R_c = A_2 - (A_2^2 - 1)^{\frac{1}{2}}$. This completes the proof.

## 3. THE KOEBE DISK

The Koebe disk for $\hat{K}$ is the largest disk centered at the origin which is contained in the range of each function in $\hat{K}$. The following theorem determines the size of this disk.

Theorem 3.1. The range of every function $f \in \hat{K}$ contains the disk $\{w : |w| < \pi/8\}$. The radius $\pi/8$ is best possible.

Proof. We will show that if a function $f$ in $\hat{K}$ omits the point $w$, then $|w| \geqslant \pi/8$. Let

$$d = \inf\{|f(\zeta)| : f \in \hat{K}, |\zeta| = 1\}.$$

From Theorem A, it suffices to minimize $|f(\zeta)|$ where $f$ is either a half-plane mapping or else $f$ is the transform of a strip mapping. Since $\hat{K}$ is rotationally invariant, we may assume that either $f(z) = z(1 - z)^{-1}$ or else $f$ is generated by a vertical strip mapping.

If $f(z) = z(1 - z)^{-1}$, then $\min\{|\zeta(1 - \zeta)^{-1}| : |\zeta| = 1\} = \tfrac{1}{2}$.

Now suppose f is obtained from a vertical strip mapping. These strip mappings in K are given by

(3.1) $g(z,x) = [1/(2i \sin x)]\log[(1 + e^{ix}z)/(1 + e^{-ix}z)], \quad 0 < x < \pi,$

and so f takes the form $f_\eta(z) = g(\eta,x)g(z,x)/[g(\eta,x) - g(z,x)]$, where $g(\eta)$, $|\eta| = 1$, is finite. Let

$$m = \inf \{|f_\eta(\zeta)| : |\zeta| = |\eta| = 1, x \in (0,\pi) .$$

Then

(3.2) $$d = \min \{\tfrac{1}{2}, m\}.$$

Observe that

$$m \geqslant \inf_{0<x<\pi} \inf \{|g(\zeta,x)g(\eta,x)/[g(\eta,x) - g(\zeta,x)]| : |\zeta| = |\eta| = 1\},$$

which implies that $m^{-1} \leqslant \sup h(x) : 0 < x < \pi\}$, where

$$h(x) = \sup\{|1/g(\zeta,x) - 1/g(\eta,x)| : |\zeta| = |\eta| = 1\}.$$

Since g is a vertical strip mapping, for each fixed x, the boundary of the range of its reciprocal consists of two circles $C_1$ and $C_2$ of radii $(\sin x)/x$ and $(\sin x)/(\pi - x)$, and centered at $w_1 = (\sin x)/x$ and $w_2 = -(\sin x)/(\pi - x)$, respectively. These two circles are symmetric with respect to the real axis and they intersect at the origin.

Thus

$$\begin{aligned} h(x) &= \text{diameter of } C_1 + \text{diameter of } C_2 \\ &= 1/g(1,x) - 1/g(-1,x) \\ &= 2\pi(\sin x)/[x(\pi - x)]. \end{aligned}$$

The derivative of $\ell(x) = (\sin x)/[x(\pi - x)]$ in $(0,\pi)$ vanishes only at $x = \pi/2$. Since $\ell(\pi/2) = 4/\pi^2$ is greater than $\lim_{x\to 0} \ell(x) = 1/\pi = \lim_{x\to\pi} \ell(x)$, we deduce that $\ell$ attains its absolute maximum at $x = \pi/2$. Thus $h(x) \leqslant 8/\pi$, that is, $m \geqslant \pi/8$.

However, if we choose g to be the symmetric vertical strip mapping in (3.1), that is,

(3.3) $$g(z) = g(z,\pi/2) = (1/2i)\log[(1 + iz)/(1 - iz)],$$

then $g(1) = \pi/4$ and $g(-1) = -\pi/4$. So $f(z) = g(1)g(z)/[g(1) - g(z)]$ omits the point $-\pi/8$ at $z = -1$. This shows that $m \leqslant \pi/8$, and hence $m = \pi/8$.

From (3.2) we conclude that $d = \pi/8$. The function g in (3.3) immediately gives sharpness of our result. This completes the proof.

## 4. A GROWTH THEOREM

For each convex function f in K, it is well-known that

(4.1) $$r(1 + r)^{-1} \leqslant |f(z)| \leqslant r(1 - r)^{-1}, \quad |z| = r < 1,$$

with equality occurring only for functions which are half-plane mappings, that is, functions of the form $f(z) = z(1 - e^{i\theta}z)^{-1}$, $\theta \in R$.

We derive sharp upper and lower bounds for $|f(z)|$, $f \in \hat{K}$.

Theorem 4.1. Let r, $0 < r < 1$, be fixed. For x in $(0, \pi)$, let

$$h(x, r) = \frac{[x - 2\arg(1 + re^{ix})] \sin x}{x \arg(1 + re^{ix})}$$

and

$$H(x, r) = \frac{[x + 2\arg(1 - re^{-ix})] \sin x}{x \arg(1 - re^{ix})}$$

Then for each $f \in \hat{K}$,

$$m(r) \leqslant |f(z)| \leqslant M(r), \quad |z| = r < 1,$$

where

$$[M(r)]^{-1} = \min\{h(x, r): \ 0 < x < \pi\} < (1 - r)r^{-1},$$

and

$$[m(r)]^{-1} = \max\{H(x, r): \ 0 < x < \pi\} > (1 + r)r^{-1}.$$

$|f(z)| = m(r)$ occurs for the functions $e^{-i\alpha}f(e^{i\alpha}z)$, $\alpha \in R$, where $f(z) = g(1, x_1)g(z, x_1)/[g(1, x_1) - g(z, x_1)]$ and g is the vertical strip mapping as defined in (3.1), and H attains its maximum at $x_1$. Similarly, $|f(z)| = M(r)$ occurs for the functions of the form given above, except that $g = g(z, x_2)$, where h attains its minimum at $x_2$.

The proof is rather lengthy, so we will break it into several parts. Notice that from Theorem A, it suffices to extremize $|f(z)|$ where f is either a half-plane mapping or else f is the transform of a strip mapping. Although this is clearly a major step in determining the extremal values, as occasionally happens in these type of problems, determining the explicit values still requires some work.

The bounds for the modulus of half-plane mappings are given by (4.1). Since $\hat{K}$ is rotationally invariant, we may assume that f is generated by a vertical strip mapping. Thus f has the form

$$f(z) = g(\eta, x)g(z, x)/[g(\eta, x) - g(z, x)] \tag{4.2}$$

where $g(\eta, x)$, $|\eta| = 1$, is finite and g is given by (3.1). We first establish the following lemma.

Lemma 4.2. For each fixed x,

$$\arg\left(1 - g(z, x)\,\frac{\sin x}{x}\right) < \arg\left(\frac{zg'(z, x)}{g(z, x)}\right) < \arg\left(1 + g(z, x)\,\frac{\sin x}{\pi - x}\right)$$

whenever $\operatorname{Im}\{z\} > 0$ in D.

Here it is understood that the argument function vanishes at $z = 0$.

Proof. We will prove the left assertion; the right assertion is proved analogously.

Specifically, we shall show that

$$(4.3)\qquad \begin{aligned} -\infty &< \limsup_{z\to\zeta} \arg(1 - g(z, x)x^{-1}\sin x) \\ &\leq \liminf_{z\to\zeta} \arg(zg'(z, x)/g(z, x)) < \infty \end{aligned}$$

for each point $\zeta$ on the boundary $\partial D^+$, where $D^+$ is the upper half-disk. An application of the generalized maximum principle for harmonic functions [3, p.254] will then complete the proof.

For each fixed x in $(0, \pi)$, Re $g < x/(2 \sin x)$ in D, so $\text{Re}\{1 - gx^{-1}\sin x\} > \frac{1}{2}$. Since g is a convex function, $\text{Re}\{zg'/g\} > \frac{1}{2}$ [4, p.73]. Thus the harmonic function $\arg(1 - gx^{-1}\sin x)$ is uniformly bounded by $\pi/2$ and continuous in $\bar{D}$, while $\arg(zg'/g)$ is bounded by $\pi/2$ and continuous in $\bar{D}$ except at $z = e^{i(\pi \pm x)}$. Therefore it is sufficient to compare radial limits almost everywhere in (4.3).

If $\zeta$ is real in $\partial D^+$, then $\zeta g'(\zeta, x)/g(\zeta, x)$ and $1 - g(\zeta, x)x^{-1}\sin x$ are both real, so (4.3) holds.

If $\zeta = e^{i(\pi-x)}$, then $\lim_{r\to 1} \arg(1 - g(re^{i(\pi-x)}, x)x^{-1}\sin x) = -\pi/2$. Since $\arg(zg'/g) > -\pi/2$ in D, (4.3) again holds.

It remains to show the validity of (4.3) for nonreal $\zeta$ in $\partial D^+$ with $\zeta \neq e^{i(\pi-x)}$. In this case, if

$$\log \frac{1 + e^{ix}z}{1 + e^{-ix}z} = u(r, \theta) + iv(r, \theta),$$

then for $z = re^{i\theta}$,

$$(4.4)\qquad \lim_{r\to 1} \arg(zg'(z, x)/g(z, x)) = \arctan \frac{u(1, \theta)}{v(1, \theta)},$$

and

$$(4.5)\qquad \lim_{r\to 1} \arg(1 - g(z, x)x^{-1}\sin x) = \arctan \frac{u(1, \theta)}{2x - v(1, \theta)}.$$

The identity

$$\frac{1 + e^{i(\theta+x)}}{1 + e^{i(\theta-x)}} = \frac{\cos((x + \theta)/2)}{\cos((x - \theta)/2)} e^{ix}$$

yields

$$(4.6.1)\qquad v(1, \theta) = \begin{cases} x, & \theta \in (0, \pi-x) \\ -(\pi-x), & \theta \in (\pi-x, \pi) \end{cases}$$

$$(4.6.2)\qquad u(1,\theta) < 0, \quad \theta \in (0,\pi).$$

So from (4.4) and (4.5), the validity of (4.3) for $\zeta = e^{i\theta}$, $\theta \in (0,\pi)$ and $\theta \neq \pi - x$, is equivalent to

$$\frac{[2x - 2v(1,\theta)]u(1,\theta)}{[2x - v(1,\theta)]v(1,\theta)} \geqslant 0.$$

But (4.6) implies the above inequality, and completes the proof of the lemma.

We will now use the above lemma to prove the following result.

Lemma 4.3. Let g be defined as in (3.1), and fix r, $0 < r < 1$. Then for $z = re^{i\theta}$, $0 \leq \theta \leq \pi$,

$$\frac{1}{g(r,x)} - \frac{\sin x}{x} \leq \left| \frac{1}{g(z,x)} - \frac{\sin x}{x} \right| \leq \frac{\sin x}{x} - \frac{1}{g(-r,x)},$$

and

$$-\frac{1}{g(-r,x)} - \frac{\sin x}{\pi - x} \leq \left| \frac{1}{g(z,x)} + \frac{\sin x}{\pi - x} \right| \leq \frac{1}{g(r,x)} + \frac{\sin x}{\pi - x}.$$

Proof. As before, we will only prove the first assertion since the other assertion follows similarly. For convenience, we write $g(z,x) = g(z)$.

For $z = re^{i\theta}$, $0 < r < 1$,

$$\frac{\partial}{\partial\theta} |h|^2 = 2\mathrm{Re}\{iz\bar{h}h'\} \text{ where } h(z) = \frac{1}{g(z)} - \frac{\sin x}{x}.$$

Differentiation yields

$$\frac{\partial}{\partial\theta} |h|^2 = |g|^{-2}\mathrm{Im}\{zg'(z)/g(z)\} - \mathrm{Im}\{zg'(z)/g^2(z)\}x^{-1}\sin x$$

$$= |g|^{-2}\ell(\theta)$$

where $\ell(\theta) = [1-(x^{-1}\sin x)\mathrm{Re}g(z)]\mathrm{Im}\{zg'(z)/g(z)\}+x^{-1}\sin x(\mathrm{Im}g(z))\mathrm{Re}\{zg'(z)/g(z)\}$.

Since $\mathrm{Re}\{zg'/g\}$ and $1-(x^{-1}\sin x)\mathrm{Re}g$ are positive, $\ell(\theta) > 0$ is equivalent to

$$\frac{\mathrm{Im}\{zg'(z)/g(z)\}}{\mathrm{Re}\{zg'(z)/g(z)\}} > \frac{-(x^{-1}\sin x)\mathrm{Im}g(z)}{1-(x^{-1}\sin x)\mathrm{Re}g(z)} = \frac{\mathrm{Im}\{1-g(z)x^{-1}\sin x\}}{\mathrm{Re}\{1-g(z)x^{-1}\sin x\}}.$$

This inequality is equivalent to

$$\arg(zg'(z)/g(z) > \arg(1-g(z)x^{-1}\sin x).$$

Applying Lemma 4.2, we conclude that $\ell(\theta) > 0$ in $(0,\pi)$. This completes the proof.

We now proceed with the proof of the theorem.

Proof of Theorem 4.1. Let

$$\delta(r) = \inf\{|f(z)| : f \in \hat{K}, |z| = r\},$$

and

$$\Delta(r) = \sup\{|f(z)| : f \in \hat{K}, |z| = r\}.$$

As observed earlier, it suffices to consider f where either $f(z) = z(1-z)^{-1}$ or else f is given by (4.2).

If $f(z) = z(1-z)^{-1}$, then f attains its maximum at $z = r$ and minimum at $z = -r$. Thus

$$r(1+r)^{-1} \leq |f(z)| \leq r(1-r)^{-1}.$$

Next let f be given by (4.2). Let

$$M(r) = \sup \{|f(z)| : |z| = r\},$$

and

$$m(r) = \inf \{|f(z)| : |z| = r\}.$$

Then by considering the reciprocal of f , we deduce that

$$[M(r)]^{-1} \geqq \inf_{0<x<\pi} \min_{\theta} \min_{\psi} \{|1/g(re^{i\theta},x) - 1/g(e^{i\psi},x)|\}, \tag{4.7}$$

and

$$[m(r)]^{-1} \leqq \sup_{0<x<\pi} \max_{\theta} \max_{\psi} \{|1/g(re^{i\theta},x) - 1/g(e^{i\psi},x)|\}. \tag{4.8}$$

In what follows, we shall show that equality is obtained in both (4.7) and (4.8), and that

$$\delta(r) = \min \{r(1 + r)^{-1}, m(r)\} = m(r),$$

and

$$\Delta(r) = \max \{r(1 - r)^{-1}, M(r)\} = M(r).$$

As observed earlier, the boundary of the range of $1/g$ consists of two circles $C_1$ and $C_2$ with centers at $x^{-1}\sin x$, $-(\pi - x)^{-1}\sin x$, and of radii $x^{-1}\sin x$, $(\pi - x)^{-1}\sin x$, respectively. Moreover, these two circles are symmetric with respect to the real axis.

Notice that the range of $1/g$ is also symmetric with respect to the real axis. Since $1/g(z,x)$ is real if and only if z is real, it suffices to consider $z = re^{i\theta}$ for $\theta$ in $[0,\pi]$. For each fixed $\theta$, $1/g(re^{i\theta},x)$ lies outside the circles $C_1$ and $C_2$. Thus the minimum distance between $1/g(re^{i\theta},x)$ and $C_1$ is given by $d_1(\theta,x)$, where

$$d_1(\theta,x) = |[g(re^{i\theta},x)]^{-1} - x^{-1}\sin x| - x^{-1}\sin x,$$

while the minimum distance between $1/g(re^{i\theta},x)$ and $C_2$ is $d_2(\theta,x)$, where

$$d_2(\theta,x) = |[g(re^{i\theta},x)]^{-1} + (\pi-x)^{-1}\sin x| - (\pi-x)^{-1}\sin x.$$

So for a fixed $\theta$ and x,

$$\min\{|[1/g(re^{i\theta},x)] - [1/g(e^{i\psi},x)]| : 0 \leqq \psi \leqq 2\pi\} = \min\{d_1(\theta,x), d_2(\theta,x)\}.$$

Similarly, for a fixed $\theta$ and x,

$$\max\{|[1/g(re^{i\theta},x)] - [1/g(e^{i\psi},x)]| : 0 \leqq \psi \leqq 2\pi\} = \max\{D_1(\theta,x), D_2(\theta,x)\},$$

where

$$D_1(\theta,x) = |[g(re^{i\theta},x)]^{-1} - x^{-1}\sin x| + x^{-1}\sin x,$$

and

$$D_2(\theta, x) = |[g(re^{i\theta}, x)]^{-1} + (\pi-x)^{-1}\sin x| + (\pi-x)^{-1}\sin x.$$

Applying Lemma 4.3, it follows that for a fixed x,

$$\min\{d_1(\theta, x) : 0 \leqq \theta \leqq \pi\} = d_1(0, x) = [1/g(r, x)] - [1/g(1, x)],$$

and

$$\min\{d_2(\theta, x) : 0 \leqq \theta \leqq \pi\} = d_2(\pi, x) = [1/g(-1, x)] - [1/g(-r, x)].$$

Also, for a fixed x

$$\max\{D_1(\theta, x) : 0 \leqq \theta \leqq \pi\} = D_1(\pi, x) = [1/g(1, x)] - [1/g(-r, x)],$$

and

$$\max\{D_2(\theta, x) : 0 \leqq \theta \leqq \pi\} = D_2(0, x) = [1/g(r, x)] - [1/g(-1, x)].$$

Now $g(r, x) = \arg(1 + re^{ix})/\sin x$, $g(1, x) = x/2\sin x$, $g(-r, x) = -\arg(1 - re^{-ix})/\sin x$, and $g(-1, x) = -(\pi - x)/2\sin x$. Since $g(r, \pi - x) = -g(-r, x)$, we see that

$$d_1(0, \pi - x) = d_2(\pi, x),$$

and

$$D_2(0, \pi - x) = D_1(\pi, x).$$

Thus it suffices to minimize $d_1(0, x)$ and to maximize $D_1(\pi, x)$. Specifically, $d_1(0, x) = h(x, r)$ and $D_1(\pi, x) = H(x, r)$, where

$$h(x, r) = [x - 2\arg(1 + re^{ix})](\sin x)/[x\arg(1 + re^{ix})],$$

and

$$H(x, r) = [x + 2\arg(1 - re^{-ix})](\sin x)/[x\arg(1 - re^{-ix})].$$

From (4.7) and (4.8),

$$[M(r)]^{-1} \geqq \inf\{h(x, r) : 0 < x < \pi\}, \tag{4.9}$$

and

$$[m(r)]^{-1} \leqq \sup\{H(x, r) : 0 < x < \pi\}. \tag{4.10}$$

Applying L'Hospital's rule, it follows that $h(0, r) = (1 - r)r^{-1} = h(\pi, r)$. A straightforward calculus argument shows that $h(\pi/2, r) = 1/\arctan r - 4/\pi$ is less than $(1 - r)r^{-1}$; hence h attains its minimum value in $(0, \pi)$.

Suppose $h(x_2) = \alpha$ is the minimum value. The proof thus far shows that the reciprocal of the function $f(z) = g(1, x_2)g(z, x_2)/[g(1, x_2) - g(z, x_2)]$ assumes the value $\alpha$ at $z = r$. Combining this with (4.9), we conclude that

$$[M(r)]^{-1} = \min\{h(x, r) : 0 < x < \pi\}.$$

Similarly, $H(0, r) = (1 + r)r^{-1} = H(\pi, r)$, and since

$$H(\pi/2, r) = \frac{4}{\pi} + \frac{1}{\arctan r} > \frac{1 + r}{r},$$

H assumes its maximum in $(0, \pi)$. Proceeding analogously as before, we conclude that

$$[m(r)]^{-1} = \max \{H(x, r) : 0 < x < \pi\}.$$

Finally, it is clear that we obtain sharpness of our result for those f as given in the statement of the theorem. This completes the proof.

Let us take a closer examination of the function h as given in Theorem 4.1. It is difficult to determine explicity the point(s) in $(0, \pi)$ at which h assumes its minimum value. So we would want to ascertain the number of zeros of $\partial h/\partial x$. Numerical evidence seems to suggest that $\partial h/\partial x$ and $\partial H/\partial x$, where H is also given in Theorem 4.1, vanish exactly once in $(0, \pi)$. Under this assumption, we give below the approximate extremal values of h and H. Note that $x_1$ and $x_2$ denote the approximate zero to $\partial H/\partial x$ and $\partial h/\partial x$, respectively.

THE EXTREMAL VALUES OF H AND h

| $r$ | $x_1$ | $x_2$ | $H(x_1,r)$ | $h(x_2,r)$ |
|---|---|---|---|---|
| 0.1 | 2.024425 | 2.140862 | 11.351998 | 8.698759 |
| 0.2 | 1.969090 | 2.202705 | 6.376196 | 3.725457 |
| 0.4 | 1.862916 | 2.336489 | 3.922480 | 1.281990 |
| 0.6 | 1.761633 | 2.490874 | 3.132843 | 0.510468 |
| 0.8 | 1.664398 | 2.686461 | 2.757472 | 0.163417 |
| 0.9 | 1.617154 | 2.821787 | 2.638368 | 0.064163 |

Corollary 4.4. For each $f \in \hat{K}$

$$\frac{r}{(1 - r^2)M(r)} \leq \left|\frac{zf'(z)}{f(z)}\right| \leq \frac{r}{(1 - r^2)m(r)}, \quad |z| = r < 1,$$

where $m(r)$ and $M(r)$ are defined by Theorem 4.1. This result is sharp.

Proof. Let $f \in \hat{K}$ and $\zeta$ in D be fixed. Let F be a Marty transformation of f, that is, $F(z) = [f((z + \zeta)/(1 + \bar{\zeta}z)) - f(\zeta)]/f'(\zeta)(1 - |\zeta|^2)$. From Theorem 4.1,

$$m(r) \leq |F(-\zeta)| \leq M(r), \quad |\zeta| = r.$$

Thus

$$m(r)(1 - r^2)/r \leq |f(\zeta)/\zeta f'(\zeta)| \leq M(r)(1 - r^2)/r.$$

By considering the Marty transformation of the extremal functions in Theorem 4.1, we obtain sharpness of our result.

REFERENCES

[1] R.W. Barnard and G. Schober, "Möbius Transformations of Convex Mappings", Complex Variables Theory Appl. 3(1984), 55 - 69.
[2] R.W. Barnard and G. Schober, "Möbius Transformations of Convex Mappings II", in print.
[3] J.B. Conway, Functions of One Complex Variable, 2nd, Ed., Springer-Verlag, New York, 1978.
[4] P.L. Duren, Univalent Functions, Springer-Verlag, New York, 1983.
[5] W.E. Kirwan, "A Note on Extremal Problems for Certain Classes of Analytic Functions", Proc. Amer. Math. Soc. 17 (1966), 1028 - 1030.

Proceedings of the Analysis Conference, Singapore 1986
S.T.L. Choy, J.P. Jesudason, P.Y. Lee (Editors)



# UNIFORM ERGODIC THEOREMS FOR OPERATOR SEMIGROUPS

Sen-Yen Shaw

Department of Mathematics, National Central University,
Chung-Li, Taiwan, Republic of China

The purpose of this article is to examine conditions for the uniform operator convergence of Cesàro means and Abel means of a locally integrable semigroup of operators on a Banach space X. We consider the case that X is a general Banach space and the case that X is a Grothendieck space with the Dunford-Pettis property.

## 1. INTRODUCTION

Let X be a Banach space, and let $\{T(t);\ t>0\}$ be a locally integrable semigroup of bounded linear operators on X. That is, $T(s+t)=T(s)T(t)$ for all $s,t>0$, and $T(\cdot)x$ is Bochner integrable over $(0,t)$ for all $t<\infty$ and $x\in X$.

Let $S(t)$, $t>0$, be the operator defined by $S(t)x:=\int_0^t T(s)x\,ds$ $(x\in X)$, and let $R_s(\lambda)$ be the operator defined by $R_s(\lambda)x:=\lim_{t\to\infty}\int_0^t e^{-\lambda u}t(u)x\,du$ $(x\in X)$ for those $\lambda$ for which the limit exists for all $x\in X$. As in [6], we denote $\sigma:=\inf\{u\in(-\infty,\infty);\ R_s(u)$ exists$\}$,

$$\sigma_a:=\inf\{u\in(-\infty,\infty);\ R_s(\lambda) \text{ is analytic for all } \lambda \text{ with } \mathrm{Re}\lambda>u\},$$

$$w_0:=\inf\{t^{-1}\log||T(t)||;\ t>0\}.$$

It is clear that $\sigma\le\sigma_a\le w_0$. For those $\lambda$ with $\mathrm{Re}\lambda>w_0$ the Bochner integrals $R(\lambda):=\lambda\int_0^\infty e^{-\lambda u}S(u)du$ exist and form a pseudo-resolvent (cf. [1, p.510]), which has $R_s(\cdot)$ as its extension to the set $\{\lambda;\ \mathrm{Re}\lambda>\sigma\}$. Thus $R_s(\cdot)$ is also a pseudo-resolvent on $\{\lambda;\ \mathrm{Re}\lambda>\sigma_a\}$.

Uniform ergodic theorems are concerned with the uniform operator convergence of the Cesàro mean $t^{-1}S(t)$ as $t\to\infty$ and of the Abel mean $\lambda R_s(\lambda)$ as $\lambda\to 0^+$. Usually one assumes "$w_0\le 0$" or the even stronger condition "$||T(t)||=o(t)$ $(t\to\infty)$" (see [1, Theorem 18.8.4], [2]).

However, as shown by an example in [6], it is possible that $\sigma_a<0<w_0$ while $T(\cdot)$ is uniformly ergodic. Our theorems in [6] assume the weaker condition "$\sigma_a\le 0$" and show in particular that in order $T(\cdot)$ to be uniformly Cesàro-ergodic it is necessary that $||T(t)R_s(1)||=o(t)$ $(t\to\infty)$, but is not that $||T(t)||=o(t)$ $(t\to\infty)$.

In section 2, we shall formulate uniform ergodic theorems in which another necessary condition "$||T(t)S(u)||=o(t)$ $(t\to\infty)$ $\forall u>0$" is used. This condition is used. This condition is slightly stronger than that "$||T(t)S(u)x||=o(t)$ $(t\to\infty)$ $\forall x\in X$, $\forall u>0$," which is necessary for the strong ergodicity (cf. [4, Corollary

3.7]). Then, under this very condition the strong Cesàro-ergodicity implies the uniform ergodicity provided that the ground space is a Grothendieck space with the Dunford-Pettis property. This is proved in section 3.

## 2. UNIFORM ERGODIC THEOREMS ON A GENERAL BANACH SPACE

For a locally integrable semigroup $T(\cdot)$ of operators on a Banach space $X$, we have the following uniform ergodic theorem.

Theorem 1. Assume that $\sigma_a \leq 0$. Then the following statements are equivalent:

(1) $T(\cdot)$ is uniformly Cesàro-ergodic.

(2) $||T(t)R_s(1)||=o(t)$ $(t\to\infty)$, and $T(\cdot)$ is uniformly Abel-ergodic.

(3) $||T(t)R_s(1)||=o(t)$ $(t\to\infty)$, and $R(R_s(1)-I)$ is closed.

(4) $||T(t)S(u)||=o(t)$ $(t\to\infty)$ for all $u>0$, and $T(\cdot)$ is unifromly Abel-ergodic.

(5) $||T(t)S(u)||=o(t)$ $(t\to\infty)$ for all $u>0$, and $R(R_s(1)-I)$ is closed.

The equivalence of (1), (2), and (3) has been proved in [6, Theorem 4]. To see that (4) and (5) are also equivalent conditions, we prove the following lemma.

Lemma 2. (i) If $T(\cdot)$ is uniformly Cesàro-ergodic, then $||T(t)S(u)||=o(t)$ $(t\to\infty)$ for all $u>0$.

(ii) If $||T(t)S(u)||=o(t)$ $(t\to\infty)$ for all $u>0$, then $||T(t)R_s(\mu)||=o(t)$ $(t\to\infty)$ for all $\mu>\sigma_a$.

Proof. (i) Let $P:=u_0\text{-}\lim_{t\to\infty} t^{-1}S(t)$. Since $T(t)S(u)=S(t+u)-S(t)$ (see [4, Lemma 2.3]), we have that $u_0\text{-}\lim_{t\to\infty} t^{-1}T(t)S(u)=u_0\text{-}\lim_{t\to\infty} t^{-1}S(t+u) - u_0\text{-}\lim_{t\to\infty} t^{-1}S(t)=P-P=0$.

(ii) For $\lambda>w_0$ we have that

$$T(t)R(\lambda)=T(t)\int_0^\infty \lambda e^{-\lambda u}S(u)du=\lambda\int_0^\infty e^{-\lambda u}T(t)S(u)du$$

so that $||T(t)R(\lambda)||/t \leq \lambda\int_0^\infty e^{-\lambda u}(||T(t)S(u)||/t)du\to 0$ as $t\to\infty$, by Lebesgue's dominated convergence theorem. Since $R_s(\cdot)$ is a pseudo-resolvent on $\{\mu\in C;\ \mathrm{Re}\mu > \sigma_a\}$, we have $R_s(\mu)=R(\lambda)+(\lambda-\mu)R(\lambda)R_s(\mu)$ and so

$$||T(t)R_s(\mu)||/t \leq [1+|\lambda-\mu|\,||R_s(\mu)||]\,||T(t)R(\lambda)||/t \to 0 \quad \text{as } t\to\infty.$$

A locally integrable semigroup $T(\cdot)$ is said to be of class $(0,A)$ if $\lambda R(\lambda)$ converges strongly to $I$ as $\lambda\to\infty$. The operator $A^0: x \to \lim_{t\to 0^+} t^{-1}(T(t)-I)x$ is densely defined and closable. The closure $A$ of $A^0$ is called the infinitesimal generator of $T(\cdot)$. $T(\cdot)$ is of class $(C_0)$ if it is strongly convergent to $I$ as $t\to 0^+$. In this case we have $A=A^0$.

It is known [6, Proposition 7] that if $T(\cdot)$ is of class $(0,A)$ then $\sigma=\sigma_a$ and $R_s(\lambda)=(\lambda-A)^{-1}$ for $\lambda>\sigma$. In particular we see that $R(\lambda R_s(\lambda)-I)=R(A(\lambda-A)^{-1})=R(A)$. Since the Cesàro-ergodicity implies $||S(t)||=O(t)$ $(t\to\infty)$, and since the latter condition in turn implies $\sigma\leq 0$ (see the proof of Proposition 8 in [6]), one can easily deduce from Theorem 1 a complete chatacterization of the uniform Cesàro-ergodicity of (0,A) semigroups.

Theorm 3. Let $T(\cdot)$ be a semigroup of class (0,A). The following statements are equivalent:

(1) $T(\cdot)$ is uniformly Cesàro-ergodic.

(2) $||S(t)||=O(t)$ $(t\to\infty)$, $||T(t)R_s(1)||=o(t)$ $(t\to\infty)$, and $R(A)$ is closed.

(3) $||S(t)||=O(t)$ $(t\to\infty)$, $||T(t)S(u)||=o(t)$ $(t\to\infty)$ for all $u>0$, and $R(A)$ is closed.

(4) $||S(t)||=O(t)$ $(t\to\infty)$, $||T(t)S(u)||=o(t)$ $(t\to\infty)$ for all $u>0$, and $T(\cdot)$ is uniformly Abel-ergodic.

## 3. GROTHENDIECK SPACE WITH THE DUNFORD-PETTIS PROPERTY

A Banach space X is called a Grothendieck space if it has the property that every weakly$^*$ convergent sequence in the dual space $X^*$ is weakly convergent.

The following strong ergodic theorem is a combination of Proposition 4.2 of [4] and Theorems 1 and 2 of [5]. We state it here for use in Theorem 5.

Theorem 4. Let $T(\cdot)$ be a locally integrable semigroup of operators on a Grothendieck space X.

(i) If $||S(t)||=O(t)$ $(t\to\infty)$ and $||T(t)S(u)||=o(t)$ $(t\to\infty)$ for all $u>0$, then $P:x\to s\text{-}\lim_{t\to\infty} t^{-1}S(t)x$ is a bounded projection with $R(P)=\bigcap_{t>0} N(T(t)-I)$ and $N(P)=\overline{\text{span}}\{R(T(t)-I);\ t>0\}$, and $s\text{-}\lim_{t\to\infty} t^{-1}S^*(t)x^*$ exists for all $x^*\in X^*$.

(ii) $T(\cdot)$ is strongly Cesàro-ergodic (i.e. $D(P)=X$) if and only if it satisfies: $||S(t)||=O(t)$ $(t\to\infty)$; $||T(t)S(u)x||=o(t)$ $(t\to\infty)$ for all $x\in X$, $u>0$; $w^*\text{-}cl(R^*)=\overline{R^*}$ (or $w^*\text{-}cl(R(A^*))=\overline{R(A^*)}$ in case $T(\cdot)$ is of class $(C_0)$), where $R^*:=\text{span}\{R(T^*(t)-I);\ t>0\}$.

X is said to have the Dunford-Pettis property if $\langle x_n, x_n^*\rangle\to 0$ whenever $x_n\to 0$ weakly in X and $x_n^*\to 0$ weakly in $X^*$. $L^\infty$ is a Grothendieck space with the Dunford-Pettis property. For other examples of such spaces see [3].

It was recently proved by Lotz [3] that, on a Grothendieck space with the Dunford-Pettis property, every $(C_0)$-semigroup is uniformly continuous and every strongly ergodic discrete semigroup $\{T^n\}$ is uniformly ergodic. It has been shown in [7] that the same assertions are true for cosine operator functions. The following theorem about the ergodicity of locally integrable semigroups is of the same nature.

Theorem 5. Let $T(\cdot)$ be a locally integrable semigroup of operators on a Grothendieck space X with the Dunford-Pettis property. Suppose that $||T(t)S(u)||=o(t)$ $(t\to\infty)$ for all $u>0$. Then $T(\cdot)$ is uniformly Cesàro-ergodic if and only if it is strongly Cesàro-ergodic.

For the proof of this theorem we need

Lemma 6 ([3]). Let $V_n$ be a sequence of bounded linear operators on a Banach space X with the Dunford-Pettis property. Suppose that

(1) $w\text{-}\lim_{n\to\infty} V_n x_n = 0$ whenever $\{x_n\}$ is bounded in X;

(2) $w\text{-}\lim_{n\to\infty} V_n^* x_n^* = 0$ whenever $\{x_n^*\}$ is bounded in $X^*$.

Then $||V_n^2|| \to 0$. In particular, $V_n - I$ and $V_n + I$ are invertible for large n.

Proof of Theorem 5. If $T(\cdot)$ is stronlgy ergodic, then $P \in B(X)$, $X = R(P) \oplus N(P)$ with $R(P) = \bigcap_{t>0} N(T(t)-I)$. Since $R(P)$ is fixed by $t^{-1}S(t)$ for all $t>0$, we may assume that $P=0$ without loss of generality.

Let $V_n = n^{-1}S(n)$. Then $s\text{-}\lim_{n\to\infty} V_n x = Px = 0$ for all $x \in X$, so that for any bounded sequence $\{x_n^*\}$ in $X^*$, the sequence $\{V_n^* x_n^*\}$ converges weakly$^*$ and hence weakly to 0. By (i) of Theorem 4 we have that $s\text{-}\lim_{n\to\infty} V_n^* x^*$ exists and is equal to $w^*\text{-}\lim_{n\to\infty} V_n^* x^* = P^* x^* = 0$ for all $x^* \in X^*$. Hence $\{V_n x_n\}$ converges weakly to 0 whenever $\{x_n\}$ is bounded. It follows from Lemma 6 that $V_n - I$ is invertible for large n.

Using the identity $S(t)(T(u)-I) = (T(t)-I)S(u)$, the assumption $||T(t)S(u)|| = o(t)$ $(t\to\infty)$ $\forall u>0$, and Lebesgue's dominated convergence theorem, we obtain

$$\begin{aligned} ||t^{-1}S(t)|| &= ||(V_n - I)^{-1} t^{-1} S(t)(n^{-1}S(n) - I)|| \\ &\le ||(V_n - I)^{-1}||\, n^{-1} \int_0^n ||t^{-1}S(t)(T(u)-I)||\,du \\ &= ||(V_n - I)^{-1}||\, n^{-1} \int_0^n t^{-1} ||(T(t)-I)S(u)||\,du \\ &\to 0 \quad \text{as} \quad t\to\infty. \end{aligned}$$

Hence $T(\cdot)$ is uniformly Cesàro-ergodic.

The following Corollary is deduced from Lemma 2(i), Theorems 4(ii) and 5.

Corollary 7. Let $T(\cdot)$ and X be as assumed in Theorem 5. Then $T(\cdot)$ is uniformly Cesàro-ergodic if and only if it satisfies: $||S(t)|| = O(t)$ $(t\to\infty)$; $||T(t)\cdot S(u)|| = o(t)$ $(t\to\infty)$ for all $u>0$; and $w^*\text{-}cl(R^*) = \overline{R^*}$ (or $w^*\text{-}cl(R(A^*)) = \overline{R(A^*)}$ in case $T(\cdot)$ is strongly continuous and hence uniformly continuous).

This and Theorem 3 yield the following

Corollary 8. Let $T(\cdot)$ be a uniformly continuous semigroup on a Grothendieck space with the Dunford-Pettis property. Suppose that $||S(t)|| = O(t)$ $(t\to\infty)$ and $||T(t)S(u)|| = o(t)$ $(t\to\infty)$ for all $u>0$. Then the following conditions are equivalent:

(1) $T(\cdot)$ is strongly Cesàro-ergodic.

(2) $T(\cdot)$ is uniformly Cesàro-ergodic.

(3) $T(\cdot)$ is uniformly Abel-ergodic.

(4) $R(A)$ is closed.

(5) $w^*\text{-}cl(R(A^*)) = \overline{R(A^*)}$.

This is illustrated by the following example.

For $0 \le \lambda \le 1$ let $g_\lambda$ be the function $g_\lambda(s) := is \cdot I_{[\lambda,1]}(s)$, where $I_{[\lambda,1]}$ is the characteristic function of the interval $[\lambda,1]$. The multiplication operators $T_\lambda(t): f \to \exp(tg_\lambda)f$ $(f \in L^\infty[0,1])$, $-\infty < t < \infty$, form a uniformly continuous group of isometrical isomorphisms of $L^\infty[0,1]$, and generator is the multiplication operator $A_\lambda : f \to g_\lambda f$. It is easy to see that $N(A_\lambda) = I_{[0,\lambda]} L^\infty[0,1]$ and $R(A_\lambda) = I_{[\lambda,1]} L^\infty[0,1]$ for $\lambda>0$, and $N(A_0)=\{0\}$ and $R(A_0) \neq \overline{R(A_0)} = \{f \in L^\infty[0,1];\ \lim_{s\to 0} f(s)=0\}$. Since $R(A_\lambda)$ is closed for $\lambda>0$, $T_\lambda(\cdot)$ is uniformly ergodic to the multication by $I_{[0,\lambda]}$. Since $R(A_0)$ is not closed, $T_0(\cdot)$ is not uniformly ergodic and hence not strongly ergodic. In fact, $\lim_{t\to\infty} t^{-1}\int_0^t T_0(s)f ds$ exists if and only if $\lim_{s\to 0} f(s)=0$, and in this case, the limit is the zero function. These facts can also be verified by direct computation.

REFERENCES

[1] Hille, E. and Phillips, R.S., Functional Analysis and Semigroups (Amer. Math. Soc. Colloq. Publ., vol. 31, Amer. Math. Soc., Providence, R.I., 1957).

[2] Lin, M., On the uniform ergodic theorem. II, Proc. Amer. Math. Soc. 46 (1974), 217-225.

[3] Lotz, H.P., Tauberian theorems for operators on $L^\infty$ and similar spaces, Functional Analysis: Surveys and Recent Results III, (1984), 117-133.

[4] Shaw, S.-Y., Ergodic properties of operator semigroups in general weak topologies, J. Funct. Anal. 49(1982), 152-169.

[5] Shaw, S.-Y., Ergodic theorems for semigroups of operators on a Grothendieck space, Proc. Janpan Acad. Ser. A 59(1983), 132-135.

[6] Shaw, S.-Y., Uniform ergodic theorems for locally integrable semigroups and pseudo-resolvents, Proc. amer. Math. Soc. 98(1986), 61-67.

[7] Shaw, S.-Y., On w*-continuous cosine operator functions, J. Funct. Anal. 66(1986), 73-95.

Proceedings of the Analysis Conference, Singapore 1986
S.T.L. Choy, J.P. Jesudason, P.Y. Lee (Editors)




# WEIGHTED NORM INEQUALITIES FOR SOME MAXIMAL FUNCTIONS

Wang Silei

Department of Mathematics, Hangzhou University, Hangzhou,
People's Republic of China

The main purpose of this paper is to study the weighted norm inequalities for some maximal functions. The results are sharp.

## §1. INTRODUCTION.

Let $\mu_{\lambda,\gamma}(f)(x)$ denote the following maximal functions defined by

$$(1.1) \quad \mu_{\lambda,\gamma}(f)(x) = \sup_{y>0} \left( \int_{\mathbb{R}^n} |u(x-t,y)|^\gamma \, y^{-n} \left(\frac{y}{|t|+y}\right)^{n\lambda} dt \right)^{1/\gamma},$$

where $1 < \lambda \leq \gamma < \infty$, $x = (x_1, x_2, \ldots, x_n)$ and $t = (t_1, t_2, \ldots, t_n)$ are points in n-dimensional Euclidean space $\mathbb{R}^n$, $u(x,y) = u(f)(x,y)$, $y > 0$ is the Poisson integral of $f \in L^p(\mathbb{R}^n)$ for some $p \geq 1$. The purpose of this paper is to study weighted norm inequalities for these maximal functions which are the generalizations of the "maximal function" $\mu_\lambda(f)(x)$ defined by E. M. Stein [S1, p.236]. Originally Stein introduced the "maximal function" $\mu_\lambda(f)(x)$

$$\mu_\lambda(f)(x) = \sup_{y>0} \left( \int_{\mathbb{R}^n} |u(x-t, y)|^2 \, y^{-n} \left(\frac{y}{|t|+y}\right)^{n\lambda} dt \right)^{1/2}$$

in the 1-dimensional periodic set-up [S2], where these functions play an important role in questions related to Fourier series.

Before we state our results, we consider the measures with respect to which norms are taken have the form $d\mu(x) = W(x)dx$. A weight $W(x)$ is said to satisfy condition $A_p$ for some p, $1 < p < \infty$ if $W(x)$ is nonnegative, locally integrable function which satisfies

$$(A_p) \quad \left(\frac{1}{|Q|} \int_Q W(x)dx\right) \left(\frac{1}{|Q|} \int_Q W(x)^{-1/(p-1)} dx\right)^{p-1} \leq C$$

for all n-dimensional cubes Q with sides parallel to the coordinate planes, $|Q|$ being the volume of Q, and C a constant independent of Q. When $p = 1$, $W(x)$ is said to satisfy condition $A_1$, if $W(x)$ is nonnegative, locally integrable and

$(A_1)\quad (MW)(x) \le C\cdot W(x),$

where $(MW)(x)$ is the Hardy-Littlewood maximal function of $W$, i.e.,

$$(MW)(x) = \sup_{Q\ni x} \frac{1}{|Q|}\int_Q W(z)dz.$$

We will write $W \in A_p$, or $W \in A_1$ for such $W$. These classes were introduced in Muckenhoupt [M] and in an equivalent form by Rosenblum [R].

For a measurable set $S$, we will use the notation

$$W\{S\} = \int_S W(x)dx$$

for the $W$-measure of $S$. The following well known properties, which we shall need about $A_p$, are listed below. For proofs see [M].

Let $(Mf)(x)$ denote the Hardy-Littlewood maximal function of $f$; then

$$(1.2)\qquad \int_{\mathbb{R}^n} (Mf(x))^p\, W(x)\, dx \le C\int_{\mathbb{R}^n} |f(x)|^p\, W(x)\, dx$$

for $1 < p < \infty$ if and only if $W \in A_p$. The $A_p$ condition is also necessary and sufficient for the weak type inequality

$$(1.3)\qquad W\{(Mf)(x) > \alpha\} \le C\,\alpha^{-p}\int_{\mathbb{R}^n} |f(x)|\, W(x)\, dx$$

for $1 \le p < \infty$.

Moreover, $W \in A_p$ implies that $W \in A_q$ $(q \ge p)$; conversely, if $p > 1$, $W \in A_p$, then

$$(1.4)\qquad W(x) \in A_{p-\varepsilon}$$

for some $\varepsilon > 0$ depending on $W$.

We now state our results.

THEOREM 1. Let $1 < \lambda \le \gamma < \infty$ and $p_o = \gamma/\lambda$, $p > \mu > p_o$.

(i) If $f \in L^p(Wdx)$, $W \in A_{p/p_o}$, then there is a constant $C = C_{\lambda,\gamma,\mu,n}$ so that

$$(1.5)\qquad \mu_{\lambda,\gamma}(f)(x) \le C((M_\mu f)(x))^{1-p_o/\gamma}(M(Mf)^{p_o})^{1/\gamma}(x),$$

where $(M_\mu f)(x) = ((M|f|^\mu)(x))^{1/\mu}$.

(ii) If $p_o < p < \infty$, $W \in A_{p/p_o}$ and $f \in L^p(Wdx)$, then the mapping $f \to \mu_{\lambda,\gamma}(f)$ from $L^p(Wdx)$ to $L^p(Wdx)$ is of strong type $(p,p)$, i.e.,

$$(1.6)\qquad \int_{\mathbb{R}^n} (\mu_{\lambda,\gamma}(f)(x))^p\, W(x)\, dx \le C \int_{\mathbb{R}^n} |f(x)|^p\, W(x)\, dx,$$

the constant C being independent of f.

THEOREM 2. Let $1 < \lambda \le \gamma < \infty$, $p_0 = \gamma/\lambda$. If $W \in A_1$ and $f \in L^{p_0}(Wdx)$, then the mapping $f \to \mu_{\lambda,\gamma}(f)$ is of weak type $(p_0,p_0)$ from $L^{p_0}(Wdx)$ to $L^{p_0}(Wdx)$, i.e.,

$$(1.7)\qquad W\{\, \mu_{\lambda,\gamma}(f)(x) > \alpha \,\} \le C\, \alpha^{-p_0} \int_{\mathbb{R}^n} |f(x)|^{p_0}\, W(x)\, dx \qquad (\alpha > 0),$$

C being a constant independent of f.

Theorems 1 and 2 are proved in §2 and §3.

In comparison with weighted norm inequalities for Hardy-Littlewood maximal functions, it is natural to ask whether the conditions "$W \in A_{p/p_0}$" and "$W \in A_1$" in Theorem 1(ii) and Theorem 2 can be replaced by more weaker conditions "$W \in A_p$" and "$W \in A_{p_0}$" respectively. The results obtained in §4 are as follows.

THEOREM 3.

(i) In general, inequality (1.6) does not hold for $W \in A_p$.

(ii) Suppose that $W(x) = |x|^{\alpha}$. Then (1.6) holds true if and only if $W \in A_{p/p_0}$.

THEOREM 4.

(i) In general, (1.7) does not hold for $W \in A_{p_0}$.

(ii) If $W(x) = |x|^{\alpha}$, then the condition $W \in A_1$ is also necessary and sufficient for the inequality (1.7).

Some further properties are investigated in §5. First, we prove that the weak type estimates for mapping $f \to \mu_{\lambda,\gamma}(f)$ cannot be strengthened to strong estimates in the case $p = p_0 = \gamma/\lambda$. More precisely the following theorem is true.

THEOREM 5. Under the assumptions of Theorem 2, the weak inequality cannot be strengthened by a strong type inequality. In fact, there exists a function $g \in L^{p_0}(Wdx)$ with $W \in A_1$ such that the strong type inequality (1.6) for g does not hold true.

As for $p < \gamma/\lambda$, we have

THEOREM 6. Let $1 \le p < p_0 = \gamma/\lambda$. Then there exist a weight $W(x) \in A_1$ and $f \in L^p(Wdx)$ so that $\mu_{\lambda,\gamma}(f)(x) \equiv \infty$ everywhere.

§2. The following lemmas will be used in the proof of Theorem 1 and 2.

LEMMA 1. Let $p \ge \mu \ge 1$. If $M_p f(x)$ is finite for some $x \in \mathbb{R}^n$, then the Poisson integral $u(x,y)$ of $f$ is finite everywhere in $\mathbb{R}^{n+1}_+$, and

$$(2.1)\quad |u(x-t,y)| \le C_n \left(1 + \frac{|t|}{y}\right)^n Mf(x),$$

and more generally

$$(2.2)\quad |u(x-t,y)| \le C_{n,\mu}\left(1 + \frac{|t|}{y}\right)^{n/\mu} M_\mu f(x),$$

where

$$(2.3)\quad M_\mu f(x) = (M(M|f|^\mu)(x))^{1/\mu}$$

PROOF. First we note that the finiteness of $M_p f(x)$ for some $x$ leads to the existence of the integral

$$(2.4)\quad \int_{\mathbb{R}^n} \frac{|f(t)|}{1+|t|^{\lambda n}}\, dt < \infty \qquad (\lambda > 1).$$

This condition is equivalent to the finiteness of the Poisson integral $u(x,y)$ of $f$ at all points $(x,y) \in \mathbb{R}^{n+1}_+$.

Thus,

$$|u(x,y)| = \left| \frac{c}{y^n} \int_{\mathbb{R}^n} f(x-t) \frac{1}{(1 + |\frac{t}{y}|^2)^{(n+1)/2}}\, dt \right|$$

$$(2.5)\quad \le C \sum_{k=-\infty}^{\infty} \sup_{2^k y \le |t| < 2^{k+1} y} \frac{1}{(1 + |\frac{t}{y}|^2)^{(n+1)/2}} \int_{|t|<2^{k+1}y} |f(x-t)|\, dt$$

$$\le C\cdot Mf(x) \int_{\mathbb{R}^n} \frac{dz}{(1 + |z|^2)^{(n+1)/2}}.$$

Now, by virtue of inequality (2.5), the argument used to prove Lemma 4 in [S1, p.92] also works for the proof of (2.1) and (2.2).

LEMMA 2. If $1 < p < \infty$ and $W \in A_p$ with constant $K$, then there is a constant $C$, depending only on $p$ and $K$, such that for every cube $Q$ and its complement $Q^c$, we have

$$(2.6)\quad |Q|^p \int_{Q^c} \frac{W(x)dx}{|x-x_0|^{np}} \le C \int_Q W(x)dx \quad (x_0 \text{ being the center of } Q).$$

This lemma is known in the special case $n = 1$, see [HMW, p.232]. Their proof does not work in the general case $n > 1$. Here we shall give an alternative proof which covers both the cases $n = 1$ and $n > 1$.

PROOF OF LEMMA 2

Let $2Q$ be a cube with the same center as $Q$, but with twice as large a side. Then

$$(2.7)\qquad |Q|^p \int_{Q^c} \frac{W(x)}{|x-x_o|^{np}}\,dx = |Q|^p \Big(\int_{(2Q)^c} + \int_{2Q-Q^c}\Big) \frac{W(x)}{|x-x_o|^{np}}\,dx = I + II$$

say. Suppose that the length of $Q$ is $d$; then clearly for $x \in Q^c$, ($Q^c$ being the complement of $Q$)

$$|x-x_o| \geq d$$

so

$$|x-x_o|^{np} \geq |Q|^p.$$

Hence it follows that

$$(2.8)\qquad II \leq \int_{2Q-Q^c} W(x)dx \leq W\{2Q\} \leq C\cdot W[Q] = C\int_Q W(x)dx,$$

since the $A_p$ condition implies the "doubling condition".

As for I, we consider a suitable maximal function as follows. For a given $Q$ and $x \in (2Q)^c$, let $Q_x$ be a cube whose center is $x$ and length $d$ is equal to $2\sup_{y\in Q}|x-y|$. Then it is easy to verify that $Q \subset Q_x$ and

$$|Q_x| = (2d)^n \leq C\,|x-x_o|^n \qquad \text{(being the center of } Q\text{)},$$

since $C_1|x-y| \leq |x-x_o| \leq C_2|x-y|$ if $x \in (2Q)^c$ and $y$ is any point of $Q$.

Now, suppose that $x \in (2Q)^c$. By definition of the Hardy-Littlewood maximal function,

$$(M\chi_Q)(x) \geq \frac{1}{|Q_x|}\int_{Q_x} \chi_Q(\xi)d\xi = \frac{|Q|}{|Q_x|} \geq C\,\frac{|Q|}{|x-x_o|^n}\,.$$

As an application of (1.2), it follows that

$$C|Q|^p \int_{(2Q)^c} \frac{W(x)}{|x-x_o|^{np}}\,dx \leq \int_{(2Q)^c} (M\chi_Q(x))^p\,W(x)\,dx$$

$$\leq C\int_{\mathbb{R}^n} (\chi_Q(x))^p\,W(x)\,dx \leq C\int_Q W(x)\,dx,$$

that is

$$I \le C \int_Q W(x)dx. \tag{2.9}$$

Combining (2.8) with (2.9), this completes the proof of Lemma 2.

PROOF OF THEOREM 1.

Before we prove (i), we make some comments. We note that (1.2) shows that

$$\int_{\mathbb{R}^n} (M_\mu f(x))^p W(x)dx = \int_{\mathbb{R}^n} ((M|f|^\mu)(x))^{p/\mu} W(x)dx \le C \int_{\mathbb{R}^n} |f|^p W(x)dx < \infty$$

and

$$\int_{\mathbb{R}^n} ((M(Mf)^{p_o})(x))^{p/p_o} W(x)dx \le C \int_{\mathbb{R}^n} (Mf(x))^p \cdot W(x)dx \le C \int_{\mathbb{R}^n} |f|^p W(x)dx < \infty$$

under the assumption of Theorem 1. In particular, $f \in L^p(Wdx)$ implies $M_\mu f(x)$ and $M((Mf)^{p_o}(x)$ are finite for almost all x. Now we come to

(i). Using (2.2), we get

$$\int_{\mathbb{R}^n} |u(x-t,y)|^\gamma \left(\frac{y}{|t|+y}\right)^{n\lambda} y^{-n}\, dt \tag{2.10}$$

$$= \int_{\mathbb{R}^n} |u(x-t,y)|^{\gamma-p_o} \left(\frac{y}{|t|+y}\right)^{(\gamma-p_o)n/\mu} |u(x-t,y)|^{p_o} \left(\frac{y}{|t|+y}\right)^{(\lambda-(\gamma-p_o)/\mu)n} y^{-n}\, dt$$

$$\le C_{\mu,\lambda,\gamma,n} (M_\mu f(x))^{\gamma-p_o} \int_{\mathbb{R}^n} (Mf(t))^{p_o} \left(\frac{y}{|x-t|+y}\right)^{(\lambda-(\gamma-p_o)/\mu)n} y^{-n}\, dt.$$

Now the same method used in proving (2.5) gives the estimate of the last integral:

$$\int_{\mathbb{R}^n} (Mf(t))^{p_o} \left(\frac{y}{|x-t|+y}\right)^{(\lambda-(\gamma-p_o)/\mu)n} y^{-n}\, dt \tag{2.11}$$

$$= \frac{1}{y^n} \int_{\mathbb{R}^n} (Mf(x-t))^{p_o} \frac{1}{(1+\frac{|t|}{y})^{(\lambda-(\gamma-p_o)/\mu)n}}\, dt \le M((Mf)^{p_o})(x),$$

since the assumption $\mu > p_0$ implies the condition $\lambda - (\gamma - p_0)/\mu > 1$. Combining (2.6) and (2.11), we get (1.5).

(ii) For $p > p_0$, choose $\mu$ such that $p > \mu > p_0$ and $W \in A_{p/\mu}$. This is possible, since (1.4) holds. For $p > \mu > p_0$, we apply (1.5) and Holder's inequality and we get

$$\int_{\mathbb{R}^n} (\mu_{\lambda,\gamma}(f)(x))^p \cdot W(x)dx \le C \int_{\mathbb{R}^n} (M_\mu f(x))^{p(\gamma-p_0)/\gamma} (M(Mf)^{p_0}(x))^{p/\gamma} \cdot W(x)dx$$

$$\le C \left( \int_{\mathbb{R}^n} (M_\mu f(x))^p \cdot W(x)dx \right)^{(\gamma-p_0)/\gamma} \left( \int_{\mathbb{R}^n} (M(Mf)^{p_0})^{p/p_0}(x) \cdot W(x)dx \right)^{p_0/\gamma}$$

$$\le C \left( \int_{\mathbb{R}^n} |f|^p \cdot W(x)dx \right)^{(\gamma-p_0)/\gamma} \left( \int_{\mathbb{R}^n} (Mf(x))^p \cdot W(x)dx \right)^{p_0/\gamma}$$

$$\le C \int_{\mathbb{R}^n} |f|^p \, W(x)dx,$$

in which we have used property (1.2) and condition $W \in A_{p/\mu}$. This completes the proof of Theorem 1.

§3. DECOMPOSITION LEMMA. Let $f \in L^{p_0}(Wdx)$, $W \in A_{p_0}$, and let $\alpha > 0$ be given. There is a collection $\{Q_j\}$ of pairwise disjoint cubes with the following properties:

(3.1) $$\sum_j W\{Q_j\} \le C \cdot \alpha^{-p_0} \int_{\mathbb{R}^n} |f(x)|^{p_0} \cdot W(x)dx$$

(3.2) $$|f(x)| \le C\,\alpha \qquad \text{for } x \notin \Omega = \bigcup_j Q_j.$$

(3.3) $$\frac{1}{W\{Q_j\}} \int_{Q_j} |f(x)|^{p_0} W(x)dx \le C\,\alpha^{p_0} \qquad \text{for each } Q_j.$$

(3.4) For any cube $Q_j$ of the collection, let $(2Q_j)$ be a cube with the same center as $Q_j$, but with twice as large a side. Then no point of $\mathbb{R}^n$ lies in more than N of the cubes $(2Q_j)$.

(3.5) Two distinct cubes $Q_1$ and $Q_2$ of $\{Q_j\}$ are said to touch, if their boundaries have a common point. The following results are known. Suppose that $Q \in \{Q_j\}$. Then there at most N cubes in $\{Q_j\}$ which touch Q. Furthermore, if $Q_1$ and $Q_2$ touch, then

$$C_1 \operatorname{diam}(Q_2) \leq \operatorname{diam}(Q_1) \leq C_2 \operatorname{diam}(Q_2).$$

For proofs see [GC, p.143] and [S1, p.169].

PROOF OF THEOREM 2. Let $f \in L^{p_o}(Wdx)$, $W \in A_1$ and $\alpha > 0$ be given. First we assume that $p_o > 1$ and we have to show that

$$(3.6) \qquad W\{\mu_{\lambda,\gamma}(f)(x) > C\alpha\} \leq C\cdot\alpha^{-p_o}\int_{\mathbb{R}^n}|f(x)|^{p_o}\cdot W(x)dx \qquad (\alpha > 0)$$

with C independent of f and $\alpha$.

Apply the decomposition lemma to f and $\alpha$, to obtain a collection $\{Q_j\}$ of cubes, satisfying (3.1) through (3.5) above. Define a function g(x) on $\mathbb{R}^n$ by

$$g(x) = \begin{cases} f(x) & (x \notin \Omega = \bigcup_j Q_j), \\ \dfrac{1}{W\{Q_j\}}\displaystyle\int_{Q_j} f(x)\cdot W(x)dx & (x \in Q_j) \end{cases}$$

Setting $b(x) = f(x) - g(x)$, we obtain a decomposition with the following properties:

$$(3.7) \qquad |g(x)| \leq C\cdot\alpha \quad \text{a.e., and} \quad \int_{\mathbb{R}^n}|g(x)|^{p_o}W(x)dx \leq C\int_{\mathbb{R}^n}|f(x)|^{p_o}W(x)dx;$$

(3.8) $b(x)$ is supported on $\Omega$;

$$(3.9) \qquad \frac{1}{W\{Q_j\}}\int_{Q_j}|b(x)|^{p_o}W(x)dx \leq C\cdot\alpha^{p_o} \qquad (j = 1,2,\ldots)$$

$$(3.10) \qquad \int_{Q_j} b(x)\,W(x)dx = 0 \qquad (j = 1,2,\ldots).$$

Let $p > p_o$. Then Property (3.7) clearly implies that

$$(3.11) \quad \int_{\mathbb{R}^n} |g(x)|^p \, W(x)dx \le C\cdot\alpha^{p-p_o} \int_{\mathbb{R}^n} |g(x)|^{p_o} \, W(x)dx.$$

By Theorem 1(ii), $\mu_{\lambda,\gamma}(f)$ is a bounded operator on $L^p(Wdx)(p>p_o)$, so that by the Chebyshev inequality and (3.11),

$$(3.12) \quad W\{\mu_{\lambda,\gamma}(g)(x) > \alpha\} \le \frac{C}{\alpha^p} \int_{\mathbb{R}^n} |g(x)|^p \, W(x)dx \le \frac{C}{\alpha^{p_o}} \int_{\mathbb{R}^n} |f(x)|^{p_o} \, W(x)dx.$$

On the other hand, $\mu_{\lambda,\gamma}(f)(x) \le \mu_{\lambda,\gamma}(g)(x) + \mu_{\lambda,\gamma}(b)(x)$. So in order to prove Theorem 2, it is sufficient to prove that

$$(3.13) \quad W\{\mu_{\lambda,\gamma}(b)(x) > C\,\alpha\} \le \frac{C}{\alpha^{p_o}} \int_{\mathbb{R}^n} |f(x)|^{p_o} \, W(x)dx.$$

If $x \in \mathbb{R}^n$ and $Q_j$ is a cube from the collection, by $x \sim Q_j$ we mean that $x$ belongs to a cube $Q_\ell$ (also from the collection) which touches or coincides with $Q_j$. Note that for fixed $x$, $x \sim Q_j$ holds at most N Whitney cubes; and that if $x \notin \Omega$ then $x \sim Q_j$ never holds. Now, let

$$(3.14) \quad b_j(x) = b(x) \cdot \chi_{Q_j}(x),$$

where $\chi_E(x)$ denotes the characteristic function of the set E, and let $u_j(x,y)$ denote the Poisson integral of $b_j(x)$. By definition,

$$(3.15) \quad \mu_{\lambda,\gamma}(b)(x) = \sup_{y>0} \Big( \int_{\mathbb{R}^n} \Big(\frac{y}{|x-t|+y}\Big)^{n\lambda} y^{-n} \Big|\sum_j u_j(t,y)\Big|^\gamma dt\Big)^{1/\gamma}$$

$$\le \mu_{\lambda,\gamma}^{(1)}(b)(x) + \mu_{\lambda,\gamma}^{(2)}(b)(x),$$

where

$$(3.16) \quad (\mu_{\lambda,\gamma}^{(1)}(b)(x))^\gamma = \sup_{y>0} \int_{\mathbb{R}^n} \Big(\frac{y}{|x-t|+y}\Big)^{n\lambda} y^{-n} \Big|\sum_{t \not\sim Q_j} u_j(t,y)\Big|^\gamma dt,$$

$$(3.17) \quad (\mu_{\lambda,\gamma}^{(2)}(b)(x))^\gamma = \sup_{y>0} \int_{\mathbb{R}^n} \Big(\frac{y}{|x-t|+y}\Big)^{n\lambda} y^{-n} \Big|\sum_{t \sim Q_j} u_j(t,y)\Big|^\gamma dt.$$

Now

$$(3.18)\quad |\sum_{t\not\approx Q_j} u_j(t,y)| \le \sum_{t\not\approx Q_j} \int_{Q_j} |b(z)| \frac{y\,dz}{(|t-z|^2+y^2)^{(n+1)/2}}$$

$$\le \sum_{t\not\approx Q_j} \sup_{z\in Q_j} \frac{y}{(|t-z|^2+y^2)^{(n+1)/2}} \int_{Q_j} |b(z)|\,dz.$$

By Holder's inequality, the $A_{p_o}$ condition and (3.9),

$$(3.19)\quad \int_{Q_j} |b(z)|\,dz$$

$$\le (|Q_j|^{-1} \int_{Q_j} |b(z)|^{p_o} W(z)\,dz)^{1/p_o} (|Q_j|^{-1} \int_{Q_j} W(z)^{-1/(p_o-1)}\,dz)^{1-1/p_o} |Q_j|$$

$$\le C\,((W\{Q_j\})^{-1} \int_{Q_j} |b(z)|^{p_o} W(z)\,dz)^{1/p_o} \cdot |Q_j| \le C\cdot\alpha\cdot|Q_j|. \qquad \text{(by (3.9))}.$$

From (3.18) and (3.19), one verifies that

$$(3.20)\quad |\sum_{t\not\approx Q_j} u_j(t,y)| \le C\cdot\alpha \sum_{t\not\approx Q_j} \sup_{z\in Q_j} \frac{y}{(|t-z|^2+y^2)^{(n+1)/2}} |Q_j|$$

$$\le C\cdot\alpha \sum_{t\not\approx Q_j} \int_{Q_j} \frac{y}{(|t-z|^2+y^2)^{(n+1)/2}}\,dz$$

$$\le C\cdot\alpha \int_{\mathbb{R}^n} \frac{y\,dz}{(|t-z|^2+y^2)^{(n+1)/2}} \le C\cdot\alpha$$

since

$$|Q_j| \sup_{z\in Q_j} \frac{y}{(|t-z|^2+y^2)^{(n+1)/2}} \le C \int_{Q_j} \frac{y}{(|t-z|^2+y^2)^{(n+1)/2}}\,dz$$

for any cube $Q_j$ satisfying $t \not\approx Q_j$ by property (3.5). Therefore, by (3.16) and (3.20), we get

$$(3.21)\quad (\mu_{\lambda,\gamma}^{(1)}(b)(x))^{\gamma} \le C\cdot\alpha^{\gamma} \sup_{y>0} \int_{\mathbb{R}^n} (\frac{y}{|x-t|+y})^{n\lambda} y^{-n}\,dt \le C\cdot\alpha^{\gamma}.$$

So, by virtue of (3.15) and (3.21), to prove (3.13) we need only show that

$$W\{\mu_{\lambda,\gamma}^{(2)}(b)(x) > C\,\alpha\} \leq C\cdot\alpha^{-p_o}\int_{\mathbb{R}^n}|f(x)|^{p_o}\,W(x)dx.$$

Let $\Omega^* = \bigcup_j(2Q_j)$ and recall that

$$W\{\Omega^*\} \leq C\sum_j W\{Q_j\} \leq C\cdot\alpha^{-p_o}\int_{\mathbb{R}^n}|f(x)|^{p_o}\,W(x)\,dx,$$

so it will be enough to prove that

$$\text{(3.22)}\quad W\{x \in \mathbb{R}^n-\Omega^*;\ \mu_{\lambda,\gamma}^{(2)}(b)(x) > C\,\alpha\} \leq C\cdot\alpha^{-p_o}\int_{\mathbb{R}^n}|f(x)|^{p_o}\,W(x)dx.$$

Now $t \notin \Omega$ implies that $\sum_{t\sim Q_\ell} u_\ell(t,y)$ is an empty sum, so

$$\text{(3.23)}\quad (\mu_{\lambda,\gamma}^{(2)}(b)(x))^\gamma = \sup_{y>0}\sum_j\int_{Q_j}\left(\frac{y}{|x-t|+y}\right)^{n\lambda} y^{-n}\Big|\sum_{t\sim Q_\ell}u_\ell(t,y)\Big|^\gamma dt$$

$$\leq \sup_{y>0}\sum_j\frac{1}{|x-t_j|^{n\lambda}}\int_{Q_j} y^{\lambda n-n}\Big|\sum_{t\sim Q_\ell}u_\ell(t,y)\Big|^\gamma dt,$$

since $C_1|x-t_j| \leq |x-t| \leq C_2|x-t_j|$ where $t_j$ is the center of $Q_j$ and $t$ is any point in $Q_j$.

We will invoke two estimates on $u_\ell(t,y) = b_\ell * P(t,y)$ with $b_\ell = b\cdot\chi_{Q_\ell}$ (see (3.14)):

$$\text{(3.24)}\quad |u_\ell(t,y)| \leq C\cdot y^{-n}\int_{Q_\ell}|b(x)|dx \leq C\cdot\alpha\cdot y^{-n}|Q|_\ell,$$

and

$$\text{(3.25)}\quad \int_{Q_j}|u_\ell(t,y)|^s\,dt \leq C\cdot\alpha^s|Q_j|$$

for any $Q_\ell$ which touches $Q_j$, and $1 \leq s \leq p_o$.

In fact,

$$|u_\ell(t,y)| = |\int_{\mathbb{R}^n} b_\ell(z)P(t-z,y)dz| \leq \|P(\cdot, y)\|_{L^\infty} \|b_\ell\|_1 \leq C\, y^{-n} \|b_\ell\|_1 .$$

So (3.24) follows from (3.19).

On the other hand, the condition $W(t) \in A_1$ implies that

$$\text{(3.26)} \quad \underset{t \in Q}{\text{ess sup}} \frac{1}{W(t)} \leq C\, |Q| / \int_Q W(t)dt.$$

Therefore, by Holder's inequality, (2.5), and (1.2),

$$\int_{Q_j} |u_\ell(t,y)|^s\, dt \leq \left(\int_{Q_j} |u_\ell(t,y)|^{p_o}\, dt\right)^{s/p_o} |Q_j|^{1-s/p_o}$$

$$\leq \left(\underset{t \in Q_j}{\text{ess sup}} \frac{1}{W(t)} \int_{Q_j} |u_\ell(t,y)|^{p_o}\, W(t)\, dt\right)^{s/p_o} |Q_j|^{1-s/p_o}$$

$$\leq \left(\frac{1}{W\{Q_j\}} \int_{\mathbb{R}^n} ((Mb_\ell)(t))^{p_o}\, W(t)\, dt\right)^{s/p_o} |Q_j|$$

$$\leq C \left(\frac{1}{W\{Q_j\}} \int_{Q_\ell} (b_\ell(t))^{p_o}\, W(t)\, dt\right)^{s/p_o} |Q_j|$$

$$< C \left(\frac{1}{W\{Q_\ell\}} \int_{Q_\ell} (b_\ell(t))^{p_o}\, W(t)\, dt\right)^{s/p_o} |Q_j|$$

$$\leq C \cdot \alpha^s\, |Q_j|,$$

in which we have used the following inequalities

$$C_1\, W\{Q_\ell\} \leq W\{Q_j\} \leq C_2\, W\{Q_\ell\}$$

which follow from the condition that $Q_\ell$ touches $Q_j$, so (3.5) holds, and Wdx is a doubling measure. This completes the proof of (3.24) and (3.25).

Now we come to the last integral of (3.23). First note that for $t \in Q_j$,

$$\sum_{t \sim Q_\ell} u_\ell(t,y) = \sum_{t_j \sim Q_\ell} u_\ell(t,y) \qquad (t_j \text{ being the center of } Q_j).$$

Moreover, by (3.5) the number of terms in $\sum_{t_j \sim Q_\ell} u_\ell(t,y)$ are bounded for all $t_j (j = 1,2,\ldots)$. Hence, using (3.24) and (3.25) the last integral in (3.23) is less than

$$(3.27)\quad \int_{Q_j} y^{n(\lambda-1)} \Big|\sum_{t\sim Q_\ell} u_\ell(t,y)\Big|^\gamma \, dt$$

$$\le C\, y^{n(-1+\gamma/p_0)} \sum_{t_j\sim Q_\ell} \int_{Q_j} |u_\ell(t,y)|^{-1+\gamma/p_0} |u_\ell(t,y)|^{\gamma+1-\gamma/p_0} \, dt$$

$$\le C \sum_{t_j\sim Q_\ell} \alpha^{-1+\gamma/p_0} |Q_\ell|^{-1+\gamma/p_0} \int_{Q_j} |u_\ell(t,y)|^{\gamma+1-\gamma/p_0} \, dt$$

$$\le C\, \alpha^\gamma \sum_{t_j\sim Q_\ell} |Q_\ell|^{-1+\gamma/p_0} |Q_j|$$

$$\le C\, \alpha^\gamma |Q_\ell|^{-1+\gamma/p_0} |Q_j|$$

$$\le C\, \alpha^\gamma |Q_j|^{\gamma/p_0}.$$

The last two steps follow from (3.5) again. Thus, by (3.23) and (3.27) we get

$$(3.28)\quad (\mu^{(2)}_{\lambda,\gamma}(b)(x))^\gamma \le C\, \alpha^\gamma \sum_j \frac{|Q_j|^{\gamma/p_0}}{|x-t_j|^{n\lambda}} .$$

So in order to complete the proof of inequality (3.22), and with it, that of Theorem 2, we have only to prove that

$$(3.29)\quad \int_{\mathbf{R}^n - \Omega^*} \sum_j \frac{|Q_j|^{\gamma/p_0}}{|x-t_j|^{n\lambda}} W(x)\,dx \le C\, \alpha^{-p_0} \int_{\mathbf{R}^n} |f(x)|^{p_0} W(x)\,dx.$$

But if we invoke Lemma 2, we immediately get

$$\int_{\mathbf{R}^n - \Omega^*} \sum_j \frac{|Q_j|^{\gamma/p_0}}{|x-t_j|^{n\lambda}} W(x)\,dx = \sum_j \int_{\mathbf{R}^n - \Omega^*} \frac{|Q_j|^\lambda}{|x-t_j|^{n\lambda}} W(x)\,dx$$

$$\leq \sum_j \int_{(Q_j)^c} \frac{|Q_j|^\lambda}{|x-t_j|^{n\lambda}} W(x)\, dx \leq C \sum_j \int_{Q_j} W(x)\, dx$$

$$\leq C \sum_j W\{Q_j\} \leq C\cdot\alpha^{-p_o} \int_{\mathbb{R}^n} |f(x)|^{p_o} W(x)\, dx.$$

This completes the proof of (3.29), and with it, that of Theorem 2.

REMARK. We thus complete the proof of Theorem 2 in the case $p_o > 1$. When $p_o = 1$, the argument also works if (3.26) is used to replace Holder's inequality.

§4. We begin by a lemma.

**LEMMA 3.** Let $1 < \lambda, \gamma < \infty$ and let $0 < |x|$, $0 < y_o \leq |x|/20$. Then there exists a constant $C = C_{n,\lambda,\gamma}$, depending only on $n$, $\lambda$, $\gamma$ such that

$$\mu_{\lambda,\gamma}(f)(x) \geq C\, |x|^{-n\lambda/\gamma}\, y_o^{n\lambda/\gamma}\, |u(0,y_o)|, \tag{4.1}$$

where

$$u(0,y_o) = (f * P)(0,y_o).$$

PROOF OF LEMMA 3. First we observe that if $|t| < |x|/10$, $0 < y_o < |x|/10$, then $|x-t| + y < 6|x|/5$.
Hence

$$\left(\frac{y}{|x-t|+y}\right)^{n\lambda} > C_{n,\lambda}\left(\frac{y}{|x|}\right)^{n\lambda}$$

Therefore we have

$$(\mu_{\lambda,\gamma}(f)(x))^\gamma \geq \sup_{y_o/2\leq y\leq 3y_o/2} \int_{|t|<|x|/10} |u(t,y)|^\gamma \left(\frac{y}{|x-t|+y}\right)^{n\lambda} y^{-n}\, dt \tag{4.2}$$

$$\geq \frac{C_{n,\lambda}}{|x|^{n\lambda}} \sup_{y_o/2\leq y\leq 3y_o/2} \int_{|t|<|x|/10} |u(t,y)|^\gamma\, y^{n(\lambda-1)}\, dt.$$

On the other hand, denoting $B = B((0,y_o), y_o/2)$ the ball at center $(0,y_o) \in \mathbb{R}_+^{n+1}$ with radius $y_o/2$, by a lemma of Hardy-Littlewood [FS, p.172], we get

$$|u(0,y_0)|^\gamma \le C_n y_0^{-n-1} \iint_B |u(t,y)|^\gamma dtdy \le C_n y_0^{-n-1} \int_{y_0/2}^{3y_0/2} dy \int_{|t|<|x|/10} |u(t,y)|^\gamma dt$$

$$\le C_{n,\lambda}\, y_0^{-n\lambda-1} \int_{y_0/2}^{3y_0/2} y^{n(\lambda-1)} dy \int_{|t|<|x|/10} |u(t,y)|^\gamma dt$$

(4.3)

$$\le C_{n,\lambda}\, y_0^{-n\lambda-1} \int_{y_0/2}^{3y_0/2} \sup_{y_0/2\le y\le 3y_0/2} \left( \int_{|t|<|x|/10} |u(t,y)|^\gamma y^{n(\lambda-1)} dt\right) dy$$

$$= C_{n,\lambda}\, y_0^{-n\lambda} \sup_{y_0/2\le y\le 3y_0/2} \int_{|t|<|x|/10} |u(t,y)|^\gamma\, y^{n(\lambda-1)}\, dt.$$

From (4.2) and (4.3) we obtain the result (4.1).

We now come to prove Theorem 3.

PROOF OF THEOREM 3. It is clear that we only have to prove part (ii) of Theorem 3. Let $W(x) = |x|^\beta$. Then $W \in A_p$ if and only if $-n < \beta < n(p-1)$.

Now we consider the function

(4.4) $$f_\delta(x) = \begin{cases} |x|^{-\delta} & (|x|\le 1) \\ 0 & (|x|>1) \end{cases} \qquad (0<\delta<(n+\beta)/p).$$

Suppose that

(4.5) $$n(p/p_0-1) < \beta < n(p-1).$$

Then it can be easily verified that $f_\delta \in L^p(Wdx)$, i.e.,

(4.6) $$\int_{\mathbb{R}^n} |f_\delta(x)|^p W(x)dx = \int_{|x|\le 1} |x|^{-p\delta+\beta}\, dx < \infty$$

since $0 < \delta < (n+\beta)/p$ and $-n < \beta < n(p-1)$ imply that $-p\delta + \beta > -n$.

On the other hand,

$$(4.7)\qquad u(f_\delta)(0,y_0) = C\int_{\mathbb{R}^n} \frac{1}{|x|^\delta}\frac{1}{(|x|^2+y_0^2)^{(n+1)/2}}\,dx$$

$$\geq \frac{C}{y_0^{\,n}}\int_{|x|<y_0} |x|^{-\delta}dx = C\, y_0^{-\delta}.$$

By (4.1), we have

$$\mu_{\lambda,\gamma}(f_\delta)(x) \geq C\,|x|^{-n/p_0}\, y_0^{\,n/p_0-\delta} \qquad (0 < y_0 \leq |x|/20).$$

Now if (1.6) were true for $f_\delta$ and $W = |x|^\beta \in A_p$, there would be

$$(4.8)\qquad C\int_{|x|>20} |x|^{-np/p_0+\beta}dx \leq \int_{\mathbb{R}^n} (\mu_{\lambda,\gamma}(f_\delta)(x))^p|x|^\beta dx \leq C\int_{\mathbb{R}^n}|f_\delta|^p|x|^\beta dx < \infty.$$

However, the latter is equivalent to

$$(4.9)\qquad \beta < n(p/p_0 - 1)$$

which contradicts (4.5). In other words, (1.6) will never be true, if $|x|^\beta \notin A_{p/p_0}$. On the other hand, Theorem 1 shows that (1.6) is true if $|x|^\beta \in A_{p/p_0}$. This completes the proof of Theorem 3.

§5. PROOF OF THEOREM 4. The same argument used to prove Theorem 3 can also be applied to prove Theorem 4, by setting $f = f_\delta(x)$ with $0 < \delta < (n+\beta)/p_0$ and $W = |x|^\beta$ with $0 < \beta < n(p_0-1)$, $p_0 > 1$.

The details are omitted.

§6. PROOF OF THEOREM 5. Set $g(x) = f_\delta(x)$ with $0 < \delta < (n+\beta)/p_0$, and $W(x) = |x|^\beta$ with $-n < \beta < 0$. Then $W(x) \in A_1$ and

$$(6.1)\qquad \int_{\mathbb{R}^n} |g(x)|^{p_0}\, W(x)dx = \int_{|x|<1} |x|^{-p_0\delta+\beta}dx < \infty$$

i.e.,

$$g \in L^{p_0}(Wdx).$$

On the other hand, applying (4.1) of Lemma 3, we have

$$(6.2)\quad \mu_{\lambda,\gamma}(g)(x) \geq C\,|x|^{-n/p_0}\, y_0^{n/p_0-\delta} \qquad (0<y_0\leq |x|/20).$$

Hence

$$(6.3)\quad \int_{\mathbb{R}^n} (\mu_{\lambda,\gamma}(g)(x))^{p_0}\, W(x)dx \geq C \int_{|x|<1} |x|^{-n+\beta}dx = +\infty$$

This completes the proof of Theorem 5.

PROOF OF THEOREM 6. Let $1 \leq p < p_0 = \gamma/\lambda$. Choose $-n < \beta < 0$ such that $n/p_0 < (n+\beta)/p$. Define function $g(x) = f_\delta(x)$ with $n/p_0 < \delta < (n+\beta)/p$. Then clearly $|x|^\beta \in A_1$ $(-n < \beta < 0)$ and

$$(6.4)\quad g = f_\delta \in L^p(|x|^\beta dx).$$

On the other hand, if $x \neq 0$, by (4.1)

$$\mu_{\lambda,\gamma}(g)(x) \geq \frac{C}{|x|^{n/p_0}}\, y_0^{n/p_0-\delta} \to \infty \qquad (y_0 \downarrow 0,\ x \neq 0)$$

i.e.,

$$(6.5)\quad \mu_{\lambda,\gamma}(g) = +\infty \qquad (|x|>0).$$

If $x = 0$, then it is easily verified that

$$(6.6)\quad \mu_{\lambda,\gamma}(g)(0) \geq (Mg)(0) = +\infty.$$

This completes the proof of Theorem 6.

## REFERENCES


[CF] Coifman, R. and Fefferman, C., Weighted norm inequalities for maximal functions and singular integrals, Studia Math., 51(1974) 241-50.

[FS] Fefferman, C. and Stein, E. M., $H^p$ spaces of several variables, Acta Math., 129 (1972), 137-193.

[GC] Gercia-Cuerva, J. and Rubio de Francia, J., Weighted norm inequalities and related topics, 1985, North Holland.

[HMW] Hunt, R., Muckenhoupt, B. and Wheeden, R., Weighted norm inequalities for the conjugate function and Hilbert transform, Trans. Amer. Math. Soc. 176 (1973), 227-51.

[M] Muckenhoupt, B., Weighted norm inequalities for the Hardy maximal function, Trans. Amer. Math. Soc., 165 (1972), 207-26.

[R] Rosenblum, M., Summability of Fourier series in $L^p(d\mu)$, Trans. Amer. Math. Soc., 105 (1962), 32-42.

[S1] Stein, E. M., Singular integrals and differentiability properties of functions, 1970, Princeton.

[S2] Stein, E. M., A maximal function with applications to Fourier series, Ann. of Math., 68 (1958), 584-603

Proceedings of the Analysis Conference, Singapore 1986
S.T.L. Choy, J.P. Jesudason, P.Y. Lee (Editors)




# THE SECOND DUALS OF THE NONABSOLUTE CESARO SEQUENCE SPACES

Wu Bo-Er
South China Normal University, China

Liu Yu-Qiang
South China Normal University, China

Lee Peng-Yee
National University of Singapore, Singapore

We characterize completely the second duals of the nonabsolute Cesaro sequence spaces.

## 1. Introduction

Let X be a real sequence space. We write $x = \{x_k\}$ and define

$$X^\alpha = \{y = \{y_k\}: \sum_{k=1}^{\infty} |x_k y_k| \quad \text{converges for all } x \in X\},$$

$$X^\beta = \{y = \{y_k\}: \sum_{k=1}^{\infty} x_k y_k \quad \text{converges for all } x \in X\},$$

$$X^\gamma = \{y = \{y_k\}: \sup_{n\geq 1} \left|\sum_{k=1}^{n} x_k y_k\right| < +\infty \quad \text{for all } x \in X\}.$$

The sequences spaces $X^\alpha$, $X^\beta$ and $X^\gamma$ are called the $\alpha$-, $\beta$- and $\gamma$- duals of X respectively. We denote $(X^\zeta)^\eta$, the second dual of X, by $X^{\zeta\eta}$, where $\zeta, \eta = \alpha$, $\beta$ or $\gamma$. In what follows, we shall always write $x = \{x_k\}$, $y = \{y_k\}$ and $z = \{z_k\}$.

The non-absolute Cesaro sequence spaces $X_p$, $1 \leq p \leq \infty$, are defined in [2] as follows:

$$X_p = \{x: Cx \in \ell_p\},$$

with a norm $||x|| = ||Cx||_{\ell_p}$, where $C = (c_{nk})$ is the Cesaro matrix given by $c_{nk} = 1/n$ when $1 \leq k \leq n$ and zero otherwise. In [2], Ng and Lee characterized $X_p^\beta$, the $\beta$-dual of $X_p$, by

$$X_p^\beta = \{y: \{ky_k\} \in \ell_\infty, \{k(y_k - y_{k+1})\} \in \ell_q\}, \ 1 \leq p \leq \infty, \ \frac{1}{p} + \frac{1}{q} = 1.$$

In [3], Wu and Lee determined the $\alpha$- and $\gamma$- duals of $X_p$ for $1 \leq p \leq \infty$, showing that $X_p^\alpha = \{y: \{ky_k\} \in \ell_q\}$ where $\frac{1}{p} + \frac{1}{q} = 1$, and that $X_p^\beta = X_p^\gamma$. In particular, $X_1^\alpha = X_1^\beta = X_1^\gamma = \{y: \{ky_k\} \in \ell_\infty\}$.

In this paper, we discuss the second duals of $X_p$.

## 2. The second duals of $X_p$

We remark [1] that

i) $X^\alpha \subset X^\beta \subset X^\gamma$;

ii) if $X \supset Y$, then $X^\zeta \subset Y^\zeta$, for $\zeta = \alpha$, $\beta$ or $\gamma$;

iii) $X \subset X^{\zeta\zeta}$ for $\zeta = \alpha$, $\beta$ or $\gamma$;

iv) $X^\zeta = X^{\zeta\zeta\zeta}$, where $\zeta = \alpha$, $\beta$ or $\gamma$;

v) $X^{\beta\alpha} = X^{\gamma\alpha} \subset X^{\gamma\beta} \subset X^{\beta\beta} \subset X^{\beta\gamma} = X^{\gamma\gamma} \subset X^{\alpha\alpha} = X^{\alpha\beta} = X^{\alpha\gamma}$.

According to Remark (v) above, since $X_p^\beta = X_p^\gamma$ we have

$$X_p^{\beta\alpha} = X_p^{\gamma\alpha} \subset X_p^{\gamma\beta} = X_p^{\beta\beta} \subset X_p^{\beta\gamma} = X_p^{\gamma\gamma} \subset X_p^{\alpha\alpha} = X_p^{\alpha\beta} = X_p^{\alpha\gamma}.$$

It remains to consider the four second duals

$$X_p^{\beta\alpha} \subset X_p^{\beta\beta} \subset X_p^{\beta\gamma} \subset X_p^{\alpha\alpha}$$

First, we have

**Theorem 1.** $X_p^{\alpha\alpha} = \{z: \{z_k/k\} \in \ell_p\}$ for $1 \le p \le \infty$

The proof is easy.

**Theorem 2.** $X_p^{\beta\alpha} = \{z: \{z_k/k\} \in \ell_1\}$ for $1 \le p < \infty$.

**Proof.** If $z \in X_p^{\beta\alpha}$, then $\sum_{k=1}^\infty |y_k z_k|$ is convergent for each $y \in X_p^\beta$. We take $\{y_k\} = \{1/k\} \in X_p^\beta$, $1 \le p < \infty$. (Note that $\{1/k\} \notin X_\infty^\beta$.) Thus $\sum_{k=1}^\infty |z_k/k|$ is convergent. The reverse inclusion is obvious.

**Corollary.** $X_1^{\zeta\eta} = \{z: \{z_k/k\} \in \ell\}$ for $\zeta, \eta = \alpha$, $\beta$ or $\gamma$.

**Theorem 3.** (i) $X_\infty^{\beta\beta} = X_\infty^{\beta\gamma} = X_\infty$.

(ii) $X_\infty^{\beta\alpha} = ces_\infty$, where $ces_\infty = \{x: C|x| \in \ell_\infty\}$.

**Proof.** First note that from [2],

$$X_\infty^\beta = \{y: ky_k \to 0 \text{ as } k \to \infty \text{ and } \sum_{k=1}^\infty k|y_k - y_{k+1}| < \infty\}.$$

Now $ky_k \to 0$ as $k \to \infty$ implies that $y_k \to 0$ as $k \to \infty$. Also, if $\sum_{k=1}^\infty k|y_k - y_{k+1}| < \infty$ and $y_k \to 0$ as $k \to \infty$, then

$$|ky_k| \le \sum_{j=k}^\infty k|y_j - y_{j+1}| \le \sum_{j=k}^\infty j|y_j - y_{j+1}|$$

and, letting $k \to \infty$, $ky_k \to 0$. Thus,

$$X_\infty^\beta = \{y: y_k \to 0 \text{ as } k \to \infty \text{ and } \sum_{k=1}^{\infty} k|y_k - y_{k+1}| < \infty\}$$

(i) According to Remark (iii) above, we have $X_\infty \subset X_\infty^{\beta\beta}$.

Conversely, we define a norm of $y \in X_\infty^\beta$ by

$$\|y\| = \|\{y_k\}\|_{c_0} + \|\{k(y_k - y_{k+1})\}\|_{\ell_1}.$$

Then $X_\infty^\beta$ is a BK space (see [5] for definition of Bk spaces).

By Theorem 7.2.7 of [5], which states that for a BK space Z containing all finite sequences, $Z^\beta \subset Z^\gamma \subset Z^f$ where $Z^f = \{\{f(e^k)\} : f \in Z^*\}$ and $e^k$ is the sequence whose only non-zero term is 1 in the k-th place, we obtain $X_\infty^{\beta\beta} \subset X_\infty^{\beta\gamma} \subset X_\infty^{\beta f}$.

Let $f \in (X_\infty^\beta)^*$; by Theorem 4.4.1 of [5], $f = F + g \circ A$ with $F \in C_0^* = \ell_1$ and $g \in \ell_1^* = \ell_\infty$. Hence $f(y) = F(y) + \sum_{k=1}^{\infty} k(y_k - y_{k+1})\, g_k$ for $y \in X_\infty^\beta$ and, in particular, $f(e^k) = F(e^k) + kg_k - (k-1)g_{k-1}$ for each k with $g_0 = 0$. Let $z_k' = F(e^k)$ and $z_k'' = kg_k - (k-1)g_{k-1}$, then $f(e^k) = z_k' + z_k''$. Since $F = \{z_k'\} \in \ell_1 \subset \ell_\infty \subset X_\infty$ and $\{z_k''\} \in X_\infty$. Therefore $\{f(e^k)\} \in X_\infty$ and $X_\infty^{\beta\beta} \subset X_\infty^{\beta\gamma} \subset X_\infty^{\beta f} \subset X_\infty$. The proof is complete.

(ii) First, we show that $ces_\infty \subset X_\infty^{\beta\alpha}$. Let $z \in ces_\infty$ and $s_k = \frac{1}{k}\sum_{i=1}^{k} |z_i|$, then $\{s_k\} \in \ell_\infty$.

Use Abel's summation formula, we get

$$\sum_{k=1}^{n} |y_k z_k| = \sum_{k=1}^{n-1} k(|y_k| - |y_{k+1}|)\, s_k + n|y_n|\, s_n.$$

If $y \in X_\infty^\beta$, then $ky_k \to 0$ as $k \to \infty$ and the last term in the above equality tends to zero as $n \to \infty$. Since $\{k(y_k - y_{k+1})\} \in \ell_1$ implies $\{k(|y_k| - |y_{k+1}|\} \in \ell_1$, the series $\sum_{k=1}^{\infty} |y_k z_k|$ is convergent; this yields $z \in X_\infty^{\beta\gamma}$.

To prove the other inclusion, let $\{z_k\} \in X_\infty^{\beta\alpha}$ but $\{z_k\} \notin ces_\infty$, then $\{\frac{1}{k}\sum_{i=1}^{k} |z_i|\} \notin \ell_\infty$. Thus, there exists $\{c_k\} \in \ell_1$ such that $\sum_{k=1}^{\infty} |c_k|(\frac{1}{k}\sum_{i=1}^{k} |z_i|) = \infty$. We may assume that $c_k \geq 0$ for all k. Rearranging the terms of the series and writing $y_k = \sum_{i=k}^{\infty} c_i/i$, we have

$$\sum_{k=1}^{m} |z_k| y_k = \infty.$$

Note that $\{c_k/k\} \in \ell_1$ and $\{z_k\} \in X_\infty^{\beta\alpha}$, thus $\{y_k\} \notin X_\infty^\beta$. Since $y_k \to 0$ as $k \to \infty$ and $\{y_k\} \notin X_\infty^\beta$ implies that $\{K(y_k-y_{k+1})\} \notin \ell_1$. But $k(y_k-y_{k+1}) = c_k$ for all k, which contradicts the fact that $\{c_k\} \in \ell_1$. The proof is complete.

Next, we determine $(X_p^\zeta)^*$, the continuous duals of $X_p^\zeta$, where $\zeta = \alpha$, $\beta$ or $\gamma$.

**Theorem 4.** The sequence space $X_p^\alpha$ is a Banach space for $1 \leq p \leq \infty$ with norm $\|y\| = \|\{ky_k\}\|_{\ell_q}$, and $(X_p^\alpha)^*$ is isometric to $\ell_p$, when $1 < p \leq \infty$. Furthermore, $(X_1^\alpha)^*$ is isometric to $\ell_\infty^*$.

**Proof.** We can prove that the mapping T: $X_p^\alpha \to \ell_q$, given by $\{y_k\} \to \{ky_k\}$ is a linear isometry. Thus $X_p^\alpha \cong \ell_q$ and the conclusion follows.

Note that since $X_1^\alpha = X_1^\beta = X_1^\gamma$, we have also found the continuous duals of $X_1^\beta$ and $X_1^\gamma$.

**Theorem 5.** Let $X_p^\beta(= X_p^\gamma)$ have norm $\|y\| = \|\{k(y_k-y_{k+1})\}\|_{\ell_q}$. Then

(i) $X_\infty^\beta$ is a Banach space and $(X_\infty^\beta)^*$ is equivalent to $\ell_\infty$.

$X_p^\beta$ is not complete for $1 < p < \infty$; the completion $\hat{X}_p^\beta$ of $X_p^\beta$ is the space of all null sequences y such that $\{k(y_k-y_{k+1})\} \in \ell_q$ where $1 < p < \infty$ and $\frac{1}{p} + \frac{1}{q} = 1$.

(ii) $(X_p^\beta)^*$ is isometric to $\ell_p$ for $1 < p < \infty$.

**Proof.** (i) Let $1 < p \leq \infty$, $\frac{1}{p} + \frac{1}{q} = 1$ and $Y = \{y$: $y_k \to 0$ as $k \to \infty$, $\{k(y_k-y_{k+1})\} \in \ell_q\}$. First, we prove that $T_1 : Y \to \ell_q$, given by $\{y_k\} \to \{k(y_k-y_{k+1})\}$, is a linear isometry. The map $T_1$ is clearly linear and norm preserving. Since the mapping is injective, we need only to show that it is surjective. If $z \in \ell_q$, then the series $\sum_k z_k/k$, converges since $\{1/k\} \in \ell_p$. Let $y = \{y_k\}$ be defined by $y_k = \sum_{i=k}^\infty z_i/i$ for each k, then $y \in Y$ and $T_1y = z$. We conclude that $T_1$ is an isometric isomorphism of Y onto $\ell_q$ and thus Y is a Banach space.

Note that $Y = X_\infty^\beta$ when $q = 1$. Therefore $X_\infty^\beta$ is a Banach space and $(X_\infty^\beta)^*$ is equivalent to $\ell_\infty$.

To show that Y is the completion of $X_p^\beta$ for $1 < p < \infty$, it is enough to show that $X_p^\beta \subset Y$ and $X_p^\beta$ is dense in Y. Choose t such that $1 + \frac{1}{q} < t < 2$ and let $y_k = \sum_{i=k}^\infty i^{-t}$ then $y_k \to 0$ as $k \to \infty$ and

$$\sum_{k=1}^{\infty} |k(y_k - y_{k+1})|^q = \sum_{k=1}^{\infty} \frac{1}{k^{(t-1)q}} \text{ converges.}$$

Hence $y = \{y_k\} \in Y$. But $ky_k \geq k \sum_{i=k+1}^{2k} \frac{1}{i^t} \geq k^2(2k)^{-t} = 2^{-t} k^{2-t} \to \infty$ as $k \to \infty$ and hence $y \notin X_p^\beta$. This implies that $X_p^\beta \subset Y$ and $X_p^\beta \neq Y$.

To finish the proof we show that $X_p^\beta$ is dense in $Y$. For any $y = \{y_k\} \in Y$, there exists $y^{(N)} = (y_1 - y_{N+1}, \ldots, y_N - y_{N+1}, 0, 0, \ldots) \in X_p^\beta$, such that

$$\| y - y^{(N)} \| = \left( \sum_{k=N+1}^{\infty} | k(y_k - y_{k+1}) |^q \right)^{\frac{1}{q}} \to 0 \quad \text{as} \quad N \to \infty.$$

The proof is complete.

(ii) By (i), $\hat{X}_p^\beta$ is congruent to $\ell_q$ for $1 < p < \infty$ and $\frac{1}{p} + \frac{1}{q} = 1$. Thus $(\hat{X}_p^\beta)^*$ is congruent to $\ell_p$ for $(1 < p < \infty)$. To prove that $(X_p^\beta)^* \cong (\hat{X}_p^\beta)^*$, we recall that if $M$ is a subspace of $X$ and $M^\perp$ is the annihilator of $M$, then $M^\perp$ is a closed subspace of $X^*$ and the conjugate space $M^*$ is congruent to $X^*/M^\perp$. (see [6; p93]). Since $X_p^\beta$ is a dense subspace of $\hat{X}_p^\beta$, the annihilator of $X_p^\beta$ is the zero vector in $(\hat{X}_p^\beta)^*$. Therefore, $(X_p^\beta)^* \cong (\hat{X}_p^\beta)^*/(X_p^\beta)^\perp \cong (\hat{X}_p^\beta)^*$.

Theorem 6. $X_p^{\beta\beta} = X_p^{\beta\gamma} = X_p^{\beta\alpha} + X_p$ for $1 < p < \omega$.

Proof. Note that since $X_p^{\beta\alpha} \subset X_p^{\beta\beta}$ and $X_p \subset X_p^{\beta\beta}$, we have $X_p^{\beta\alpha} + X_p \subset X_p^{\beta\beta}$. Thus, it is enough to show that $X_p^{\beta\gamma} \subset X_p^{\beta\alpha} + X_p$.

Let $A = (a_{nk})$ be defined by $a_{nk} = n$ when $n = k$, $a_{nk} = -n$ when $n = k - 1$, and $a_{nk} = 0$ otherwise. Then $X_p^\beta$ is represented as an intersection of two sequence spaces, i.e.

$$X_p^\beta = X_1^\alpha \cap \{y : Ay \in \ell_q\}$$

We define a norm of $y \in X_p^\beta$ by

$$\| y \| = \| \{ky_k\} \|_{\ell_\infty} + \| \{ k(y_k - y_{k+1}) \} \|_{\ell_q}.$$

Then $X_p^\beta$ is a BK space.

Again, by Theorem 7.2.7 of [5],

$$X_p^{\beta\beta} \subset X_p^{\beta\gamma} \subset X_p^{\beta f}.$$

Let $f \in (X_p^\beta)^*$; by Theorem 4.4.1 of [5], $f = F + g \circ A$ with $F \in (X_1^\alpha)^*$ and $q \in \ell_q^* = \ell_p$. Hence $f(y) = F(y) + \sum_{k=1}^{\infty} k(y_k - y_{k+1})g_k$ for $y \in X_p^\beta$ and in part--icular,

$$f(e^k) = F(e^k) + kg_k - (k-1)g_{k-1}$$

for each k with $g_0 = 0$. In view of the proof of Theorem 4, we have $F = F_1 \circ T$, where $F \in (X_1^\alpha)^*$ and $F_1 = F \circ T^{-1} \in \ell_\infty^*$. Hence $F(e^k) = F_1(ke^k) = kF_1(e^k)$. Since $\ell_\infty^* = \ell_1 \oplus c_0^\perp$ (see [6; p. 426]), $F_1 = u + v$ with $u = \{u_k\} \in \ell_1$, $v = \{v_k\} \in c_0^\perp$. Thus, $F_1(e^k) = u_k$ for each k. Let $z_k' = kF_1(e^k) = ku_k$ and $z_k'' = kg_k - (k-1)g_{k-1}$; then $f(e_k) = z_k' + z_k''$ where $z' = \{z_k'\} \in X_p^{\beta\alpha}$ and $z'' = \{z_k''\} \in X_p$. Therefore

$X_p^{\beta\beta} \subset X_p^{\beta\gamma} \subset X_p^{\beta f} \subset X_p^{\beta\alpha} + X_p$.

The proof is complete.

## REFERENCES

1. D.J.H. Garling, The β- and γ-duality of sequence spaces, Proc. Camb. Phil. Soc. 63(1967) 963-981.
2. Ng Peng-Nung and Lee Peng Yee, Cesaro sequence spaces of a nonabsolute type, Comm. Math. 20 (1978), 429-433.
3. Lee Peng Yee, Cesaro sequence spaces, Math. Chronicle 13 (1984), 29-45.
4. Wu Bo Er and Lee Peng Yee, The duals of some sequence spaces of a non-absolute type, SEA Bull. Math. 9(1985), 77-80.
5. Albert Wilansky, Summability through functional analysis (1984).
6. Gottfried Köthe, Topological Vector Spaces I (1983).

Proceedings of the Analysis Conference, Singapore 1986
S.T.L. Choy, J.P. Jesudason, P.Y. Lee (Editors)



# BANACH REDUCIBILITY OF DECOMPOSABLE OPERATORS

Xu Feng
Northeast Normal University, Changchun, China

Zou Chenzu
Jilin University, Changchun, China

In this paper, we continue the study of Banach reducibility of decomposable operators. The notion of complete Banach reducibility for bounded operators is introduced, and various facts are proved about completely Banach reducible weak decomposable operators.

Throughout this paper, $X$ is a complex Banach space and $B(X)$ is the Banach algebra of all bounded linear operators on $X$. If $T \in B(X)$ has the single-valued extension property, $\sigma_T(x)$ denotes the local spectrum of $T$ at $x$. $\mathbb{C}$ denotes the complex plane. For every $E \subset \mathbb{C}$, $X_T(E) = \{x | \sigma_T(x) \subset E\}$ denotes a spectral manifold of $T$.

In our previous paper [1], we have investigated the Banach reducibility of decomposable operators. In this paper we will investigate further the Banach reducibility of decomposable operators.

**DEFINITION 1.** $T \in B(X)$ is said to have complete Banach reducibility if for every invariant subspace $M$ of $T$, there exists another invariant subspace $N$ of $T$ such that $X = M \dot{+} N$, where $\dot{+}$ denotes the direct sum.

K. Tanahashi [2] proved that: if $H$ is a complex Hilbert space, $T$ is a bounded linear operator on $X$ with the single-valued extension property and $E \subset \mathbb{C}$, then $\sigma(T|\overline{H_T(E)}) \subset E$ if and only if $H_T(E)$ is closed. Clearly, the above result is right when $H$ is a Banach space.

**PROPOSITION.** Let $T \in B(X)$ have the single-valued extension property and $E \subset C$. Then

$$\sigma(T|\overline{X_T(E)}) \subset E$$

if and only if $X_T(E)$ is closed.

**LEMMA.** Let $T \in B(X)$ have the single-valued extension property and complete Banach reducibility. Then, for every closed set $F$ of $\mathbb{C}$,

$$\sigma_{T|\overline{X_T(F)}}(x) = \sigma_T(x), \quad x \in \overline{X_T(F)}.$$

**PROOF.** Since $\overline{X_T(F)}$ is a reducible subspace for $T$, there exists $Z$, an invariant subspace of $T$, such that

$$X = \overline{X_T(F)} \dotplus Z.$$

Suppose P is the projection of X onto $\overline{X_T(F)}$ along Z; then TP = PT. Obviously,

$$\sigma_{T|\overline{X_T(F)}}(x) \supset \sigma_T(x), \quad x \in \overline{X_T(F)}.$$

Now for arbitrary $x \in \overline{X_T(F)}$, we show $\sigma_T(x) \subset \sigma_{T|\overline{X_T(F)}}(x)$. Let $f(\lambda)$ : $\rho_T(x) \to X$ be analytic. Then

$$(\lambda I-T)f(\lambda) = x, \quad \lambda \in \sigma_T(x).$$

Suppose $f_1(\lambda) = P(f(\lambda))$, then $f_1(\lambda) : \rho_T(x) \to \overline{X_T(F)}$ is analytic and

$$P(\lambda I-T)f(\lambda) = Px = x, \quad \lambda \in \rho_T(x).$$

$$\begin{aligned} P(\lambda I-P) &= (\lambda P-PT)f(\lambda) = (\lambda I|\overline{X_T(F)}-TP)f(\lambda) \\ &= (\lambda I|\overline{X_T(F)} - T|\overline{X_T(F)})Pf(\lambda) \\ &= (\lambda I|\overline{X_T(F)} - T|\overline{X_T(F)})f_1(\lambda). \end{aligned}$$

So

$$(\lambda I|\overline{X_T(F)} - T|\overline{X_T(F)})f_1(\lambda) = x, \quad \lambda \in \rho_T(x).$$

The above fact shows

$$\rho_T(x) \subset \rho_{T|\overline{X_T(F)}}(x).$$

Thus, we have obtained

$$\rho_T(x) = \rho_{T|\overline{X_T(F)}}(x).$$

**THEOREM 1.** Let $T \in B(X)$ be a completely Banach reducible weak decomposable operator. Then T is a quasi-decomposable operator.

**PROOF.** Since T is a weak decomposable operator, T has the single-valued extension property. First, we show that for an arbitrary closed subset F of C,

$$\sigma(T|\overline{X_T(F)}) \subset F.$$

Let $\lambda_0 \notin F$. Then there exists an $\varepsilon > 0$ such that

$$G_1 = \{\lambda | |\lambda-\lambda_0| < 2\varepsilon\}, \quad \overline{G_1} \cap F = \emptyset.$$

Let

$$G_2 = \{\lambda | |\lambda-\lambda_0| > \varepsilon\}.$$

Then $\{G_1, G_2\}$ forms an open covering of $\sigma(T)$. Since T is a weak decomposable operator, there exist spectral maximal subspaces $Y_1$ and $Y_2$ of T such that

$$X = \overline{Y_1+Y_2}, \quad \sigma(T|Y_i) \subset G_i, \quad i = 1,2.$$

Moreover,T is a completely Banach reducible operator and $\overline{X_T(F)}$ is an invariant subspace of T, so there exists a projection P on X such that

$$PX = \overline{X_T(F)} \quad \text{and} \quad TP = PT.$$

We obtain

$$PX = P(\overline{Y_1+Y_2}) \subset \overline{PY_1 + PY_2}.$$

Obviously, $Y_1$ is a hyper-invariant subspace of T, so $PY_1 \subset Y_1$ and hence $PY_1 \subset Y_1 \cap \overline{X_T(F)}$.

Since T is a completely Banach reducible operator, there exists an invariant subspace $Z_1$ of T such that $X = Y_1 \dotplus Z_1$. Let $P_1$ be the projection of X onto $Y_1$ along $Z_1$. Then $TP_1 = P_1T$. Thus, for arbitrary $x \in X_T(F)$, we have $\sigma_T(x) \subset F$ and

$$\sigma_T(P_1x) \subset \sigma_T(x) \cap \sigma(T|Y_1) \subset F \cap G_1 = \emptyset.$$

Thus, $\sigma_T(P_1x) \subset F \cap G_1 = \phi$, and hence $P_1x = 0$. Therefore

$$P_1(\overline{X_T(F)}) = 0.$$

Then $PY_1 = 0$, and hence

$$\overline{X_T(F)} \subset \overline{PY_1 + PY_2} = \overline{PY_2} \subset Y_2.$$

By the lemma, for every $x \in \overline{X_T(F)}$,

$$\sigma_{T|\overline{X_T(F)}}(x) = \sigma_T(x) \subset \sigma(T|Y_2) \subset G_2.$$

Therefore

$$\sigma(T|\overline{X_T(F)}) \subset G_2.$$

We have $\lambda_0 \notin \sigma(T|\overline{X_T(F)})$, thus

$$\sigma(T|\overline{X_T(F)}) \subset F.$$

By the proposition, $X_T(F)$ is closed, hence T is a quasi-decomposable operator.

**THEOREM 2.** Let $T \in B(X)$ be a completely Banach reducible weak decomposable operator. Then T is a strongly identically decomposable operator.

**PROOF.** By Theorem 1, for a closed subset $F \subset C$, $X_T(F)$ is an invariant subspace of T. On the other hand, let $G \supset F$ be an open subset. Then $X_T(F) \subset X_T(\overline{G})$, and there exists an invariant subspace N of T such that

$$X = X_T(\overline{G}) \dotplus N.$$

Since $X/X_T(\overline{G})$ and N are topologically isomorphic by R. Lange [3], we have

$$\sigma(T^{X_T(\overline{G})}) \subset G^c,$$

where $T^{X_T(\overline{G})}$ denotes the coinduced operator by T on the quotient space

$X/X_T(\overline{G})$. Hence

$$\sigma(T|N) = \sigma(T^{X_T(\overline{G})}) \subset G^c \subset F^c.$$

Therefore

$$N \subset \overline{X_T(F^c)}.$$

Since $\overline{X_T(F^c)}$ is an invariant subspace of T, there exists another invariant subspace Z of T such that

$$X = \overline{X_T(F^c)} \dotplus Z.$$

Thus, $Z \subset X_T(\overline{G})$, so we have

$$Z \subset \bigcap_{G \supset F} X_T(\overline{G}) = X_T(F),$$

and hence

$$X = X_T(F) + \overline{X_T(F^c)}.$$

By [4], T is a strongly identically decomposable operator.

**THEOREM 3.** Let X be a weakly complete Banach space and $T \in B(X)$ be a completely Banach reducible weak decomposable operator with the property (B). Then T is a spectral operator.

**PROOF.** By Theorem 2, any closed subset $F \in \phi_1(T)$, and since T has the property (A, B, C) (see [5]), therefore $F \in \phi_1(T)$. Thus, we obtain that $\phi(T)$ is the family of Borel subsets of C. By [5], T is a spectral operator, where $\phi_1(T)$ and $\phi(T)$ were introduced by Dunford and Schwartz [5].

## REFERENCES

[1] Xu Feng and Zou Chenzu, Banach reducibility of decomposable operators, Dongbei Shida Xuebao 4(1983), 61-69.

[2] K. Tanahashi, Reductive weak decomposable operators are spectral, Proc. Amer. Math. Soc., 87(1983), 44-46.

[3] R. Lange, Equivalent conditions for decomposable operators, Proc. Amer. Math. Soc., 82(1981), 401-406.

[4] Liu Guanyu, Decomposable operators with finite order and their properties, J. Nanjing Univ., 3(1982), 598-606.

[5] N. Dunford and J. Schwartz, Linear Operators (III), New York, 1971.

Proceedings of the Analysis Conference, Singapore 1986
S.T.L. Choy, J.P. Jesudason, P.Y. Lee (Editors)



# THERE CAN BE NO LIPSCHITZ VERSION OF MICHAEL'S SELECTION THEOREM

DAVID YOST
Mathematics Department, I.A.S., Australian National University,
G.P.O. Box 4, Canberra, A.C.T. 2601, AUSTRALIA

Given a real Banach space E, let H(E) denote the family of closed, bounded, convex nonempty subsets of E. We equip H(E) with the Hausdorff metric : for A,B ∈ H(E), set $d_H(A,B) = \sup(\{d(x,A) : x \in B\} \cup \{d(x,B) : x \in A\})$. Let X be a metric space and $\psi : X \to H(E)$ a continuous map. Then, as a special case of Michael's selection theorem [M, Theorem 3.2"], $\psi$ admits a continuous selection. This means that there is a continuous map $f : X \to E$ satisfying $f(x) \in \psi(x)$ for all x in X.

Michael's theorem is actually somewhat more general. It states that $\psi$ admits a continuous selection, assuming only that $\psi$ is lower semicontinuous, and allowing the values of $\psi$ to be all the nonempty, closed, convex subsets of E (not necessarily bounded). We will not concern ourselves with the more general result, or with the various other selection theorems proved by Michael ([M], and the references therein). Michael's theorem has had applications to numerous areas of mathematics, including infinite-dimensional topology [E,J], functional analysis [C,N,V,W], approximation theory [D,O,X], mathematical economics [G], differential equations and inclusions [H,K], optimization theory [R] and vector bundles [T]. This list of references is not meant to be comprehensive; it is just an arbitrary selection. (Not all of these papers require the full strength of Michael's selection theorem. For the topological properties of H(E), at least in the finite dimensional case, we refer to [Q].)

Naturally, various mathematicians (see, for example [I,Y]) have thought about the following problem : if $\psi : X \to H(E)$ is Lipschitz continuous, is it possible to choose $f : X \to E$ to be Lipschitz continuous also? This is easily seen to be equivalent to the following problem : does there exist, for a given Banach space E, a Lipschitz map $f : H(E) \to E$ satisfying the identity $f(A) \in A$ ?

Bressan [B] and Przeslawski [P] showed that this is possible if E is finite dimensional. (Thanks are due to Alicja Sterna-Karwat for bringing [P] to our attention.) The existence of Lipschitz selections in some other special cases is proved in [L],[U] and [Z]. But in general, as we now show, the preceding problems have a negative solution.

For a counterexample we take E = D, the set of bounded functions from [0,1] into $\mathbb{R}$ which are continuous at every irrational point in (0,1), and right continuous with left limits at every rational point. Equipped with the supremum norm, D is a separable Banach space. The nonexistence of a Lipschitz selection from H(D) to D follows by combining some results from [A] and [X] . However, we think it is worthwhile to give a self-contained argument.

For $f \in D$, denote by Pf the set of best approximants to f in the subspace C of continuous functions. That is, $Pf = \{g \in C : \| f - g \| = d(f,C)\}$. Let $Jf : [0,1] \to \mathbb{R}$ denote the "jump function" of f, $Jf(t) = f(t) - f(t-)$. (For convenience we may set $f(0-) = f(0)$.) Obviously Jf is a bounded function. Our next result is a special case of [S,7.5.6].

**Lemma 1** For each $f \in D$, we have $Pf \in H(C)$ and $d(f,C) = 1/2 \| Jf \|$.

**Proof** Obviously Pf is a closed, bounded, convex subset of C ; we must show it is nonempty. For any $g \in C$ and any t, we have either $| g(t) - f(t) | \geq 1/2 | f(t) - f(t-) |$ or $| g(t-) - f(t-) | \geq 1/2 | f(t) - f(t-) |$. Taking the supremum over t , we conclude that $\| g-f \| \geq 1/2 \| Jf \|$.

Next we will exhibit a $g \in C$ satisfying $\| g - f \| \leq 1/2 \| Jf \|$. This will simultaneously show that $d(f,C) = 1/2 \| Jf \|$ and that $g \in Pf$. Define $f_0, f_1 : [0,1] \to \mathbb{R}$ by $f_0(t) = \max\{f(t), f(t-)\}$ and $f_1(t) = \min\{f(t), f(t-)\}$. Clearly, $f_0$ is upper semicontinous, $f_1$ is lower semicontinuous, and $f_0(t) - 1/2 \| Jf \| \leq f_1(t) + 1/2 \| Jf \|$ for all t. From Michael's theorem, we obtain a $g \in C$ which satisfies the identity $f_0(t) - 1/2 \| Jf \| \leq g(t) \leq f_1(t) + 1/2 \| Jf \|$. Clearly $\| g - f \| \leq 1/2 \| Jf \|$.

Now let us define $Qf = f - Pf$ for each $f \in D$. It is clear that $Qf \in H(D)$ and that $Qf = Qg \iff f - g \in C$. The next result could be deduced from [X, Lemma 1.1 and Corollary 2.3], but the following argument is simpler.

**Lemma 2** For all $f,g \in D$, we have (i) $d_H(Pf, Pg) \leq 2 \| f - g \|$
and (ii) $d(f - g,C) \leq d_H(Qf, Qg) \leq 3d(f - g,C)$.

**Proof** (i) Set $\gamma = \| f - g \|$ and $\delta = d(g,C)$. By symmetry, it suffices to show that, given $a \in Pf$, we can find $b \in Pg$ with $\| a - b \| \leq 2\delta$. Now $\| g - a \| \leq \| g - f \| + d(f,C) \leq 2 \| g - f \| + d(f,C) = 2\gamma + \delta$. Thus $g - a - \delta \leq 2\gamma$ and $-2\gamma \leq g - a + \delta$. Lemma 1 now gives us some $c \in P(g-a) = C \cap B(g-a,\delta)$. Thus $g-a-\delta \leq c \leq g-a + \delta$. If we put $b(t) = \max\{-2\gamma, \min\{2\gamma, c(t)\}\} + a(t)$, then $b(t) = \min\{2\gamma, \max\{-2\gamma, c(t)\}\} + a(t)$ and so $g-a-\delta \leq b-a \leq g-a + \delta$. Clearly $b \in C$ and $\| g-b \| \leq \delta$, whence $b \in Pg$. Obviously $\| b-a \| \leq 2\gamma$.

(ii) For any $h \in Pf$, we have $d(f-g,C) = d(f-h,g+C) \leq d(f-h,Qg) \leq d_H(Qf,Qg)$. This proves the first inequality. To prove the second inequality, note that, for any $h \in C$,

$$\begin{aligned} d_H(Qf,Qg) &= d_H(f + h - P(f + h), g - Pg) \\ &\leq \| f + h - g \| + d_H(P(f + h), Pg) \leq 3 \| f - g + h \|. \end{aligned}$$

Lemma 2 tells us that the quotient space D/C is Lipschitz equivalent to a subset of H(D) . Aharoni and Lindenstrauss [A, Remark (ii)] showed that there is no Lipschitz lifting from D/C to D . This gives us our main result.

**Theorem 1** There is no Lipschitz continuous selection from $H(D)$ to $D$.
**Proof** (after [A]) Suppose that such a selection exists. Then there certainly exists a Lipschitz selection $\psi : Q(D) \to D$, where $Q(D) = \{Qf : f \in D\} \subset H(D)$.

Let us equip $Q(D)$ with the equivalent metric

$$d(Qf, Qg) = d(f - g, C) = {}^1/_2 \, \| J(f - g) \|.$$

By Lemma 2, $\psi$ is also Lipschitz with respect to $d$, so we may work exclusively with $d$ henceforth. Let $K$ be the best Lipschitz constant for $\psi$. Choose $f_1$ and $f_2$ in $D$ so that

$$\| \psi Qf_1 - \psi Qf_2 \| / d(Qf_1, Qf_2) > K - {}^1/_5 .$$

Since $\psi$ is Lipschitz, we assume without loss of generality that $d(Qf_1, Qf_2) = 1$. Choose an irrational $t_0$ so that

$$| (\psi Qf_1)(t_0) - (\psi Qf_2)(t_0) | > K - {}^1/_5 .$$

By continuity at $t_0$, there is an open interval $I$ and a rational $q \in I$ such that

$$| (\psi Qf_1)(t) - (\psi Qf_2)(t) | \geq K - {}^1/_5 , \text{ for all } t \in I ,$$

$${}^1/_2(\psi Qf_1 + \psi Qf_2) \text{ varies by less than } {}^1/_5 \text{ on } I , \text{ and}$$

$$| J(f_1 - f_2)(q) | \leq 1 .$$

Let $s = \pm 1$ be the sign of $J(f_1 + f_2)(q)$, and put $g = {}^1/_2(f_1 + f_2 + s\chi_{[0,q)})$. Here, of course, $\chi_{[0,q)}$ denotes the characteristic function of the interval $[0,q)$. Now for any $t \neq q$, we have

$$| J(f_1 - g)(t) | = {}^1/_2 | J(f_1 - f_2)(t) | \leq d(Qf_1, Qf_2) = 1 \text{ . Also,}$$

$$| J(f_1 - g)(q) | = {}^1/_2 | J(f_1 - f_2)(q) + s | \leq 1 \text{ . Thus}$$

$$d(Qf_1, Qg) = {}^1/_2 \| J(f_1 - g) \| \leq {}^1/_2 \text{ . Similarly}$$

$$d(Qf_2, Qg) \leq {}^1/_2 \text{ , and so}$$

$$\| \psi Qf_i - \psi Qg \| \leq {}^1/_2 K, \text{ for } i = 1,2 .$$

Then, for every $t \in I$, we have

$$| {}^1/_2(\psi Qf_1 + \psi Qf_2)(t) - (\psi Qg)(t) |$$

$$= \max_{i=1,2} | (\psi Qf_i)(t) - (\psi Qg)(t) | - {}^1/_2 \, | (\psi Qf_1)(t) - (\psi Qf_2)(t) |$$

$$\leq \max_{i=1,2} \| \psi Qf_i - \psi Qg \| - {}^1/_2 \, (K - {}^1/_5)$$

$$\leq {}^1/_{10} \, .$$

Simple calculations then show that $\psi Qg$ varies by less than ${}^2/_5$ on $I$ . In particular $| (J\psi Qg)(q) | \leq {}^2/_5$ . But $\psi Qg \in g - Pg \subset g + C$ , whence $J\psi Qg = Jg$. Since $| (Jg)(q) | = {}^1/_2 \, | J(f_1 + f_2)(q) + s | \geq {}^1/_2$ , we have a contradiction.

If $F$ is a closed linear subspace of a Banach space $E$ , and there is no Lipschitz selection $H(F) \to F$ , then clearly there is no Lipschitz selection $H(E) \to E$ . In particular, there is no Lipschitz selection when $E = C$ , as $C$ contains an isometric copy of every separable Banach space.

Denote by $\mathbb{R}^n$ Euclidean n-space, and by $K_n$ the smallest Lipschitz constant for selections $f : H(\mathbb{R}^n) \to \mathbb{R}^n$ . It is known [B,P] that $K_n \leq n$ . Thus, Lipschitz selections exist, but the Lipschitz constants might not be uniformly bounded as $n$ becomes larger. Quite recently, Krysztof Przeslawski [personal communication] proved that $K_n$ does go to infinity with $n$ , and hence that there is no Lipschitz selection $H(E) \to E$ whenever $E$ is an infinite dimensional Hilbert space.

But now, let us recall Dvoretzky's theorem [F] : for any infinite dimensional Banach space E, any $n \in \mathbb{N}$ , and any $\varepsilon > 0$ , there is an n-dimensional subspace of $E$ which is $\varepsilon$-isometric to $\mathbb{R}^n$ . From this, and Przeslawski's result, the following can be proved.

**Theorem 2** There exists a Lipschitz selection from $H(E)$ to $E$ if and only if the Banach space $E$ is finite dimensional.

Theorem 2 was stated as a conjecture when this paper was originally presented. It lies much deeper than Theorem 1, and full details will appear elsewhere.

## REFERENCES


[A] I.Aharoni and J.Lindenstrauss, *Uniform equivalence between Banach spaces*, Bull. Amer. Math. Soc. **84** (1978) 281-283.

[B] A.Bressan, *Misure di curvatura e selezioni lipschitziane*, preprint.

[C] A.Clausing and S.Papadopolou, *Stable convex sets and extremal operators*, Math. Ann. **231** (1978) 193-203.

[D] F.Deutsch, *A survey of metric selections*, Contemporary Math. **18** (1983) 49-71.

[E] K.D.Elworthy, *Embeddings, isotopy and stability of Banach spaces*, Compositio Math. **24** (1972) 175-226.

[F] T.Figiel, *A short proof of Dvoretzky's theorem on almost spherical sections of convex bodies*, Compositio Math. **33** (1976) 297-301.

[G] R.Guesnerie and J.-J.Laffont, *Advantageous reallocation of initial resources*, Econometrica **46** (1978) 835-841.

[H] G.Haddad and J.M.Lasry, *Periodic solutions of functional differential inclusions and fixed points of σ-selectionable correspondences*, J. Math. Anal. Appl. **96** (1983) 295-312.

[I] A.D.Ioffe, *Single-valued representation of set-valued mappings II; Applications to differential inclusions*, SIAM J. Control. Optim **21** (1983) 641-651.

[J] K.John and V.Zizler, *Weak compact generating in duality*, Studia Math **55** (1976) 1-20.

[K] W.G.Kelley, *Periodic solutions of generalized differential equations*, SIAM J. Appl. Math. **30** (1976) 70-74.

[L] S.Lojasiewicz Jr, A.Plis and R.Suarez, *Necessary conditions for a nonlinear control system*, J. Diff. Eqns. **59** (1985) 257-265.

[M] E.Michael, *Continuous selections. I*, Ann. of Math. **63** (1956) 361-382.

[N] R.R.Nelson, *Pointwise evaluation of Bochner integrals in Marcinkiewcz spaces*, Nederl. Akad. Wetensch. Proc. Ser. A. **85** (1982) 365-379.

[O] C.Olech, *Approximation of set-valued functions by continuous functions*, Colloq. Math. **19** (1968) 285-293.

[P] K.Przeslawski, *Linear and Lipschitz continuous selectors for the family of convex sets in Euclidean vector spaces*, Bull. Pol. Acad. Sci. **33** (1985) 31-33.

[Q] J.Quinn, S.B.Nadler Jr and N.M.Stavrakas, *Hyperspaces of compact, convex sets*, Pacific J. Math **39** (1971) 439-469.

[R] R.T.Rockafellar, *Integrals which are convex functionals, II*, Pacific J. Math. **39** (1971) 439-469.

[S] Z.Semadeni, *Banach spaces of continuous functions*, PWN, Warsaw (1971).

[T] A.J.Tromba, *The Euler characteristic of vector fields on Banach manifolds and a globalization of Leray-Schauder degree*, Adv. in Math. **28** (1978) 148-173.

[U] V.A.Ubhaya, *Lipschitz condition in minimum norm problems on bounded functions*, J. Approx. Theory **45** (1985) 201-218.

[V] M.Valadier, *Closedness in the weak topology of the dual pair* $L_1$, $C$, J. Math. Anal. Appl. **69** (1979) 17-34.

[W] D.Werner, *Extreme points in function spaces*, Proc. Amer. Math. Soc. **89** (1983) 598-600.

[X] D.Yost, *Best approximation and intersections of balls in Banach spaces*, Bull. Austral. Math. Soc. **19** (1979) 285-300.

[Y] D.Yost, *Best approximation operators in functional analysis*, Proc. Centre Math. Anal. Austral. Nat. Univ. **8** (1984) 249-270.

[Z] P.Zecca and G.Stefani, *Multivalued differential equations on manifolds with application to control theory*, Illinois J. Math. **24** (1980) 560-575.

Proceedings of the Analysis Conference, Singapore 1986
S.T.L. Choy, J.P. Jesudason, P.Y. Lee (Editors)



# A NEW SMOOTHNESS OF BANACH SPACES

Wenyao ZHANG

Department of Mathematics
Liaoning Normal University
Dalian, Liaoning
China*

If a Banach space $X$ is weakly very smooth, then every monotone basis on $X$ is shrinking.

Throughout this paper, $X$ is a Banach space, $S = \{x \in X, \|x\| = 1\}$, $S^* = \{x^* \in X^*, \|x^*\| = 1\}$. Let $x \in X$, $x \neq 0$, if $f \in S^*$, $f(x) = \|x\|$, then $f$ is called a supporting functional of $x$. If each point in $S$ has a unique supporting functional, then $X$ is called smooth. It is known (see, e.g., [1]) that $X$ is smooth if and only if for any $x \in S$, $f_n \in S^*$, $f_n(x) \longrightarrow 1$ implies that $\{f_n\}$ converges weak* to a supporting functional of $x$. A Banach space $X$ is very smooth if for any $x \in S$, $Jx$ has a unique supporting functional in $X^{***}$. It was proved [1] that $X$ is very smooth if and only if for any $x \in S$, $f_n \in S^*$, $f_n(x) \longrightarrow 1$ implies that $\{f_n\}$ converges weakly to a supporting functional of $x$. Now we introduce a concept of smoothness which is weaker than very smoothness. Let $X$ be a Banach space, if for any $x \in S$, $f_n \in S^*$, $f_n(x) \longrightarrow 1$ implies that $\{f_n\}$ has a subsequence which converges weakly in $X^*$, then $X$ is called weakly very smooth.

Obviously a very smooth Banach space is weakly very smooth, and every reflexive Banach space is weakly very smooth, thus there are weakly very smooth Banach spaces which are not very smooth. On the other hand, it is routine to show that if a weakly very smooth Banach space is smooth, then it is very smooth.

It is known that if $X^*$ is very smooth, then $X$ is reflexive. By the well known James' characterization of reflexive spaces it follows immediately that if $X^*$ is weakly very smooth, then $X$

*Present address: Department of Mathematics, The University of Iowa, Iowa City, Iowa, 52242, U.S.A.

is reflexive.

Let $X$ be a Banach space with a basis $\{e_n\}$, we define $P_n\left[\sum_{i=1}^{\infty} a_i e_i\right] = \sum_{i=1}^{n} a_i e_i$ the natural projections associated to $\{e_n\}$ and $f_n\left[\sum_{i=1}^{\infty} a_i e_i\right] = a_n$ the biorthogonal functional associated to $\{e_n\}$. If for any $x^* \in X^*$, $\lim_n \sup\left\{\left|x^*(x)\right| : x = \sum_{i=n+1}^{\infty} a_i e_i,\ \|x\| \le 1\right\} = 0$, then $\{e_n\}$ is called a shrinking basis. For other concepts and results appearing in this paper, we refer to [2] and [3].

THEOREM: Let $X$ be a Banach space with a monotone basis $\{e_n\}$, if $X$ is weakly very smooth, then the biorthogonal functional $\{f_n\}$ associated to $\{e_n\}$ is a basis of $X^*$.

Proof: Let $\{P_n\}$ be the associated natural projections of $\{e_n\}$. To show that $\{f_n\}$ is a basis of $X^*$, we need to show that $\overline{\operatorname{span}\{f_n\}} = X^*$. By Bishop-Phelps Theorem [4], it remains to show that $\{x^* \in X^* :$ there exists an $x \in S$ such that $x^*(x) = \|x^*\|\} \subset \overline{\operatorname{span}\{f_n\}}$.

Let $x_0^* \in X^*$, $x_0^*(x_0) = \|x_0^*\|$ with $x_0 \in S$, Without loss of generality, we assume that $\|x_0^*\| = 1$. Since $\{e_n\}$ is monotone, $\|P_n\| = 1$ for every $n$, thus we have $\|P_n^*\| = 1$. On the other hand, for any $\epsilon > 0$, there exists a natural number $N$, such that $\|P_n x_0 - x_0\| < \epsilon$ whenever $n > N$. Hence

$$\|x_0^*\| = x_0^*(x_0) \le |x_0^*(P_n x_0)| + \epsilon \le (P_n^* x_0^*)(x_0) + \epsilon \le \|P_n^* x_0^*\| + \epsilon.$$

We conclude that $\lim\|P_n^* x_0^*\| = \|x_0^*\| = 1$, and so there exists a natural number $N_1$, such that $\|P_n^* x_0^*\| > 0$ whenever $n > N_1$. Thus $\frac{P_n^* x_0^*}{\|P_n^* x_0^*\|} \in S^*$ for $n > N_1$, and $\frac{P_n^* x_0^*}{\|P_n^* x_0^*\|}(x_0) = \frac{x_0^*(P_n x_0)}{\|P_n^* x_0^*\|} \longrightarrow 1$ $(n \longrightarrow \infty)$. Now $X$ is weakly very smooth, so $\left\{\frac{P_n^* x_0^*}{\|P^* x^*\|}\right\}$ has a subsequence $\left\{\frac{P^*_{n_i} x^*_0}{\|P^*_{n_i} x^*_0\|}\right\}$ such that $\frac{P^*_{n_i} x^*_0}{\|P^*_{n_i} x^*_0\|} \xrightarrow{w} y^* \in X^*$. But for any $x \in X$, $P^*_{n_i} x^*_0 (x) = x^*_0(P_{n_i} x) \longrightarrow x^*_0(x)$, hence $P^*_{n_i} x^*_0 \xrightarrow{w^*} x^*_0$. Since $\lim\|P^*_n x^*_0\| = 1$, it follows that $y^* = x^*_0$, and $\frac{P^*_{n_i} x^*_0}{\|P^*_{n_i} x^*_0\|}$ $\xrightarrow{w} x^*_0$.

Observe that $\frac{P^*_{n_i} x^*_0}{\|P^*_{n_i} x^*_0\|} = \frac{1}{\|P^*_{n_i} x^*_0\|} \left[\sum_{k=1}^{n_i} x^*_0(e_k) f_k\right] \in \overline{\text{span}\{f_n\}}$ and that $\overline{\text{span}\{f_n\}}$ is weakly closed, we have $x^*_0 \in \overline{\text{span}\{f_n\}}$. Hence $\{f_n\}$ is a basis of $X^*$. Q.E.D.

If $X^*$ is locally uniformly convex, then $X$ is very smooth, hence weakly very smooth; thus we have

COROLLARY 1. Let $X$ be a Banach space with a monotone basis, if $x^*$ is locally uniformly convex, then the basis is shrinking.

REMARK. There exists a Banach space $X$ with a basis, but $X^*$ is separable and $X^*$ does not have basis [3]. For such a Banach space $X$, we can define a new equivalent norm on $X$ under which $X^*$ is locally uniformly convex [2, p. 118]. Under this new norm $X^*$ also does not have basis, hence any basis of $X$ under this norm is not monotone. Thus there are weakly very smooth Banach spaces which have bases but which do not have a monotone basis. For other examples of Banach spaces with a basis which do not have monotone basis, see [5, p. 248].

It is known that if $X$ is very smooth, then $X^*$ has Radon-Nikodym property. Since every separable dual space has Radon-Nikodym property, we have

COROLLARY 2. Let $X$ be a Banach space with a monotone basis, if $X$ is weakly very smooth, then $X^*$ has Radon-Nikodym property.

The author would like to thank his adviser, Professor Bor-Luh Lin, for valuable guidance in revising this paper.

REFERENCES

[1] Sullivan, F., Geometrical properties determined by the higher duals of a Banach space. Illinois J. Math. 21 (1977) 315-331.
[2] Diestel, J., Geometry of Banach Spaces—Selected Topics. Lecture Notes in Math. 485 (Springer-Verlag, 1975).
[3] Lindenstrass, J. and Tzafriri, L., Classical Banach spaces I (Springer-Verlag, 1977).
[4] Bishop, E. and Phelps, R.R., The support functionals of a convex set, Convexity, Proceedings of Symposia in Pure Math., AMS 7 (1963).
[5] Singer, I., Bases in Banach spaces I (Springer-Verlag, 1970).
[6] James, J., Characterizations of reflexivity, Studia Math. 23 (1964) 205-216.